CAMBRIDGE TRACTS IN MATHEMATICS

General Editors

170   Polynomials and Vanishing Cycles

POLYNOMIALS AND VANISHING CYCLES

MIHAI TIBĂR

Université des Sciences et Technologies de Lille, France

CAMBRIDGE
UNIVERSITY PRESS

CAMBRIDGE UNIVERSITY PRESS
Cambridge, New York, Melbourne, Madrid, Cape Town, Singapore, São Paulo

Cambridge University Press
The Edinburgh Building, Cambridge CB2 8RU, UK

Published in the United States of America by Cambridge University Press, New York

www.cambridge.org
Information on this title: www.cambridge.org/9780521829205

First published 2007

Printed in the United Kingdom at the University Press, Cambridge

A catalog record for this publication is available from the British Library

ISBN 978-0-521-82920-5 hardback

For My Mother, *in Memoriam*

Contents

Preface

Vanishing cycles appear naturally in the picture when studying families of hypersurfaces, usually regarded as singular fibrations. The behaviour of vanishing cycles seems to be the cornerstone for understanding the geometry and topology of such families of spaces. There is a large literature, mostly over the last 40 years, showing the various ways in which vanishing cycles appear. For instance, we may associate to a holomorphic function f its sheaf of vanishing cycles, encoding information about the singularities and the monodromy of f.

Although quite sophisticated information is available (e.g. in Hodge theoretic terms, see the survey [Di2]), there are many open questions on the geometry of vanishing cycles (see for instance Donaldson's paper [Don] for an intriguing one).

This book proposes a systematic geometro-topological approach to the vanishing cycles appearing especially in *nonproper fibrations*. In such fibrations, some of the vanishing cycles do not correspond to the singularities on the space. Nevertheless, if the fibration extends to a proper one, then new singularities appear at the boundary and their relation to the original context may explain the presence of those vanishing cycles. The study of this type of problem in the setting of singular spaces and stratified singularities started notably with the works of Goresky and MacPherson, Hamm and Lê.

The situations where nonproper fibrations appear fall into two types are:
1. fibration on a noncompact space X, which is the restriction of a fibration over a given compact space Y such that $X = Y \setminus V$ for some subspace $V \subset Y$;
2. fibration on a noncompact X, which can be extended, nonuniquely, to a proper fibration on a larger space.

In case 1, the singularities of the given extension on the space Y are studied and then the information for the fibration on X are extracted. In case 2, first a 'good' candidate for the extension over some space Y should be found, and pursued as in case 1. For instance, let $f : \mathbb{C}^{n+1} \to \mathbb{C}$ be any polynomial function. This defines a nonproper fibration and can be extended to a meromorphic function $\tilde{f}/x_0^d : \mathbb{P}^{n+1} \dashrightarrow \mathbb{P}^1$, where $\tilde{f}$ is the homogenization of f of degree $d = \deg f$, by the new variable x_0. Here, the embedding is $\mathbb{C}^{n+1} \subset \mathbb{P}^{n+1}$ and $\tilde{f}/x_0^d$ restricts to f on $\mathbb{C}^{n+1} = \mathbb{P}^{n+1} \setminus \{x_0 = 0\}$, but we may consider other embeddings.

The leading idea of this monograph is to bring into new light a bunch of topics – holomorphic germs, polynomial functions, pencils on quasi-projective spaces – conceiving them as aspects of a single theory with vanishing cycles at its core. A synthetic table with the topics and their relations is given in Figure 9. The new and highly general branches – meromorphic functions and non-generic Lefschetz pencils – complete and extend the landscape.

Parts I and II focus on complex polynomial functions f and discuss recent results in connection to the 'vanishing cycles at infinity' introduced in [ST2]. (Some aspects are discussed in real variables in Part I.) The specificity of the situation is the loss of 'information' toward infinity (e.g. singular points, curvature of fibres, vanishing cycles) and the aim is to explain the phenomena and to quantify this loss whenever possible. Roughly, the strategy is to compactify the family of fibres of f, study the proper extension of f especially at its singularities at infinity, and then derive the consequences of this study for the original affine setting.

Some evidence for the crucial importance of singularities at infinity in understanding the behaviour of polynomials is the famous unsolved Jacobian Conjecture. In two complex variables, an equivalent formulation of this conjecture is the following, cf. [LêWe,ST2]: *If $f : \mathbb{C}^2 \to \mathbb{C}$ has no critical points but has singularities at infinity, then, for any polynomial $h : \mathbb{C}^2 \to \mathbb{C}$, the zero locus $Z(\mathrm{Jac}(f, h))$ of the Jacobian is not empty.* Corollary 3.3.3 will show that, if the polynomial f has no critical points and no singularities at infinity, then all the fibres of f are CW-complexes with trivial homotopy groups, hence contractible. In this situation the Abhyankar–Moh–Suzuki theorem [AM, Suz] tells that f is linearizable, so the case left is indeed the one of singularities at infinity, as stated in the above conjecture.

Counting the vanishing cycles is an important issue and relates to enumerative geometry. In the complex setting, this is managed by an omnipresent character, the *polar curve*, to the role of which is dedicated Part II. Intersecting with the polar curve opens the way to counting points with multiplicities, which yields several invariants of the affine varieties, up to the embedding: CW-complex structure, relative homology groups, Euler obstruction, Chern–MacPherson cycle. Numerical polar invariants may control, under 'reasonable' circumstances, the behaviour of families of affine hypersurfaces or of polynomials: equisingularity at infinity, topological triviality, the integral of the curvature, the Gauss–Bonnet defect, etc. The geometry of polar curves enters into the study of the various aspects of the monodromy of f.

Part III studies the topology of pencils of hypersurfaces (or *meromorphic functions*) on stratified complex spaces. The context is a general one: *non-generic pencils*, which means pencils of hypersurfaces that may have singularities in the base locus (axis). This represents a unitary viewpoint on the Lefschetz–Morse–Zariski–Milnor theory, which is concerned with the change of topology when singularities occur while scanning the space by the levels of a function. Here, 'singularities' also means singularities at the boundary (whenever the space is open) and singularities in the axis of the pencil. This new standpoint, issued from [Ti8, Ti9, Ti12, Ti10], yields an extension of the classical context of *generic pencils*, also called *Lefschetz pencils*.

This book relies on the research I have done over the past 12 years, some of which was joint work. I owe very special thanks to Dirk Siersma. Several chapters stem from our joint papers [ST1-8] and handwritten notes, over which we have spent an immeasurable amount of time in Utrecht, in Lille, as RiP-ers in Oberwolfach and in many other places. I warmly thank my collaborators Alberto Verjovsky, Anatoly Libgober, José Seade, Alexandru Zaharia, Jörg Schürmann, Clément Caubel and Arnaud Bodin. Many results of our common papers were integrated into the book structure.

The monograph is intended for researchers and graduate students. The idea was to give transparent proofs, such that also nonspecialists can follow and get to grips with the literature. A list of exercises is provided at the end of almost every chapter, with a few hints at the appendix. I have privileged the self-containedness to the abundance of results. Besides the new presentation, there are also a few new results (Theorem 3.1.2, the determinacy scheme in Figure 1.1, Section 3.3, Proposition 4.1.5), improvements of some older statements, and a couple of new proofs in larger generality (e.g. Theorem 3.2.1, the global geometric monodromy in case of t-singularities §8.1).

As prerequisites, a good idea of differential and algebraic topology (homology, homotopy), and the basics on analytic and algebraic geometry are required. For singularity theory, some familiarity with Milnor's classical book [Mi2] is assumed. Reference is made to the appropriate literature whenever more involved results are needed for specialized topics. A list of some relevant textbooks and monographs is given at the end.

During the preparation of this manuscript since 2000, I benefited from the hospitality of several research institutes, to which I express my whole gratitude: Newton Institute (Cambridge), Institute for Advanced Study (Princeton), Mathematisch Forschungsinstitut Oberwolfach, Centre de Recerca Matemàtica

(Barcelona), Institutul de Matematică al Academiei Române (Bucureşti), International Centre for Theoretical Physics (Trieste), Banach Center (Warsaw).

Finally, it is a great pleasure to acknowledge the full non-mathematical loving support of my wife Teodorina, my children Alexandra and Ştefan, my parents Maria and Mircea and my brother Mălin. This book is dedicated to them.

January 2007
Lille, France

PART I

Singularities at infinity of polynomial functions

1

Regularity conditions at infinity

1.1 Atypical values

Let $\mathbb{K}$ be either the real field $\mathbb{R}$ or the complex field $\mathbb{C}$. Let $f : \mathbb{K}^n \to \mathbb{K}$ be a polynomial function of $n \geq 2$ variables. We denote by $\mathrm{Sing}f$ the singular locus of f, that is the set of points $x \in \mathbb{K}^n$ such that the gradient $\mathrm{grad}f(x) = (\frac{\partial f}{\partial x_1}, \ldots, \frac{\partial f}{\partial x_n})(x)$ is equal to zero. Alternately, $\mathrm{Sing}f$ is the set of points $x \in \mathbb{K}^n$ where f is not a submersion. It then follows by the implicit function theorem (submersion theorem) that f is differentially isotopic to a trivial fibration in some small neighbourhood of x. On a compact Riemannian manifold M (with or without boundary), the Ehresmann theorem tells us that we can get a fibration which is locally trivial on the target: if $v \in \mathbb{R}^p$ is a regular value of a function $h : M \to \mathbb{R}^p$, then this function is a trivial fibration over a small enough neighbourhood of v. In the situation of our polynomial function, this result cannot be applied. The reason is that we cannot control the trivialization in the 'neighbourhood of infinity'. This justifies the following definition.

Definition 1.1.1 We say that f is *topologically trivial at* $t_0 \in \mathbb{K}$ if there is a neighbourhood D of $t_0 \in \mathbb{K}$ such that the restriction $f_| : f^{-1}(D) \to D$ is a topologically trivial fibration. If t_0 does not satisfy this property, then we say that t_0 is an *atypical value* and that $f^{-1}(t_0)$ is an *atypical fibre*. We shall denote by $\mathrm{Atyp}f$ the set of atypical fibres of f.

It is not so difficult to prove that the set $f(\mathrm{Sing}f)$ of critical values of f is a finite subset of $\mathbb{K}$. In the complex setting we have the inclusion (see Exercise 1.4):

$$f(\mathrm{Sing}f) \subset \mathrm{Atyp}f. \tag{1.1}$$

The inclusion can be strict even in very easy examples, such as the following one (given by Broughton [Br2]).

Example 1.1.2 $f : \mathbb{K}^2 \to \mathbb{K}$, $f(x,y) = x^2 y + x$. We can quickly see that $\mathrm{Sing} f = \emptyset$. We have $f^{-1}(\varepsilon) = \{ y = (\varepsilon - x)/x^2 \}$ for $\varepsilon \ne 0$ and $f^{-1}(0) = \{ x(xy + 1) = 0 \}$. Therefore, $f^{-1}(\varepsilon)$ is homeomorphic to $\mathbb{K}^* := \mathbb{K} \setminus \{0\}$, whereas $f^{-1}(0)$ is homeomorphic to the disjoint union $\mathbb{K} \sqcup \mathbb{K}^*$. We obtain $\mathrm{Atyp} f \supset \{0\}$. This inclusion is actually an equality (Exercise 1.3).

It turns out that the set of atypical values of f is also finite. We shall give a proof in Corollary 1.2.13 after discussing several issues.[1] However the proofs of the finiteness of the set $\mathrm{Atyp} f$ are not constructive.

Among the natural problems that occur are the following ones:

to find a procedure to decide whether a noncritical value is atypical or not;
to describe how the topology of fibres changes at such a value.

In order to answer the first question, we would try to produce a trivialization at infinity as in Definition 1.1.1 by integrating the gradient vector field $\mathrm{grad} f$. It may then happen that some integral curves 'disappear' at infinity. This is due to the fact that $\mathrm{grad} f$ may tend to zero along nonbounded sequences of points. This phenomenon is well known in nonlinear analysis: we say that f does not satisfy the *Palais–Smale condition* [PaSm].

We shall explain two regularity conditions at infinity that go beyond the Palais–Smale condition: ρ-*regularity* and t-*regularity*. The former depends on the choice of a proper non negative C^1-function ρ, which defines a codimension one foliation in the neighbourhood of infinity. The latter condition depends on a compactification of f, but allows us to apply algebro-geometric tools, particularly efficient in the complex setting.

Triviality at infinity. Let $f : \mathbb{K}^n \to \mathbb{K}$ be a polynomial function of degree d. We consider the following algebraic subset of $\mathbb{P}^n_{\mathbb{K}} \times \mathbb{K}$:

$$\mathbb{X}_{\mathbb{K}} := \{ \tilde{f}(x_0, x) - t x_0^d = 0 \} \subset \mathbb{P}^n_{\mathbb{K}} \times \mathbb{K}, \qquad (1.2)$$

where $\tilde{f}$ denotes the projectivization of f by the new variable x_0. Let:

$$\tau : \mathbb{X}_{\mathbb{K}} \to \mathbb{K}$$

be the projection to $\mathbb{K}$ and let us denote by $\mathbb{X}^\infty_{\mathbb{K}} := \mathbb{X}_{\mathbb{K}} \cap \{ x_0 = 0 \}$ the part at infinity of $\mathbb{X}_{\mathbb{K}}$.

Note 1.1.3 In case $\mathbb{K} = \mathbb{C}$ our set $\mathbb{X}_{\mathbb{C}}$ is precisely the closure in $\mathbb{P}^n_{\mathbb{K}} \times \mathbb{K}$ of the graph $\{ (x,t) \in \mathbb{K}^n \times \mathbb{K} \mid f(x) = t \}$ of f and the part at infinity $\mathbb{X}^\infty_{\mathbb{C}}$ is a divisor of $\mathbb{X}_{\mathbb{C}}$. It is clearly not so in the real case (example: $f(x,y) = x^4 + y^2$). This is one of the reasons why we shall stick to the complex setting later on.

We may identify $\mathbb{K}^n$ to $\mathbb{X} \setminus \mathbb{X}^\infty$ via the canonical map $x \overset{i}{\mapsto} ([x : 1], f(x))$, which fits into the commuting diagram:

$$\begin{array}{ccc} \mathbb{K}^n & \overset{i}{\longrightarrow} & \mathbb{X} \\ {\scriptstyle f} \searrow & & \swarrow {\scriptstyle \tau} \\ & \mathbb{K} & \end{array} \qquad (1.3)$$

In this way we get an extension of f which is a proper function.[2]

Definition 1.1.4 (Local triviality at infinity)

Let $y \in \mathbb{X}^\infty$ and let us denote by $B_\varepsilon \subset \mathbb{P}_{\mathbb{K}}^n \times \mathbb{K}$ an open ball of radius ε centred at y and by $D_\delta \subset \mathbb{K}$ an open disk of radius δ centred at $\tau(y)$. (In the real setting D_δ means just a symmetric interval).

We say that f is *locally trivial at infinity, at* $y \in \mathbb{X}^\infty$, if there exists $\varepsilon_0 > 0$ for which the following condition holds: for any $0 < \varepsilon \le \varepsilon_0$, there is a $\delta > 0$ such that the restriction:

$$\tau_| : (\mathbb{X} \setminus \mathbb{X}^\infty) \cap B_\varepsilon \cap \tau^{-1}(D_\delta) \to D_\delta \qquad (1.4)$$

is a trivial topological fibration.

Definition 1.1.5 (Topological triviality at infinity)

We say that f is *topologically trivial at infinity at the value* $t_0 \in \mathbb{K}$ if there exists a compact set $K \subset \mathbb{K}^n$ and a disk D_δ centred at t_0 such that the restriction:

$$f_| : (\mathbb{K}^n \setminus K) \cap f^{-1}(D_\delta) \to D_\delta \qquad (1.5)$$

is a trivial topological fibration.

If we replace D_δ by D_δ^* in the fibrations (1.4) and (1.5), then we can show that these are locally trivial fibrations, without any condition, see Appendix A1.1 and Theorem 3.1.6 respectively. A more precise notion, the ϕ-*controlled topological triviality*, will be given in Definition 3.1.8.

1.2 ρ-regularity and t-regularity

We introduce a regularity condition which is based on a control function. Let $K \subset \mathbb{K}^n$ be some compact (eventually empty) set and let:

$$\rho : \mathbb{K}^n \setminus K \to \mathbb{R}_{\ge 0}$$

be a proper C^1-submersion.

Definition 1.2.1 (ρ-**regularity at infinity**)
We say that f is ρ-*regular at* $y \in \mathbb{X}^\infty$ if there is an open ball $B_\varepsilon \subset \mathbb{P}^n_\mathbb{K} \times \mathbb{K}$ centred at y and some disk $D_\delta \subset \mathbb{K}$ at $\tau(y)$ such that either $f^{-1}(D_\delta) \cap B_\varepsilon = \emptyset$ or, for all $c \in D_\delta$, the fibre $f^{-1}(c) \cap B_\varepsilon$ intersects all the levels of the restriction $\rho_{|B_\varepsilon \cap \mathbb{K}^n}$ and this intersection is transversal.

We say that the fibre $f^{-1}(t_0)$ is ρ-*regular at infinity* if f is ρ-regular at all points $y \in \mathbb{X}^\infty \cap \tau^{-1}(t_0)$. In this case we also say that t_0 is a ρ-regular value of f.

The transversality of the fibres of f to the levels of a control function recalls the well-known control functions ('fonction tapissante' or 'rug function', a notion due to Thom [Th1, Th2]) used by Thom and Mather in their First Isotopy Theorem along Whitney stratifications. More recently, the Isotopy Theorem has been proved for stratifications that are *(c)-regular*, a weaker regularity condition developed by Bekka [Be], which is also based on control functions.

If we use the Euclidean norm ρ_E in place of the function ρ in the above definition, then the ρ_E-regularity is a large-scale version[3] of the *transversality to small spheres*, the condition used by Milnor in the local study of holomorphic functions [Mi2, §4,5], see 3.1.

Remark 1.2.2 The fact that $f^{-1}(t_0)$ is ρ-regular at infinity is independent on the proper extension of f, since it is equivalent to the following: for any sequence $(x_k)_{k \in \mathbb{N}} \subset \mathbb{K}^n$, $|x_k| \to \infty$, $f(x_k) \to t_0$, there exists some $k_0 = k_0((x_k)_{k \in \mathbb{N}})$ such that, if $k \geq k_0$, then f is transversal to ρ at x_k.

Example 1.2.3 $\rho : \mathbb{K}^n \to \mathbb{R}_{\geq 0}$, $\rho(x) = (\sum_{i=1}^n |x_i|^{2p_i})^{1/2p}$, where $(w_1, \dots, w_n) \in \mathbb{N}^n$, $p = \text{lcm}\{w_1, \dots, w_n\}$ and $w_i p_i = p$, $\forall i$. This is a control function which can be used especially for polynomials f which are quasihomogeneous of type $(w_1, \dots, w_n)$, see Exercise 1.7.

Proposition 1.2.4 If the fibre $f^{-1}(t_0)$ is ρ-regular at infinity, then f is topologically trivial at infinity at t_0.

Proof The set of points at infinity $\overline{f^{-1}(t_0)} \cap H^\infty$ is a compact set and therefore we can cover it by a finite number of balls B_i as in Definition 1.2.1. Let N be the union of these balls. Let D_i be the disk centred at t_0 which corresponds to the ball B_i in Definition 1.2.1, and let D be the smallest of those disks.

So N is a neighbourhood of $\tau^{-1}(t_0) \cap \mathbb{X}^\infty$ and we propose to show that the restriction $f_| : N \cap f^{-1}(D) \to D$ is a trivial fibration. Notice that Thom–Mather's Isotopy Theorem does not apply since the function $f_|$ is not proper. We use here ρ as global control function and construct a lift of the (real or complex) vector field $\partial/\partial t$ on D to a (real or complex) vector field $\mathbf{w}$ without zeros on $N \cap \mathbb{K}^n$, which is tangent to the levels $\rho = $ constant. We then get our topologically trivial fibration[4] by integrating $\mathbf{w}$. For all the details of this type of construction we may refer the reader to Verdier's proof [Ve, Theorem 4.14] of Thom–Mather's Isotopy Theorem. □

Remark 1.2.5 In view of Definition 3.1.8, the proof of Proposition 1.2.4 yields the following more precise statement: *If the fibre $f^{-1}(t_0)$ is ρ-regular at infinity, then f is ρ-controlled topologically trivial at t_0.*

Corollary 1.2.6 If the fibre $f^{-1}(t_0)$ is nonsingular and ρ-regular at infinity, then $f^{-1}(t_0)$ is not an atypical fibre, i.e. $t_0 \notin \mathrm{Atyp} f$.

Proof From the proof of Proposition 1.2.4, we have a vector field $\mathbf{w}$ without zeros on a neighbourhood of infinity $N \cap \mathbb{K}^n$ and which is a lift of the unit vector field $\partial / \partial t$ on D.

Since the fibre $f^{-1}(t_0)$ is nonsingular, the gradient $\mathrm{grad} f$ has no zeros on $f^{-1}(t_0)$. Moreover, for any large ball $B \subset \mathbb{K}^n$ there exists a (small enough) disk D centred at t_0 such that $\mathrm{grad} f$ has no zeros on $B \cap f^{-1}(D)$. Therefore the vector field $\mathrm{grad} f$ is nowhere zero on $B \cap f^{-1}(D)$.

We then take a ball B such that $B \cap N \cap f^{-1}(D)$ is open and such that $(B \cup N) \cap f^{-1}(D) = f^{-1}(D)$. By a partition of unity, we glue the vector field $\mathbf{w}$ to the vector field $\mathbf{u} := \frac{\mathrm{grad} f}{\| \mathrm{grad} f \|^2}$. The result is a vector field defined on $f^{-1}(D)$ which has the properties that it has no zeros and it is a lift of $\partial / \partial t$, since both vector fields $\mathbf{w}$ and $\mathbf{u}$ have these two properties. We then get a global trivialization over D by integrating this vector field. □

The relative conormal. Let $\mathcal{X} \subset \mathbb{K}^N$ be a $\mathbb{K}$-analytic variety. In the real case, assume that $\mathcal{X}$ contains at least a regular point. Let $U \subset \mathbb{K}^N$ be an open set and let $g : \mathcal{X} \cap U \to \mathbb{K}$ be $\mathbb{K}$-analytic and not constant. One calls *relative conormal of g* the subspace of the restriction of the cotangent bundle $T^*(\mathbb{K}^N)_{|\mathcal{X} \cap U}$ defined as follows[5]:

$$T^*_{g|\mathcal{X} \cap U} := \mathrm{closure}\{(y, \xi) \in T^*(\mathbb{K}^N) \mid y \in \mathcal{X}^0 \cap U, \xi(T_y(g^{-1}(g(y)))) = 0\},$$

where $\mathcal{X}^0 \subset \mathcal{X}$ is the open dense subset of the regular points of $\mathcal{X}$ where g is a submersion. The relative conormal is *conical*, which means the following:

$$(y, \xi) \in T^*_{g|\mathcal{X} \cap U} \Rightarrow (y, \lambda \xi) \in T^*_{g|\mathcal{X} \cap U}, \forall \lambda \in \mathbb{K}^*.$$

Let $\pi : T^*_{g|\mathcal{X} \cap U} \to \mathcal{X} \cap U$ denote the canonical projection and let $(T^*_{g|\mathcal{X} \cap U})_x := \pi^{-1}(x)$ for some $x \in \mathcal{X} \cap U$ such that $g(x) = 0$. We show that $(T^*_{g|\mathcal{X} \cap U})_x$

depends on the germ of g at x only up to multiplication by a unit in the analytic germ algebra $\mathcal{O}_{\mathcal{X},x}$.

Lemma 1.2.7 Let $\gamma : (\mathcal{X}, x) \to \mathbb{K}$ be $\mathbb{K}$-analytic and such that $\gamma(x) \neq 0$. Then $(T^*_{g|\mathcal{X}\cap U})_x = (T^*_{\gamma g|\mathcal{X}\cap U})_x$.

Proof Suppose first that $(\mathcal{X}, x)$ is nonsingular. We have $\operatorname{grad} \gamma g = \gamma \operatorname{grad} g + g \operatorname{grad} \gamma$, hence:

$$\frac{\operatorname{grad} \gamma g}{\|\operatorname{grad} g\|} = \gamma \frac{\operatorname{grad} g}{\|\operatorname{grad} g\|} + \operatorname{grad} \gamma \frac{g}{\|\operatorname{grad} g\|}.$$

Since γ is analytic, $\|\operatorname{grad} \gamma\|$ and γ are bounded within some neighbourhood of x. We have the following inequality due to Łojasiewicz [Łoj1]:

$$\|\operatorname{grad} g\| \geq |g|^\theta, \quad \text{for some } 1 > \theta > 0,$$

which is valid in some small enough neighbourhood of x. Since $g(x) = 0$ we get that $\frac{g(y)}{\|\operatorname{grad} g(y)\|}$ tends to zero as the point y tends to x. Therefore, along any sequence of points tending to x, we have $\lim \frac{\operatorname{grad} \gamma g}{\|\operatorname{grad} g\|} = \gamma(x) \lim \frac{\operatorname{grad} g}{\|\operatorname{grad} g\|}$. This shows that the limits of the directions $\operatorname{grad} \gamma g$ and $\operatorname{grad} g$ are the same.

Let $g^{-1}(0)$ be denoted by $\mathcal{Y}$. In the general case we resolve $\mathcal{X}$ within an embedded resolution $p : \tilde{\mathcal{X}} \to \mathcal{X}$, to a smooth variety $\tilde{\mathcal{X}}$. This is an isomorphism over $\mathcal{X}_{\text{reg}} \setminus \mathcal{Y}$. Now apply the result proved above in the nonsingular case to the functions $g \circ p$ and $(\gamma \circ p)(g \circ p)$, then pull down to the conormal of $\mathcal{X}$, by the following:

$$
\begin{array}{ccccc}
(T^*\mathbb{K}^N)_{|\tilde{\mathcal{X}}\cap p^{-1}(U)} & \leftarrow & p^*(T^*\mathbb{K}^N)_{|\mathcal{X}\cap U} & \to & (T^*\mathbb{K}^N)_{|\mathcal{X}\cap U} \\
& \searrow & \downarrow & & \downarrow \pi \\
& & \tilde{\mathcal{X}} \cap p^{-1}(U) & \xrightarrow{p} & \mathcal{X} \cap U.
\end{array}
$$

$\square$

Remark 1.2.8 In the above proof, it is not necessary that γ is an analytic function, it is only important that γ and $\|\operatorname{grad} \gamma\|$ are bounded.

Let us come back to our polynomial function f and its associated space $\mathbb{X}_\mathbb{K} \subset \mathbb{P}^n \times \mathbb{K}$. The subspace $\mathbb{X}_\mathbb{K}^\infty \subset \mathbb{X}_\mathbb{K}$ is covered by the union of the affine charts $U_i \times \mathbb{K}$, where $U_i := \{x_i \neq 0\}$, for $0 < i \leq n$. In each chart, the subspace $\mathbb{X}_\mathbb{K}^\infty$ is defined by the equation $x_0 = 0$ and therefore the relative conormal $T^*_{x_0|\mathbb{X}\cap(U_i\times\mathbb{K})}$ is well defined. The equations $x_0 = 0$ differ from one chart to the other by multiplication with a rational function of type x_i/x_j.

Since this function is nonzero on $\mathbb{X}_{\mathbb{K}}^{\infty}$, we deduce from Lemma 1.2.7 that the fibre $(T_{x_0|\mathbb{X}\cap(U_i\times\mathbb{K})}^*)_y$ is independent on the chart and therefore we may write $(T_{x_0|\mathbb{X}}^*)_y$. We then have the following definition:

Definition 1.2.9 We say that (y,ξ) is a *characteristic covector at infinity* if $\xi\in(T_{x_0|(\mathbb{X}\setminus\mathbb{X}^\infty)\cap U}^*)_y$, where $y\in\mathbb{X}_{\mathbb{K}}^\infty\cap\overline{f^{-1}(\tau(y))}$ and $U\subset\mathbb{X}$ is some neighbourhood of y. We denote by $\mathcal{C}_{\mathbb{K}}^\infty$ the *subspace of characteristic covectors at infinity*.

We have proved above that $\mathcal{C}_{\mathbb{K}}^\infty$ is an analytic subspace of the restriction of the cotangent bundle $T^*(\mathbb{P}^n\times\mathbb{K})$ over $\mathbb{X}_{\mathbb{K}}^\infty$. We also note that $\mathcal{C}_{\mathbb{K}}^\infty$ is conical* and therefore we may consider its projectivization $\mathbb{P}(\mathcal{C}_{\mathbb{K}}^\infty)$. With these definitions, let us introduce the announced regularity condition.

Definition 1.2.10 (*t-regularity at infinity; t-singularities,* $\mathrm{Sing}^\infty f$)
The fibre $f^{-1}(t_0)$ (or that f) is *t-regular* at $y\in\mathbb{X}_{\mathbb{K}}^\infty\cap\overline{f^{-1}(t_0)}$ iff $(y,\mathrm{d}t)$ is not a characteristic covector at infinity, i.e. $(y,\mathrm{d}t)\notin\mathcal{C}_{\mathbb{K}}^\infty$. We also say that $f^{-1}(t_0)$ is *t-regular at infinity* if this fibre is *t*-regular at all its points at infinity.

We call *t-singularities of f at infinity* the points $y\in\mathbb{X}_{\mathbb{K}}^\infty$ where f is not *t*-regular. We denote by $\mathrm{Sing}^\infty f$ the set of *t*-singularities of f at infinity.

Remark 1.2.11 It follows from the definition that, if $y\in\overline{\mathrm{Sing}f}\cap\mathbb{X}_{\mathbb{K}}^\infty$, then f is not *t*-regular at y. The reciprocal is obviously not true (Example 1.1.2) and therefore we need to investigate more closely what is the set $\mathrm{Sing}^\infty f$.

We now relate the *t*-regularity to the *ρ*-regularity. Let us denote by ρ_E the Euclidean norm in $\mathbb{K}^n$. This is a control function and the ρ_E-regularity is well defined.

Proposition 1.2.12 If f is *t*-regular at $y\in\mathbb{X}_{\mathbb{K}}^\infty$, then f is ρ_E-regular at y. In particular, if $f^{-1}(t_0)$ is *t*-regular at infinity, then $f^{-1}(t_0)$ is ρ_E-regular at infinity.

Proof Let $d^\infty:\mathbb{X}_{\mathbb{K}}\to\mathbb{R}$ be defined by:

$$\begin{cases} d^\infty(x,f(x))=1/\rho_E^2(x), & \text{for } x\in\mathbb{K}^n\\ d^\infty(y)=0, & \text{for } y\in\mathbb{X}_{\mathbb{K}}^\infty. \end{cases}$$

By computing in local charts, we can see that d^∞ is a rational function. (After Thom [Th1], this is an example of a *rug function*.) Moreover, in the neighbourhood of some point $y\in\mathbb{X}_{\mathbb{K}}^\infty$, the functions $|x_0|^2$ and d^∞ differ by a nonzero factor and have the same zero locus, the germ of $\mathbb{X}_{\mathbb{K}}^\infty$ at y. By Lemma 1.2.7 we have:

$$(T_{d^\infty|\mathbb{X}}^*)_y=(T_{|x_0|^2|\mathbb{X}}^*)_y. \tag{1.6}$$

---

* In the complex setting, $\mathcal{C}_{\mathbb{C}}^\infty$ is also a *Lagrangean* subvariety of $T^*(\mathbb{P}^n\times\mathbb{C})_{|\mathbb{X}^\infty}$.

Let us consider the real setting first. Then $(T^*_{|x_0|^2|\mathbb{X}})_y = (T^*_{x_0|\mathbb{X}})_y$ and the latter is in turn equal to $(C^\infty_{\mathbb{R}})_y$. The condition $(y, dt) \notin (T^*_{d^\infty|\mathbb{X}})_y$ is therefore equivalent to the t-regularity at $y \in \mathbb{X}^\infty$. On the other hand, it implies that, in some neighbourhood of y intersected with $\mathbb{K}^n$, the fibres $t = $ constant are transversal to the levels of the function d^∞. These levels are the same as the levels of the function ρ_E, so this finishes the proof in the real case.

The complex case $\mathbb{K} = \mathbb{C}$ now. By the conical structure of $T^*(\mathbb{K}^n)$ we may take the projectivization $\mathbb{P}T^*(\mathbb{K}^n)$, which is the quotient space by the action of $\mathbb{K}^*$. Let us introduce the map $\iota : \mathbb{P}T^*(\mathbb{R}^{2n}) \to \mathbb{P}T^*(\mathbb{C}^n)$ between the real and the complex projectivized conormal bundles (where $\mathbb{R}^{2n}$ is the real underlying space of $\mathbb{C}^n$) defined as follows: if ξ is conormal to a hyperplane $H \subset \mathbb{R}^{2n}$, then $\iota([\xi])$ is the conormal to the unique complex hyperplane included in H. This is clearly a continuous map. We then have the following equality:

$$\mathbb{P}(C^\infty_{\mathbb{C}})_y = \iota(\mathbb{P}(T^*_{|x_0|^2|\mathbb{X}})_y),$$

since the complex tangent space $T_x\{x_0 = \text{constant}\}$ is exactly the unique complex hyperplane contained in the real tangent space $T_x\{|x_0|^2 = \text{constant}\}$. The equality follows by the fact that ι commutes with taking limits. Now $(y, dt) \notin (C^\infty_{\mathbb{C}})_y$ implies $(y, \iota^{-1}([dt])) \notin \mathbb{P}(T^*_{|x_0|^2|\mathbb{X}})_y$. This implies the ρ_E-regularity at y since the equality (1.6) is true in the complex setting too. $\square$

We may now give a proof of the finiteness of the set of nonregular values, based on Whitney stratifications of semi-algebraic sets (see §A1.1).

Corollary 1.2.13 Let $f : \mathbb{K}^n \to \mathbb{K}$ be a polynomial. The set $\tau(\text{Sing}^\infty f)$ of the values $t_0 \in \mathbb{K}$ such that $f^{-1}(t_0)$ is not t-regular at infinity is a finite set.

In particular, the set of values of f that are not ρ_E-regular at infinity is a finite set.

Proof There exists a Whitney stratification $\mathcal{W} = \{\mathcal{W}_i\}_i$ of $\mathbb{X}_{\mathbb{K}}$ with $\mathbb{K}^n$ as a stratum and with a finite number of strata, which has in addition the Thom property with respect to the function x_0 in any local chart.[*] We call it a *Thom–Whitney stratification at infinity*.[6]

If $\tau^{-1}(t_0)$ is transversal to a stratum $\mathcal{W}_i \subset \mathbb{X}^\infty$ at some point y, then $\tau^{-1}(t_0)$ is transversal to the limits of the tangents spaces to the levels of x_0, by the assumed (a_{x_0})-property of our Thom–Whitney stratification. Therefore $f^{-1}(t_0)$ is transversal to the levels of x_0 in the neighbourhood of y and hence f is t-regular at infinity at y (see also §2.2 and (2.6)).

[*] See §A1.1 and also the proof of Proposition 2.2.3.

The restriction of the projection $\tau : \mathbb{X}_{\mathbb{K}} \to \mathbb{K}$ to a stratum contained in $\mathbb{X}^{\infty}$ has a finite number of critical values: since strata are semi-algebraic, the set of critical values of τ is semi-algebraic, discrete, and has a finite number of connected components.* Since the number of strata is itself finite, this implies that the set of values t_0 such that $\tau^{-1}(t_0)$ is not transversal to all the strata it meets, is a finite set. By the above discussion, this contains the set $\tau(\text{Sing}^{\infty}f)$.

The second claim follows from Proposition 1.2.12. $\qquad\square$

Corollary 1.2.14 The set $\text{Atyp}f$ of atypical values of f is a finite set.

Proof We have the inclusion:

$$\text{Atyp}f \subset \tau(\text{Sing}f \cup \text{Sing}^{\infty}f). \qquad (1.7)$$

Indeed, this follows from Corollary 1.2.6 by the fact that t-regularity implies ρ_E-regularity (Proposition 1.2.12). Now, the set $\tau(\text{Sing}^{\infty}f)$ is finite (Corollary 1.2.13) and the set $\text{Sing}f$ is finite too (Exercise 1.2). $\qquad\square$

1.3 The Malgrange condition

We introduce another regularity condition, which is more computable and which will turn out to be equivalent to the t-regularity.[7]

Definition 1.3.1 Let $\{p_j\}_{j \in \mathbb{N}}$ be a sequence of points in $\mathbb{K}^n$ and let us consider the following properties:

(L_1) $\|p_j\| \to \infty$ and $f(p_j) \to t_0$, as $j \to \infty$.
(L_2) $p_j \to y \in \mathbb{X}_{\mathbb{K}}^{\infty}$, as $j \to \infty$.

We say that the fibre $f^{-1}(t_0)$ verifies the *Malgrange condition* if there is $\delta > 0$ such that, for any sequence of points with property (L_1), we have:

$$\|p_j\| \cdot \|\,\text{grad} f(p_j)\| > \delta. \qquad (1.8)$$

We say that f verifies the Malgrange condition at $y \in \mathbb{X}_{\mathbb{K}}^{\infty}$ if there is $\delta > 0$ such that (1.8) holds for any sequence of points with property (L_2).

Right from the definition, it follows that $f^{-1}(t_0)$ verifies the Malgrange condition if and only if f verifies the Malgrange condition at any point $y \in \tau^{-1}(t_0) \cap \mathbb{X}_{\mathbb{K}}^{\infty}$. We have the following characterization.

---

* This uses Tarki–Seidenberg theorem and Whitney's finiteness theorem, see for instance [GWPL].

Proposition 1.3.2 A polynomial $f : \mathbb{K}^n \to \mathbb{K}$ is t-regular at $y \in \mathbb{X}_{\mathbb{K}}^{\infty}$ if and only if f verifies the Malgrange condition at this point.

Proof Let $U_i \times \mathbb{K}$, where $U_i = \{x_i \neq 0\}$, for $0 \le i \le n$, be the affine charts of $\mathbb{P}_{\mathbb{K}}^n \times \mathbb{K}$. Let $y \in \mathbb{X}_{\mathbb{K}}^{\infty}$ be a point, say in the chart $\mathbb{X}_{\mathbb{K}} \cap (U_n \times \mathbb{K})$. We assume as usual that y is in the closure of some fibre of f; if not, then there is nothing to prove.

By definition, t-regular at y means that at any point $x \in \mathbb{K}^n$ in some small enough neighbourhood $\mathcal{N}_y$ of y, the levels $\{x_0 = \text{const.}\}$ are transversal to the levels $\{t = \text{const.}\}$ in $\mathbb{X}_{\mathbb{K}}$. Now from (1.2), $\mathbb{X}_{\mathbb{K}} \cap (U_n \times \mathbb{K})$ is defined by the equation $F_n := \tilde{f}(x_0, x_1, \ldots, x_{n-1}, 1) - t x_0^d = 0$. The t-regularity at y is therefore equivalent to the fact that the normal vector to $\{F_n = 0\} \cap \{x_0 = \text{constant}\}$ at $x \in \mathcal{N}_y$ has some nonzero component other than $\frac{\partial(F_n)}{\partial t}(x)$. This amounts to the following inequality:

$$\delta \left| \frac{\partial(F_n)}{\partial t} \right| < \left\| \frac{\partial(F_n)}{\partial x_1}, \ldots, \frac{\partial(F_n)}{\partial x_{n-1}} \right\|, \tag{1.9}$$

for some $\delta > 0$, in some neighbourhood of y.

Let us translate this inequality in the affine coordinates, i.e. in the chart $U_0 \times \mathbb{K}$: we divide by $|x_0^{d-1}|$ and replace x_0 by x_n^{-1}. We then get:

$$|x_n| \cdot \left\| \frac{\partial f}{\partial x_1}, \ldots, \frac{\partial f}{\partial x_{n-1}} \right\| > \delta. \tag{1.10}$$

This clearly implies Malgrange's inequality (1.8).

Reciprocally, let us assume that (1.8) is true in some neighbourhood of y. Let us first remark that $\|x\| \sim |x_n|$ in the neighbourhood of y. (The equivalence '$\sim$' means that the quotient $\frac{\|x\|}{|x_n|}$ tends to a nonzero constant as the point tends to y.) Therefore (1.8) is equivalent to the following inequality in the neighbourhood of the point y:

$$|x_n| \cdot \left\| \frac{\partial f}{\partial x_1}, \ldots, \frac{\partial f}{\partial x_{n-1}}, \frac{\partial f}{\partial x_n} \right\| > \delta', \tag{1.11}$$

for some $\delta' > 0$. We then prove that $\left\| \frac{\partial f}{\partial x_n} \right\| \le A \left\| \frac{\partial f}{\partial x_1}, \ldots, \frac{\partial f}{\partial x_{n-1}} \right\|$, for some $A > 0$, in some neighbourhood of y. This implies an inequality of type (1.10). We have seen that (1.10) is equivalent to (1.9) and hence[8] to the t-regularity at y. $\square$

Note 1.3.3 The *Łojasiewicz exponent $L_y(f)$ at y* $L_y(f)$ is defined as the smallest real number $\theta \in \mathbb{R}$ such that, for some neighbourhood U of y and some constant

$C > 0$, we have:

$$\| \operatorname{grad} f(x) \| \geq C \|x\|^{\theta}, \quad \forall x \in U \cap \mathbb{K}^n.$$

The Malgrange condition at $y \in \mathbb{X}_{\mathbb{K}}^{\infty}$ is equivalent to saying that the $L_y(f) \geq -1$. There are two other regularity conditions used in the literature similar to Malgrange condition but more restrictive: *Fedoryuk's condition* (or *tameness*, see Proposition 2.3.1), cf. [Fed, Br1, Br2], which is just the Palais–Smale condition mentioned in §1.1:

$$\| \operatorname{grad} f(x) \| > \delta, \tag{1.12}$$

which amounts to saying that $L_y(f) \geq 0$ within the neighbourhood $U \cap \mathbb{K}^n$, and *Parusiński's condition* [Pa1]:

$$\|x^{(N-1)/N}\| \cdot \| \operatorname{grad} f(x) \| > \delta, \tag{1.13}$$

which means that $L_y(f) \geq -(N-1)/N$, for some positive integer N, within $U \cap \mathbb{K}^n$. These conditions are useful for checking the Malgrange condition in particular cases. Nevertheless these two conditions are not generic: the example $f(x, y, z) = x + x^2 y + x^4 yz$ from [Pa2] shows that they do not hold at any fibre. In contrast to this, the ρ_E-regularity at infinity holds at all but a finite number of fibres of f (cf. Corollary 1.2.13).

A couple of other conditions have been used by Némethi and Zaharia [Ne2, NZ], which are between the Fedoryuk condition and ρ_E-regularity.

Comparing *t*-regularity to Whitney equisingularity. Let $y = (z, t_0) \in \mathbb{X}_{\mathbb{K}}^{\infty}$ be a point in the chart $\mathbb{X}_{\mathbb{K}} \cap (U_n \times \mathbb{K})$. The *t*-regularity at y is equivalent, by the inequality (1.9), to the following *integral closure* condition:

$$\frac{\partial F_n}{\partial t} \in \overline{\left(\frac{\partial F_n}{\partial x_1}, \cdots, \frac{\partial F_n}{\partial x_{n-1}} \right)}. \tag{1.14}$$

In the complex setting, it is known from [Te1, BrSc] that the *Whitney equisingularity* along the line $\{z\} \times \mathbb{C}$ at y is equivalent to the following integral closure criterion:

$$\frac{\partial F_n}{\partial t} \in \overline{\mathfrak{m}_y \left(\frac{\partial F_n}{\partial x_0}, \frac{\partial F_n}{\partial x_1}, \cdots, \frac{\partial F_n}{\partial x_{n-1}} \right)}, \tag{1.15}$$

where $\mathfrak{m}_y$ is the maximal ideal of the analytic local algebra $\mathcal{O}_{\mathbb{X}, y}$.

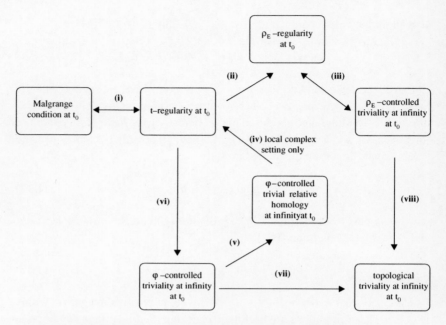

Figure 1.1. Implications between regularity and triviality conditions at $t_0 \in \mathbb{K}$

Determinacy scheme. Figure 1.1 summarizes the determinacies among several regularity conditions at infinity, at some value $t_0 \in \mathbb{K}$.

The equivalence (i) and the implication (ii) follow from the definitions and from Proposition 1.3.2, respectively 1.2.12.

We anticipate on the notions of 'ϕ-controlled topological triviality' and 'ϕ-controlled trivial relative homology at infinity' given in Definition 3.1.8. Then the implication (iii) is Proposition 1.2.4 with Remark 1.2.5, and (v) immediately follows from Definition 3.1.8. The implications (vii) and (viii) are obvious from the definitions.

The implication (vi) is the implication (d)$\Rightarrow$(a) of Theorem 2.2.5, proved in full generality. The implication (iv) holds in the *local complex setting* only and will be proved in Proposition 3.1.9.

We give below two examples. The first one shows that local ρ_E-regularity is more general than local t-regularity, namely that the converse of Proposition 1.2.12 does not hold. Another example illustrating this fact was given by Păunescu and Zaharia [PZ1].

Example 1.3.4 Let $f : \mathbb{C}^3 \to \mathbb{C}, f(x, y, z) = x + x^2 y$, as in Example 1.1.2 but viewed here as a polynomial in three variables. Let us first investigate the *t*-regularity. By Proposition 1.3.2 and the inequality (1.10) in the chart $y \neq 0$, the *t*-regularity is equivalent to the condition $|y| \cdot \|\frac{\partial f}{\partial x}\| \not\to 0$, as $\|x, y, z\| \to \infty$. But in our example, $|y| \cdot \|1 + 2xy\|$ may tend to zero (e.g. for $y \to \infty$, $x = 1/y^3 - 1/(2y)$, $z = ay$). In this way we can show (as an easy exercise) that f is not *t*-regular at any point of the line at infinity $\{x_0 = x = t = 0\} \subset \mathbb{X}^\infty$. Consequently, the only non- *t*-regular value of f is zero.

Let us now find the set of points $(p, t) \in \mathbb{X}^\infty$ where f is not ρ_E-regular. These are the points on the boundary at infinity of the set of solutions of the equality $\overline{\text{grad} f} = (\lambda x, \lambda y, \lambda z)$. This equality amounts to the system: $\|x\|^2 \bar{x} = y + 2\bar{x}\|y\|^2$, $z = 0$, which has a real algebraic set $A \subset \mathbb{C}^2$ as the solution. We can show as an exercise that $\bar{A} \cap \mathbb{X}^\infty$ consists of a single point on the line $\{x_0 = x = t = 0\}$.

The following example, due to Păunescu and Zaharia [PZ2], shows that the topological triviality at infinity at t_0 in the sense of Definition 1.1.5 does not imply *t*-regularity, and that the inclusion (1.7) is strict.

Example 1.3.5 [PZ2] The polynomial function $f : \mathbb{C}^4 \to \mathbb{C}$:

$$f(x, y, z, u) = x + y - 2x^2y^3 + x^3y^6 + zy^3 - z^2y^5 + uy^5$$

is not *t*-regular and not ρ_E-regular at the fibre $f^{-1}(0)$ (see [PZ2] for details). Nevertheless, it is a component of an automorphism of $\mathbb{C}^4$, consequently Atyp$f = \emptyset$, Sing$f = \emptyset$ but $\tau(\text{Sing}^\infty f) = \{0\}$. Indeed, we apply two successive changes of coordinates: $w = u - z^2$ and next $v = z - 2x^2 + x^3y^3 + wy^2$. Then f becomes $f(x, y, v, w) = x + y + vy^3$ and the automorphism of $\mathbb{C}^4$ may be for instance $\varphi(x, y, v, w) = (f, y, v, w)$.

Exercises

1.1 Show that every connected component of Sing$f$ is included into some fibre of f.
1.2 Prove that the critical set $f(\text{Sing} f)$ consists of finitely many values.
1.3 Show that Atyp$f = \{0\}$ in Example 1.1.2.
1.4 If $\mathbb{K} = \mathbb{C}$, then any critical value of f is atypical.
1.5 If a fibre $f^{-1}(t_0)$ of a complex polynomial is ρ-regular at infinity, then this fibre has at most isolated singularities.
1.6 Let f be a quasi-homogeneous polynomial of type $(w_1, \ldots, w_n)$. Show that, if a value $c \in \mathbb{C}$ is atypical, then c is a critical value of f. Conclude that only the value 0 can be atypical.
1.7 In case of the Euclidean norm ρ_E, find explicitly the vector field **w** constructed in the proof of Proposition 1.2.4.

1.8 By Corollary 1.2.14, the set of values $t_0 \in \mathbb{K}$ such that $f^{-1}(t_0)$ is not ρ_E-regular at infinity, is a finite set, denoted say by Λ_{f,ρ_E}. Let us choose some open disk $D_i \subset \mathbb{K}$ centered at a_i, for each $a_i \in \Lambda_{f,\rho_E}$. Show that there exists $R_0 \gg 0$ such that $f^{-1}(t)$ is transversal to the sphere S_R, for any $t \in \mathbb{K} \setminus (\cup_{i \in \Lambda_{f,\rho_E}} D_i)$ and any $R > R_0$.

2

Detecting atypical values via singularities at infinity

2.1 Polar curves

Local polar loci. Let us consider the map germ:

$$(x_0, \tau) : (\mathbb{X}_{\mathbb{K}} \cap (U_i \times \mathbb{K}), y) \to (\mathbb{K}^2, 0),$$

for some $y \in \mathbb{X}_{\mathbb{K}}^{\infty}$ in the chart $U_i \times \mathbb{K}$. Then consider the singular loci of the map (x_0, τ) and of the restriction of τ to the nonsingular subspace $\mathbb{K}^n \cap (U_i \times \mathbb{K})$ and take the analytic closure of their difference:

$$\Gamma(x_0, \tau)_y^{(i)} := \text{closure}[\text{Sing}\,(x_0, \tau)_{|\mathbb{K}^n \cap (U_i \times \mathbb{K})} \setminus \text{Sing}\,\tau_{|\mathbb{K}^n \cap (U_i \times \mathbb{K})}] \subset \mathbb{X},$$

viewed as a germ at y.

Definition 2.1.1 The germ at $y \in \mathbb{X}_{\mathbb{K}}^{\infty}$ of the analytic set $\Gamma(x_0, \tau)_y^{(i)}$ is called *local polar locus of τ with respect to x_0*. It depends on the affine chart $U_i \times \mathbb{K}$ since the function x_0 does so.

We have the following direct consequence of Definitions 2.1.1 and 1.2.10:

$$\Gamma(x_0, \tau)_y^{(i)} \neq \emptyset \implies y \text{ is a t-singularity at infinity of } f. \qquad (2.1)$$

It seems that the local polar locus is intimately related to the relative conormal of x_0, which has been defined in §1.2. Indeed, we have the following description. Let B be some small enough open ball centred at y within a fixed chart $U_i \times \mathbb{K}$. Let $\mathbb{P}T^*_{x_0|(\mathbb{X}\setminus\mathbb{X}^{\infty})\cap B} \subset \mathbb{P}T^*(U_i \times \mathbb{K})$ be the projectivized of the relative conormal of x_0. It is an analytic variety of dimension $n + 1$. Let π and p be the projections

17

from $\mathbb{P}T^*_{x_0|(\mathbb{X}\setminus\mathbb{X}^\infty)\cap B}$ to $\mathbb{X}$ and to $\mathbb{P}(\mathbb{C}^n \times \mathbb{C})^*$ respectively. We get:

Lemma 2.1.2 The polar locus $\Gamma(x_0, \tau)^{(i)}_y$ coincides with the germ at y of the analytic set $\pi(p^{-1}(dt))$.

We shall see that the set of points y such that $\Gamma(x_0, \tau)^{(i)}_y \neq \emptyset$ for some i is a finite set and that the polar loci $\Gamma(x_0, \tau)^{(i)}_y$ are curves. This will turn out to be useful in finding the set of t-singular values of τ, which is itself finite, by Corollary 1.2.13. In contrast, the set of $\mathcal{T}$-singularities may be of dimension >0 (see for instance Example 1.3.4).

The local polar curves at infinity[1] are similar to polar loci defined by Teissier but they are not generic in the sense of Teissier, since the linear form x_0 may be not in general position with respect to τ. We send to §7.1, §7.3 for more details and some historical remarks. Moreover, the whole of Part II is devoted to applications of polar curves and polar varieties.

The global polar locus relative to a linear function. Given a polynomial $f : \mathbb{K}^n \to \mathbb{K}$ and a linear function $l : \mathbb{K}^n \to \mathbb{K}$, the following analytic set:

$$\Gamma(l, f) := \text{closure}[\text{Sing}(l, f) \setminus \text{Sing} f] \subset \mathbb{K}^n$$

will be called the *affine polar locus*[2] *of f relative to l.* We shall use the notation $l_H : \mathbb{K}^n \to \mathbb{K}$ for the unique linear form which defines the linear hyperplane $H = \{l_H = 0\} \subset \mathbb{K}^n$ viewed as a point in the *dual projective space* $\check{\mathbb{P}}^{n-1}$. Let also $\overline{\Gamma}(l, f)$ denote the closure of $\Gamma(l, f)$ in $\mathbb{X}_\mathbb{K}$ and let $\overline{\Gamma}(l, f)_y$ denote its germ at some point $y \in \mathbb{X}^\infty_\mathbb{K}$.

Proposition 2.1.3 Let $y = ([x_0 : x], t_0) \in \mathbb{X}^\infty \cap \overline{f^{-1}(t_0)}$. The following equality of analytic set-germs holds:

$$\Gamma(x_0, \tau)^{(i)}_y = \overline{\Gamma}(x_i, f)_y. \tag{2.2}$$

Proof Let $y = ([0 : x_1 : \cdots : x_n], t_0)$ with $x_i \neq 0$ for some fixed i. In the chart $U_i \times \mathbb{K}$, we denote $f^{(i)} := x_i^{-d}\tilde{f}(x_0, x_1, \ldots, x_n)$. Then the polar locus $\Gamma(x_0, \tau)^{(i)}_y$ is the germ at y of the analytic set $\overline{\mathcal{G}_i} \subset \mathbb{X}_\mathbb{K}$, where:

$$\mathcal{G}_i := \{([x_0 : x], t) \in \mathbb{X} \setminus \mathbb{X}^\infty \mid \frac{\partial f^{(i)}}{\partial x_1} = \cdots = \frac{\partial f^{(i)}}{\partial x_{i-1}} = $$
$$\frac{\partial f^{(i)}}{\partial x_{i+1}} \cdots = \frac{\partial f^{(i)}}{\partial x_n} = 0\} \setminus \text{Sing} f^{(i)}.$$

On the intersection of charts $(U_0 \cap U_i) \times \mathbb{K}$, the function $\frac{\partial f^{(i)}}{\partial x_j}$ differs from $\frac{\partial f^{(0)}}{\partial x_j}$ by a nowhere zero factor, for all $j \neq 0, i$. Thus $\overline{\mathcal{G}_i}$ equals $\overline{\Gamma}(x_i, f)$ and this proves (2.2). $\qquad\square$

We need the following particular case of the polar curve theorem 7.1.2:

Lemma 2.1.4 There exists an open dense set $\Omega_f \subset \check{\mathbb{P}}^{n-1}$ (respectively, a Zariski-open set in the complex case) such that, for any $H \in \Omega_f$, the critical set $\Gamma(l_H, f)$ is either a (real, resp. complex) curve or it is empty.

Definition 2.1.5 A system of coordinates $(x_1, \ldots, x_n)$ in $\mathbb{K}^n$ is called *generic* with respect to f iff $\{x_i = 0\} \in \Omega_f, \forall i$.

It follows from Lemma 2.1.4 that such systems are generic among the systems of coordinates.

In case of nonisolated $\mathcal{T}$-singularities, the general affine polar curve might be empty, as shown in Example 1.3.4. We come now to the proof of the announced general finiteness result.

Theorem 2.1.6 *If the system of coordinates $(x_1, \ldots, x_n)$ is generic with respect to f, then there is a finite number of points $y \in \mathbb{X}^\infty$ at which $\Gamma(x_0, \tau)_y^{(i)} \neq \emptyset$. At these points, the polar loci are curves.*

Proof Fix a generic system of coordinates. By Lemma 2.1.4, the set $\Gamma(x_i, f)$ is a curve or empty. By Proposition 2.1.3, we have the equality of germs $\Gamma(x_0, \tau)_y^{(i)} = \overline{\Gamma}(x_i, f)_y$. Then the assertion follows by the fact that $\cup_{i=1}^n \overline{\Gamma}(x_i, f)$ is a finite union of algebraic curves and therefore its intersection with $\mathbb{X}^\infty$ is a finite set. $\qquad \square$

Let us show that in the complex case $\mathbb{K} = \mathbb{C}$, if we know that f is t-regular at all points in some neighbourhood of a point $y \in \mathbb{X}_\mathbb{C}^\infty$, except eventually y (we say that y *is an isolated t-singularity*), then the implication (2.1) becomes an equivalence.

Theorem 2.1.7 *Let $y \in \mathbb{X}_\mathbb{C}^\infty$ where f is either t-regular or has an isolated t-singularity at infinity. Then f is t-regular at y if and only if $\Gamma(x_0, \tau)_y^{(i)} = \emptyset$ for all i.*

Proof By Lemma 2.1.2, the polar locus $\Gamma(x_0, \tau)_y^{(i)}$ is precisely the germ at y of $\pi(p^{-1}(dt))$. The projection $p : \mathbb{P}T^*_{x_0|(\mathbb{X}\backslash\mathbb{X}^\infty)\cap B} \to \mathbb{P}(\mathbb{C}^n \times \mathbb{C})^*$ is an analytic map between varieties of dimensions $n + 1$ and n, respectively. By dimension reasons (see also the proof of Theorem 7.1.2), we have either $\dim(p^{-1}(dt)) \geq 1$ or $p^{-1}(dt) = \emptyset$. Since the projection π is one-to-one over the nonsingular part of $\mathbb{X}\backslash\mathbb{X}^\infty$, we get, via Lemma 2.1.2, that either $\dim \Gamma(x_0, \tau)_y^{(i)} \geq 1$ or the polar locus $\Gamma(x_0, \tau)_y^{(i)}$ is empty.

By hypothesis, we have that y is an isolated point where the t-regularity is eventually not fulfilled, so we have the inclusion:

$$\pi(p^{-1}(dt)) \cap \mathbb{X}^\infty \cap B \subset \{y\}. \tag{2.3}$$

Now, by definition, f is not t-regular at y if and only if $(y, dt) \in \mathbb{P}(\mathcal{C}_\mathbb{C}^\infty)_{|\bar{B} \cap \mathbb{X}^\infty} = \mathbb{P}(T^*_{x_0|(\mathbb{X} \setminus \mathbb{X}^\infty) \cap B})_{|\mathbb{X}^\infty}$ and this, if and only if we have the converse, the inclusion in (2.3). This converse inclusion is equivalent to $p^{-1}(dt) \neq \emptyset$ at (y, dt), hence to $\Gamma(x_0, \tau)_y^{(i)} \neq \emptyset$. $\qquad\square$

It follows from Theorems 2.1.6 and 2.1.7 that, if f has effectively an isolated t-singularity at some point at infinity, then the germ of the polar locus at this point is a nonempty curve.

2.2 The case of isolated singularities

We consider here four types of singularities at $\mathbb{X}^\infty$, and next focus on the case where they are isolated. In the complex setting, this abuts to precise criteria for detecting atypical values.

Sets of singularities. The singular locus of the hypersurface $\mathbb{X} \subset \mathbb{P}^n \times \mathbb{K}$ is precisely:

$$\mathbb{X}_{\text{sing}} = \Sigma \times \mathbb{K},$$

where:

$$\Sigma := \{\frac{\partial f_d}{\partial x_1} = \cdots = \frac{\partial f_d}{\partial x_n} = 0, f_{d-1} = 0\}$$

is an algebraic subset of the hyperplane at infinity $H^\infty := \{x_0 = 0\} \subset \mathbb{P}^n$. The singular locus of $\mathbb{X}^\infty$ is

$$\mathbb{X}_{\text{sing}}^\infty = W \times \mathbb{K},$$

where

$$W := \{\frac{\partial f_d}{\partial x_1} = \cdots = \frac{\partial f_d}{\partial x_n} = 0\} \subset H^\infty.$$

The singularities of f, i.e. the affine set $\text{Sing} f = Z\left(\frac{\partial f}{\partial x_1}, \cdots, \frac{\partial f}{\partial x_n}\right)$, may be identified, by diagram (1.3), with the singularities of τ on $\mathbb{X} \setminus \mathbb{X}^\infty$.

Canonical stratification at infinity. There exists, cf. [Mat, §4], the least fine Whitney stratification of $\mathbb{X}$ that contains as strata $\mathbb{X} \setminus \mathbb{X}^\infty$ and $\mathbb{X}^\infty \setminus \mathbb{X}_{\text{sing}}$. This is not the canonical Whitney stratification of $\mathbb{X}$ with the largest nonsingular open subset $\mathbb{X} \setminus \mathbb{X}_{\text{sing}}$ as one of the strata, nevertheless it is a canonical Whitney stratification with two imposed strata. We shall denote it by $\mathcal{W}$ and call it the *canonical Whitney stratification at infinity of* $\mathbb{X}$.

Singularities of a map with respect to a stratification. Let us recall the definition of stratified singularities and send to Appendix A1.2 for the stratified Morse theory.

Definition 2.2.1 Let $\mathcal{X} \subset \mathbb{K}^N$ be a $\mathbb{K}$-analytic set endowed with a $\mathbb{K}$-analytic stratification $\mathcal{S} = \{S_i\}_{i \in I}$, which satisfies the Whitney (a) property.

For a $\mathbb{K}$-analytic map $g = (g_1, \ldots, g_p) : \mathcal{X} \to \mathbb{K}^p$, the *critical locus* of g with respect to $\mathcal{S}$ is the following subset of $\mathcal{X}$ (which is closed analytic, by the assumed Whitney (a)-regularity):

$$\text{Sing}_{\mathcal{S}} g := \bigcup_{i \in I} \text{Sing} \, g_{|S_i}.$$

Let $\text{Sing} \, \tau$ be the singular set of $\tau : \mathbb{X} \to \mathbb{C}$ with respect to the canonical Whitney stratification at infinity $\mathcal{W}$ defined above and denote $\text{Sing}^{\infty} \tau := \text{Sing} \, \tau \cap \mathbb{X}^{\infty}$.

We have that the projection τ is transversal to the stratum $\mathbb{X}^{\infty} \setminus \mathbb{X}_{\text{sing}}$, and therefore, since $\mathbb{X}_{\text{sing}} = \Sigma \times \mathbb{K}$, we have (see Exercises 2.3 and 2.4):

$$\text{Sing}^{\infty} \tau \subset \Sigma \times \mathbb{K}. \tag{2.4}$$

We consider the following classes of polynomials.[3]

Definition 2.2.2

 (i) f is a $\mathcal{F}$-*type* polynomial if its compactified fibres and their restrictions to the hyperplane at infinity have at most isolated singularities.
 (ii) f is a $\mathcal{B}$-*type* polynomial if its compactified fibres have at most isolated singularities.
 (iii) f is a $\mathcal{W}$-*type* polynomial if dim $\text{Sing} \, \tau \le 0$. In this case, we say that f has *isolated $\mathcal{W}$-singularities at infinity*.
 (iv) f is a $\mathcal{T}$-*type* polynomial if dim $\text{Sing}^{\infty} f \cup \text{Sing} f \le 0$, i.e. if both its singular set $\text{Sing} f$ and the set $\text{Sing}^{\infty} f$ of t-singularities at infinity consist of isolated points.

We should stress again that having singularities of the types in the above definition depends on the chosen system of coordinates $x_1, \ldots, x_n$ on $\mathbb{C}^n$.

Proposition 2.2.3 There are the following inclusions:

$$\mathcal{F}\text{-class} \subset \mathcal{B}\text{-class} \subset \mathcal{W}\text{-class} \subset \mathcal{T}\text{-class} \qquad (2.5)$$

where the last inclusion holds in the complex setting $\mathbb{K} = \mathbb{C}$.

Proof The first inclusion is clear from the definition. Let us prove the second inclusion. It is easy to show (see Exercise 2.1) that $\mathcal{B}$-type polynomials are characterized by the condition: dim Sing$f \le 0$ and dim $\Sigma \le 0$. It then follows that the set $\mathbb{X}_{\text{sing}} = \Sigma \times \mathbb{K}$ is a finite union of lines. There is an algebraic Whitney stratification at infinity $\mathcal{W}'$ of $\mathbb{X}$ (eventually more refined than the canonical one) consisting of the following strata: $\mathbb{X} \setminus \mathbb{X}^\infty$, $\mathbb{X}^\infty \setminus \mathbb{X}_{\text{sing}}$, the lines $\cup_{y \in \Sigma}\{y\} \times \mathbb{K}$ without a finite set of points, and those points. It is easy to verify the Whitney (b)-regularity (see §A1.1) between pairs of strata of this stratification, since both $\mathbb{X} \setminus \mathbb{X}^\infty$ and $\mathbb{X}^\infty \setminus \mathbb{X}_{\text{sing}}$ are connected and included in the regular part of $\mathbb{X}$. The excepted set $\mathbb{X}_{\text{sing}}$ is a disjoint union of lines and, since the stratification is algebraic, these lines verify Whitney (b)-regularity except eventually at finitely many points.

By (2.4), the singularities of τ on the strata at infinity are included into $\mathbb{X}_{\text{sing}}$. Moreover, τ is transversal to all positive-dimensional strata at infinity of $\mathcal{W}'$ since, for any point (y, t) on such a stratum, the whole line $\{y\} \times \mathbb{K}$ is contained in it and τ is the projection on $\mathbb{K}$. Then the singularities of τ with respect the canonical stratification at infinity $\mathcal{W}$ are among the point-strata of $\mathbb{X}^\infty$. It then follows that Sing $\tau = $ Sing$f \cup $ Sing$^\infty \tau$ is a finite set, so the polynomial is of $\mathcal{W}$-type.

The third inclusion follows from the following inclusion:

$$\text{Sing}^\infty f \subset \text{Sing}^\infty \tau = \text{Sing} \tau \cap \mathbb{X}^\infty. \qquad (2.6)$$

This inclusion holds over the complex numbers and its proof goes as follows. The Whitney stratification $\mathcal{W}$ is also a Thom stratification with respect to the (a_{x_0})-regularity condition. This is proved in Appendix A1.1, Theorem A1.1.7. The fact that $y \in \mathbb{X}^\infty$ is not a $\mathcal{W}$-stratified singularity means that the map τ is transversal to the stratum of $\mathcal{W}$ to which y belongs. In turn, the transversality of τ to the Thom stratification $\mathcal{W}$ at y implies the t-regularity at y. □

Examples 2.2.4
(a) Polynomials of two variables with reduced fibres (hence with dim Sing$f \le$ 0) are all of $\mathcal{F}$-type.
(b) The polynomial $h = x^3 y + x + z^2 : \mathbb{C}^3 \to \mathbb{C}$ is of $\mathcal{W}$-type but not of $\mathcal{B}$-type, since dim $\Sigma = 1$.

(c) The polynomial $g := x^2y + x : \mathbb{C}^3 \to \mathbb{C}$ has nonisolated $\mathcal{W}$-singularities at infinity (namely in the fibre $g^{-1}(0)$). It turns out that it has nonisolated $\mathcal{W}$-singularities at infinity in any coordinates.

Theorem 2.2.5 *Let f be a complex polynomial of $\mathcal{T}$-type. Then the following are equivalent:*

(a) *f is φ-controlled topologically trivial at t_0 (see Example 3.1.7 and Definition 3.1.8 for the notation and terminology).*
(b) *The difference in Euler characteristics $\chi(f^{-1}(t_0)) - \chi(f^{-1}(t))$ is equal to $(-1)^n \mu_{f^{-1}(t_0)}$, where $\mu_{f^{-1}(t_0)}$ is the sum of the Milnor numbers of the isolated singularities of the hypersurface $f^{-1}(t_0)$.*
(c) *$\Gamma(x_0, \tau)_y^{(i)} = \emptyset$, for all $y \in \tau^{-1}(t_0) \cap \mathbb{X}^\infty$ and all $i = 1, \ldots, n$.*
(d) *$f^{-1}(t_0)$ is t-regular at infinity.*
(e) *$f^{-1}(t_0)$ is ρ_E-regular at infinity.*
(f) *$f^{-1}(t_0)$ has vanishing cycles at infinity (cf. §3.3).*

Proof (a) $\Rightarrow$ (b).[4] If (b) were not true, then the equality (3.24) from Corollary 3.3.3 shows that the 'jump at infinity' $\lambda_{f^{-1}(t_0)}$ is not zero. This implies that f is not topologically trivial at infinity at t_0 (cf. Definition 1.1.5).

(b) $\Leftrightarrow$ (c) since both conditions are equivalent to $\lambda_{f^{-1}(t_0)} = 0$ by the just-mentioned equality (3.24) and by Corollary 3.3.1, respectively.

(c) $\Leftrightarrow$ (d) is Theorem 2.1.7.

(d) $\Rightarrow$ (e) is Proposition 1.2.12, holding without the assumption about $\mathcal{T}$-singularities.

(d) $\Rightarrow$ (a). This implication occurred anticipatedly in Figure 1.1. The proof does not need the assumption "isolated $\mathcal{T}$-singularities" and holds over $\mathbb{C}$ or over $\mathbb{R}$.

The function φ from Example 3.1.7 is equal to x_0 in local charts and defines $\mathbb{X}^\infty$. Then the proof of Proposition 1.2.12 applies to the function φ instead of ρ_E and shows that $f^{-1}(t_0)$ is φ-regular at infinity. We get the φ-controlled topological triviality by using Proposition 1.2.4 and Remark 1.2.5.

(e) $\Rightarrow$ (b). If the fibre $f^{-1}(t_0)$ is ρ_E-regular at infinity, then, by the proof of Proposition 1.2.4, for some small enough disk $D \in \mathbb{C}$ centered at t_0, there is $M_0 \gg 0$, such that, for all $M \geq M_0$, the levels $\rho_E = M$ are transversal to the fibres $f^{-1}(t)$, $\forall t \in D$. We therefore get the homotopy equivalence of pairs $(f^{-1}(D), f^{-1}(t)) \simeq (B_M \cap f^{-1}(D), B_M \cap f^{-1}(t))$.

Next, at each isolated singularity of the fibre $f^{-1}(t_0)$, we take a Milnor ball B_i and provided that the disk D is small enough, we have a local Milnor fibration (cf. Theorem 3.1.1):

$$f_| : B_i \cap f^{-1}(D^*) \to D^*$$

and a trivial fibration $\partial B_i \cap f^{-1}(D) \to D$. We therefore have a trivial fibration $f_| : B_M \cap f^{-1}(D) \setminus \cup_i B_i$ and by excision we get:

$$H_*(B_M \cap f^{-1}(D), B_M \cap f^{-1}(t)) \simeq \oplus_i H_*(B_i \cap f^{-1}(D), B_i \cap f^{-1}(t)).$$

The relative homologies are concentrated in dimension n and the rank of the direct sum in dimension n is precisely $\mu_{f^{-1}(t_0)}$ by definition. Since $\chi(f^{-1}(D)) = \chi(f^{-1}(t_0))$, we get our claim.

The equivalence (f) $\Leftrightarrow$ (c) will be shown by Corollary 3.3.1. $\square$

2.3 Two variables

The complex setting. Complex polynomials in two variables $f(x, y)$ with isolated singularities (i.e. such that dim Sing$f \leq 0$) are moreover of $\mathcal{T}$-type. This is due to the fact that the projective closure of any fibre intersects the infinity line in $\mathbb{P}^2$ at finitely many points. This means in particular that the equivalences stated in Theorem 2.2.5 hold with the same proofs. We give some more equivalences below.[5]

Proposition 2.3.1 Let $f : \mathbb{C}^2 \to \mathbb{C}$ be a polynomial function and let $t_0 \in \mathbb{C}$. The following are equivalent:

(a) $f^{-1}(t_0)$ is a reduced fibre and f is φ-controlled topologically trivial at t_0.
(b) $f^{-1}(t_0)$ satisfies the Malgrange condition.
(c) f is *tame* at t_0 (i.e. there is no sequence of points $x_i \in \mathbb{C}^2$ with property (L_1) of Definition 1.3.1 and such that $|\operatorname{grad} f(x_i)| \to 0$ as $i \to \infty$.)

Proof The equivalence (a) $\Leftrightarrow$ (b) is just the equivalence (a) $\Leftrightarrow$ (d) of Theorem 2.2.5, via Proposition 1.2.4. The implication (c) $\Rightarrow$ (b) is clear since "tame" obviously implies the Malgrange condition. The implication (a) $\Rightarrow$ (c) was proved by Hà H.V. [Hà] by using the Łojasiewicz exponent, see Note 1.3.3. Another proof was given by Durfee [Du2]. $\square$

Note 2.3.2 In terms of the Łojasiewicz exponent (see Note 1.3.3), the *tameness* of f at some point $y \in \mathbb{X}^\infty$ amounts to the condition $L_y(f) \geq 0$. The implication (c) $\Rightarrow$ (a) of the above result shows in particular* that, if $L_y(f) \geq 0$, then $L_y(f) \geq -1$. This implication is no more true in dimensions $n \geq 3$ (see Example 1.3.4).

---

* This is part of Hà proof in [Hà].

Tameness is not a generic property. The example $f(x, y, z) = x + x^2 y + x^4 yz$ taken from [Pa2] shows that f is not tame at all fibres $t_0 \in \mathbb{C}$, whereas all fibres, except finitely many, are ρ_E-regular at infinity (cf. Corollary 1.2.13).

We may also consider the tameness in the sense of Broughton [Br2], namely: $\exists K > 0$ such that, for any sequence $|x_i| \to \infty$, we have $|\operatorname{grad} f(x_i)| \geq K$. Then '$f$ is tame in the sense of Broughton', obviously implies 'f is tame at t, for any $t \in \mathbb{C}$', whereas the reciprocal is not true; some examples can be found in [NZ, Du2].

Remarks on polynomials with at most one atypical value. We have seen that, if a polynomial function has no critical points and all its fibres are t-regular at infinity, then there are no atypical values. In the case of two complex variables, it follows by the Abhyankar–Moh–Suzuki theorem [AM, Suz] that f is equivalent, modulo an automorphism of $\mathbb{C}^2$, to a linear form.[6]

Assume now that f is a complex polynomial and has a single atypical fibre $F_0 := f^{-1}(0)$. Take a small disk $D_0 \subset \mathbb{C}$ centred at 0. Then the restriction $f_| : \mathbb{C}^2 \setminus f^{-1}(D_0) \to \mathbb{C} \setminus D_0$ is a locally trivial fibration, hence $\mathbb{C}^2$ retracts to $f^{-1}(D_0)$. In turn, $f^{-1}(D_0)$ retracts to a tubular neighbourhood of F_0 and we get the equivalence of Euler characteristics $\chi(F_0) = \chi(\mathbb{C}^2) = 1$.

Proposition 2.3.3 Let $f : \mathbb{C}^2 \to \mathbb{C}$ be a polynomial function with at most one atypical value, say $0 \in \mathbb{C}$, and such that F_0 is reduced.

(a) If F_0 is connected and nonsingular, then f is equivalent (modulo $\operatorname{Aut} \mathbb{C}^2$) to a linear form.
(b) If F_0 is connected and irreducible, then f is equivalent (modulo $\operatorname{Aut} \mathbb{C}^2$) to $x^p + y^q$, for some relatively prime integers $p, q \geq 1$.
(c) If F_0 is nonsingular but not connected, then f is equivalent (modulo $\operatorname{Aut} \mathbb{C}^2$) to $x(1 + xh(x, y))$, for some $h \in \mathbb{C}[x, y]$.

Proof (b) is Zaidenberg–Lin result [ZL, Theorem A]. A more complete statement is [ZL, Theorem B], treating the case where F_0 is assumed to be only connected. If F_0 is connected and nonsingular, then we get (a) by applying the same theorem. Alternately, we may use the fact that F_0 is isomorphic to $\mathbb{C}$ and conclude by the Abhyankar–Moh–Suzuki theorem. An argument for the isomorphism $F_0 \simeq \mathbb{C}$ is the following. There is a compactification of F_0 to a Riemann surface M which is also nonsingular and connected. Then:

$$1 = \chi(F_0) = \chi(M) - \chi(M \setminus F_0) = 2 - 2g - s, \tag{2.7}$$

where g is the genus of M and s is the number of points in $M \setminus F_0$. This implies $g = 0, s = 1$ and therefore $M \simeq \mathbb{P}^1$ and $F_0 \simeq \mathbb{C}$. This finishes (a).

(c). We first remark that there is at least one connected component $F_{0,i}$ of F_0 such that $\chi(F_{0,i}) = 1$. Indeed, $1 = \chi(F_0) = \sum_{i \in I} \chi(F_{0,i})$, but $\chi(F_{0,i}) \leq 1$, $\forall i \in I$, by a computation such as (2.7). Since $F_{0,i}$ is nonsingular, we may apply (a) to it and we get that f is equivalent to $x \cdot v(x, y)$, modulo Aut $\mathbb{C}^2$, and further that the polynomial v is of the form $\alpha + xh(x, y)$, with α a constant $\neq 0$ and some $h \in \mathbb{C}[x, y]$. An algebraic proof of (c) can be found in [Assi]. $\qquad\square$

A more refined classification (with a longer list of cases, representing distinct topological types), generalizing in particular the result by Zaidenberg and Lin to reducible polynomials, was given by Bodin [Bo1].

Another type of classification, without condition on the number of atypical values, was initiated by Siersma and Smeltink [SiSm]. They classify polynomials of degree 4, from the point of view of the singularities at infinity.

The study of singularities at infinity might be useful for approaching the Jacobian Conjecture in two variables, which has the following equivalent formulation, cf. [LêWe, ST2]:

Conjecture 2.3.4 (Jacobian Conjecture) Let $f : \mathbb{C}^2 \to \mathbb{C}$ be a polynomial function with smooth fibres, which has t-singularities at infinity (i.e. there exists at least a fibre of f which is not t-regular at infinity). Then, for any $g \in \mathbb{C}[x, y]$, the Jacobian Jac(f, g) cannot be a nonzero constant.

Remark 2.3.5 One can prove that a polynomial f with at most one atypical nonsingular value verifies the above conjecture. In case (c) of Proposition 2.3.3 this follows from results proved by Oka [Oka1] by using Newton polygon techniques.

The real setting. We shall characterize the atypical values of polynomials $f : \mathbb{R}^2 \to \mathbb{R}$, which are not critical values.

Families of real curves. Let $X \subseteq \mathbb{R}^n$ be a smooth noncompact algebraic surface and let $f : X \to \mathbb{R}$ be the restriction of a polynomial function $F : \mathbb{R}^n \to \mathbb{R}$. Let $I :=]a, b[\subset f(X)$ be an open interval such that $I \cap \text{Atyp} f = \emptyset$. The restriction $f : f^{-1}(I) \to I$ is a C^∞ trivial fibration and it restricts to a trivial fibration on any connected component $\mathcal{Y}$ of $f^{-1}(I)$. Denote $Y_t := X_t \cap \mathcal{Y}$ and observe that Y_t is connected, $\forall t \in]a, b[$.

Definition 2.3.6 We say that a point $p \in X$ is a *limit point* of the family $\{Y_t\}_{t \in]a,b[}$ when t tends to a if there exists a sequence of points $p_k \in \mathcal{Y}, k \in \mathbb{N}$, such that p_k tends to p and $f(p_k)$ tends to a. We denote:

$$\lim_{t \to a,\, t > a} Y_t := \{p \in X \mid p \text{ is a limit point of } \{Y_t\}_t \text{ when } t \to a\}$$

and we define $\lim\limits_{t \to b,\, t<b} Y_t$ analogously.

It follows that $\lim\limits_{t \to a,\, t>a} Y_t \subseteq f^{-1}(a)$ and $\lim\limits_{t \to b,\, t<b} Y_t \subseteq f^{-1}(b)$.

Definition 2.3.7 We say that Y_t *splits at infinity* when t tends to a, $t > a$, if $\lim\limits_{t \to a,\, t>a} Y_t$ contains at least two connected components of $f^{-1}(a)$.

We say that Y_t *vanishes at infinity* if $\lim\limits_{t \to a,\, t>a} Y_t = \emptyset$.

The following lemma is straightforward from the definitions.

Lemma 2.3.8 Let $a \in f(X)$ be a regular value of f. In the above notations, we have:

(i) The limit $\lim\limits_{t \to a,\, t>a} Y_t$ is either empty or equal to the union of some connected components of $f^{-1}(a)$.

(ii) Let $\{Y_t'\}_t$ be the family of curves corresponding to some connected component $\mathcal{Y}'$. If $\left(\lim\limits_{t \to a,\, t>a} Y_t \right) \cap \left(\lim\limits_{t \to a,\, t>a} Y_t' \right) \neq \emptyset$, then $Y_t = Y_t'$, for all $t > a$ close enough to a.

Let us consider the following conditions for the fibres X_t of f:

(B) The Betti numbers of the fibre X_t are constant for t within some neighbourhood of zero.

(E) The Euler characteristic $\chi(X_t)$ is constant for t within some neighbourhood of zero.

(nV) There is no connected component of X_t, which vanishes at infinity when t tends to 0, $t < 0$ or $t > 0$.

Theorem 2.3.9 [TZ] *Let* $0 \in f(X)$ *be a regular value of* f. *Then:*

(a) *The value zero is a typical value of* f.

(b) *(B) and (nV) hold.*

(c) *(E) and (nV) hold.*

(d) *(B) and (nS) hold.*

The above criteria are natural, since all the conditions are necessary, i.e. (a)$\Rightarrow$(b), (c), (d). The criterion (c) has a striking similarity to certain criteria in the complex case, as we explain in the following.

For a family $X_t = f^{-1}(t)$ given by the fibres of a complex polynomial function $f : \mathbb{C}^2 \to \mathbb{C}$, a particular case of Theorem 2.2.5(b), proved by Hà H.V. and Lê D.T. [HàL], yields the following criterion: *a reduced fibre* X_{t_0} *is typical if and only if its Euler characteristic* $\chi(X_{t_0})$ *is equal to the Euler characteristic of a general fibre of f.* We also have the following equivalent form of the Hà–Lê criterion, cf. §3.3 and particularly Corollary 3.3.3: *A regular fibre X_t of a complex polynomial function is typical if and only if there are no vanishing cycles at infinity corresponding to this fibre.*

We observe that in both the real and the complex setting, a criterion containing the idea of 'nonvanishing' occurs. Nevertheless, in the real case the two conditions of Theorem 2.3.9(c), namely the constancy of Euler characteristic and the nonvanishing at infinity, have to be considered together; alone, neither of them implies that X_t is atypical. This is shown by Example 2.3.11: (E) holds but (nV) fails; respectively Example 2.3.12: (nV) and (nS) hold but (B) and (E) fail. Example 2.3.13 shows that (E) + (nS) is not a sufficient criterion.

Proof of Theorem 2.3.9 (b)⇒(a). Let $\mathcal{D}$ be a connected component of $f^{-1}([-\varepsilon, \varepsilon])$, $\varepsilon > 0$. Let us prove that, if (nV) holds, then $\mathcal{D}$ contains at least one connected component of X_0. We may assume without dropping generality that $f(\mathcal{D}) \supset]0, \varepsilon]$. Then take a decreasing sequence $\{\varepsilon_k\}_{k\in\mathbb{N}} \subset]0, \varepsilon[, \varepsilon \to 0$. By the nonvanishing condition (nV), we can choose a bounded sequence $\{p_k\}_{k\in\mathbb{N}}$ of points $p_k \in X_{\varepsilon_k} \cap \mathcal{D}$. There exists a convergent sub-sequence and the limit of this sub-sequence has to be on X_0. On the other hand, it is on $\mathcal{D}$ since $\mathcal{D}$ is closed. Applying now Lemma 2.3.8(i), we are done.

We next show that the restriction $f_| : \mathcal{D} \cap f^{-1}([-\varepsilon, \varepsilon]) \to [-\varepsilon, \varepsilon]$ is a C^∞ trivial fibration, for small enough ε. In case that $\mathcal{D}$ contains a 'circle' component $K \subset X_0, K \overset{\text{diffeo}}{\simeq} S^1$, we may take an open tubular neighbourhood T of K such that $T \cap X_0 = K$. Since K is compact, we get that $X_t \cap T$ is compact, for any small enough $|t| > 0$. Therefore the restriction $f_| : T \cap f^{-1}([-\varepsilon, \varepsilon]) \to [-\varepsilon, \varepsilon]$ is a proper submersion (for small enough $\varepsilon > 0$) and we may apply Ehresmann's fibration theorem to conclude that it is a C^∞ trivial fibration. It also follows that the total space $T \cap f^{-1}([-\varepsilon, \varepsilon])$ is connected, hence it coincides with $\mathcal{D}$, for small enough ε.

Consider now the case when $\mathcal{D}$ contains a 'line' component $L \subset X_0, L \overset{\text{diffeo}}{\simeq} \mathbb{R}$. We have that $b_0(\mathcal{D} \cap X_t)$ is constant, for t in some neighbourhood of 0. Indeed, this number cannot decrease as $t \to 0$, by the nonvanishing condition (nV). The condition (B) means that the sum $\sum_\mathcal{D} b_0(\mathcal{D} \cap X_t)$ over all connected components $\mathcal{D}$ is constant, which shows that $b_0(\mathcal{D} \cap X_t)$ cannot increase either. In particular the condition (nS) is fulfilled. We claim that the Betti number $m := b_0(\mathcal{D} \cap X_t)$ has to be equal to 1. The sets $\mathcal{D} \cap f^{-1}(]0, \varepsilon])$ and $\mathcal{D} \cap f^{-1}([-\varepsilon, 0[)$ contain exactly m connected components, since the restriction of f on each of them is trivial. This gives m families $\{Y'_t\}_t$ from the positive side and another m families $\{Y''_t\}_t$ from the negative side, according to Definition 2.3.6. The limit of such a family must be a connected component of X_0, by Lemma 2.3.8(i) and the above discussion. Our fixed line component L of X_0 is the limit of a certain right-side family and also the limit of a certain left-side family. Now take the union $\mathcal{V}$ of these two families together with their limit L. Then $\mathcal{V}$ is a connected component of $f^{-1}([-\varepsilon, \varepsilon])$, since it is a closed set and disjoint from the other families and their limits (by Lemma 2.3.8(ii)). Therefore $\mathcal{V}$ coincides with $\mathcal{D}$.

It remains to show that the restriction $f : \mathcal{D} \cap f^{-1}([-\varepsilon, \varepsilon]) \to [-\varepsilon, \varepsilon]$ is a C^∞ trivial fibration. This will follow from Proposition 2.3.10 below, which covers a larger setting.

(d)$\Rightarrow$(a). The proof follows the pattern of the case (b)$\Rightarrow$(a) and we leave it to the reader.

(c)$\Rightarrow$(b). First, we note that (c)$\Rightarrow$(nS). Indeed, condition (nV) implies that no 'line' component of X_t vanishes at infinity when t tends to zero, and condition (E) means that the number of 'line' components of X_t does not depend on t within some neighbourhood of zero. Now conditions (nS) and (nV) show that the number of connected components of X_t is constant for t within a neighbourhood of zero. Together with (E), this implies that (B) is satisfied. Since the implication (b)$\Rightarrow$(c) is trivial, we get (b)$\Leftrightarrow$(c).

Proposition 2.3.10 Let $M \subseteq \mathbb{R}^n$ be a smooth submanifold of dimension $m+1$ and let $g : M \to \mathbb{R}^m$ be a smooth function. Assume that the function g has no critical values and that all the fibres $g^{-1}(t)$ are diffeomorphic to $\mathbb{R}$ and closed in $\mathbb{R}^n$. Then g is a C^∞ trivial fibration and, in particular, $M \overset{\text{diffeo}}{\simeq} \mathbb{R}^{m+1}$.

Proof It suffices to show that g is locally trivial at some point $p \in \mathbb{R}^m$. Fix a point $q \in g^{-1}(p)$. Since g is a submersion, we can find a submanifold $T \subset M$ such that it is transversal to the fibres of g, that $q \in T$, and that the restriction of g to T is a C^∞ diffeomorphism onto a small open ball $B \subset \mathbb{R}^m$ centered at p.

We can take a smooth vector field $w : g^{-1}(B) \to \mathbb{R}^n$ without zeros and tangent to the fibres of g, then normalize it to a unit vector field with respect to the Riemannian metric of $\mathbb{R}^n$. The fibres being closed and diffeomorphic to $\mathbb{R}$, this vector field defines a global flow $\psi : T \times \mathbb{R} \to g^{-1}(B)$, which is a diffeomorphism. Since T is diffeomorphic to B, it follows that $g^{-1}(B)$ is diffeomorphic to $B \times \mathbb{R}$. $\qquad\square$

Let us investigate by several examples the 'vanishing' and 'splitting' phenomena occurring in Theorem 2.3.9, following [TZ]. Let $f : \mathbb{R}^2 \to \mathbb{R}$ of the following form:

$$f(x, y) := \alpha(y)x^2 + 2\beta(y)x + \gamma(y).$$

Let $A := \{y \in \mathbb{R} \mid \alpha(y) = 0\}$ and let $\varepsilon > 0$ such that $I :=]-\varepsilon, \varepsilon[$ contains only regular values of f and:

$$\beta(y) = 0 \text{ and } |\gamma(y)| \geq \varepsilon \text{ for any } y \in A. \tag{2.8}$$

Then for any $t \in I$, the equation $f = t$ in the variable x has two complex solutions $x_{1,2}(y,t)$. Let $\Delta(y,t) = \beta^2(y) - \alpha(y)\gamma(y) + t\alpha(y)$ and let us denote:

$$\mathcal{G} := \{(y,t) \in \mathbb{R}^2 \mid \Delta(y,t) \geq 0\}, \quad \mathcal{K} := \{(y,t) \in \mathbb{R}^2 \mid \Delta(y,t) = 0\},$$

$$\mathcal{L}(s) := \{(y,t) \in \mathbb{R}^2 \mid t = s\} \text{ and } \mathcal{A} := \{(y,t) \in \mathbb{R}^2 \mid y \in A\}.$$

Then $x_{1,2}(y,t) \in \mathbb{R}$ if and only if $(y,t) \in \mathcal{G}$. Moreover, if $(y,t) \in \mathcal{G}$ and y tends to a point in A, then $|x_{1,2}(y,t)|$ tends to infinity. Note also that $\mathcal{A} \subseteq \mathcal{K}$ and $\mathcal{K} \setminus \mathcal{A} \subseteq \partial \mathcal{G}$ since $\dfrac{\partial \Delta}{\partial t} = \alpha(y) \neq 0$ for $y \notin A$.

For $t_0 \in I$, the topology of the fibre $f^{-1}(t_0)$ can be described using the projections:

$$\{(x,y,t) \in \mathbb{R}^3 \mid f(x,y) = t\} \xrightarrow{\pi} \mathbb{R}^2 \to \mathbb{R}, \quad (x,y,t) \xmapsto{\pi} (y,t) \mapsto t.$$

More precisely, the connected components of the sets $\mathcal{F}(t_0) := \mathcal{G} \cap \mathcal{L}(t_0)$ and $\mathcal{F}(t_0) \setminus \mathcal{A}$ can be only segments and isolated points. By (2.8), if P is an isolated point of $\mathcal{F}(t_0)$ such that $P \in \mathcal{A}$, then $\pi^{-1}(P) = \emptyset$. Moreover, if Q is an isolated point of $\mathcal{F}(t_0)$ such that $Q \notin \mathcal{A}$, then $\pi^{-1}(Q)$ is an isolated point of $f^{-1}(t_0)$, hence a critical point of f; but t_0 is a regular value of f.

Now, consider the one-dimensional connected components of $\mathcal{F}(t_0) \setminus \mathcal{A}$. Let $\mathcal{J}$ be such a segment, let $\overline{\mathcal{J}}$ be its closure in $[-\infty, \infty] \times \mathbb{R}$ and let $n(\mathcal{J})$ be the number of endpoints of $\overline{\mathcal{J}}$ which are contained in $\mathcal{K} \setminus \mathcal{A}$.

If $(y,t_0) \in \mathcal{J} \setminus \partial \overline{\mathcal{J}}$, then $\pi^{-1}(y,t_0)$ consists of two distinct points. Now assume that $(y,t_0) \in \mathcal{J} \setminus \partial \overline{\mathcal{J}}$ tends to an endpoint Q of $\mathcal{J}$. By studying the three possibilities: $Q \in \mathcal{K} \setminus \mathcal{A}$, $Q \in \mathcal{A}$, and $Q = (\pm\infty, t_0)$, we get the following:

  (i) If $n(\mathcal{J}) = 2$, then $\pi^{-1}(\overline{\mathcal{J}})$ is diffeomorphic to a circle.
 (ii) If $n(\mathcal{J}) = 1$, then $\pi^{-1}(\overline{\mathcal{J}})$ is diffeomorphic to a line.
(iii) If $n(\mathcal{J}) = 0$, then $\pi^{-1}(\overline{\mathcal{J}})$ is diffeomorphic to a disjoint union of two lines.

The conclusion of this analysis is that we can read the topology of the fibre $f^{-1}(t)$ from the pictures of $\mathcal{G}$, $\mathcal{K}$, and $\mathcal{A}$. Then using Theorem 2.3.9, we decide whether or not 0 is an atypical value of f.

Example 2.3.11 The polynomial:

$$f(x,y) := x^2 y^3 (y^2 - 25)^2 + 2xy(y^2 - 25)(y+25) - (y^4 + y^3 - 50y^2 - 51y + 575)$$

has the property that zero is a regular and atypical value, but the Betti numbers of the fibres $f^{-1}(t)$ are constant, for $|t|$ small enough. We shall show that all these fibres have five noncompact connected components.

Let us observe that condition (2.8) is satisfied and that the set $\mathcal{K}$ contains, besides the lines in $\mathcal{A}$, the graph of the function $y \longmapsto \varphi(y) := \frac{-(y^2 - 25)^2(y+1)}{y}$, which has two connected components separated by the vertical asymptote $\{y = 0\}$. The set $\mathcal{G}$ consists of the lines in $\mathcal{A}$ and the region of the plane situated between the two connected components of the graph of φ. The only local extrema of the function φ are two local maximums, for $y = \pm 5$, and a local minimum, between -5 and -1. For $|t|$ sufficiently small, the equation $\varphi(y) = t$ has five (complex) solutions $a_j(t), j = 1, \ldots, 5$. One of these solutions, say $a_3(t)$, is a real one, for all t, while the other four are real if and only if $t \leq 0$. Assume that:

$$\lim_{t \to 0} a_1(t) = \lim_{t \to 0} a_2(t) = -5 \quad \text{and} \quad \lim_{t \to 0} a_4(t) = \lim_{t \to 0} a_5(t) = 5 \,.$$

For $|t|$ sufficiently small and $t < 0$, the set $\mathcal{F}(t) \setminus \mathcal{A}$ has five connected components and each of them corresponds to a line component in $f^{-1}(t)$. Namely, we have:

$$\mathcal{F}(t) \setminus \mathcal{A} = ([a_1(t), -5[\times \{t\}) \cup (] -5, a_2(t)] \times \{t\}) \cup$$

$$\cup ([a_3(t), 0[\times \{t\}) \cup (]0, a_4(t)] \times \{t\}) \cup ([a_5(t), \infty[\times \{t\}) \,.$$

We also have:

$$\mathcal{F}(0) \setminus \mathcal{A} = ([-1, 0[\times \{0\}) \cup (]0, 5[\times \{0\}) \cup (]5, \infty[\times \{0\}) \,.$$

Therefore, when $t < 0$ tends to zero, the line components in $f^{-1}(t)$ corresponding to the segments $([a_1(t), -5[\times \{t\}) \cup (] -5, a_2(t)] \times \{t\})$ will vanish at infinity since $\lim_{t \to 0} a_1(t) = \lim_{t \to 0} a_2(t) = -5 \in A$. Also, each of the line components in $f^{-1}(t)$ corresponding to the segments $(]0, a_4(t)] \times \{t\}) \cup ([a_5(t), \infty[\times \{t\})$ will split into two line components for $t = 0$ since $\lim_{t \to 0} a_4(t) = \lim_{t \to 0} a_5(t) = 5 \in A$.

For $|t|$ sufficiently small and $t \geq 0$, the set $\mathcal{F}(t) \setminus \mathcal{A}$ has three connected components: one corresponds to a line component in $f^{-1}(t)$ and each of the other two corresponds to two line components in $f^{-1}(t)$. Namely, we have:

$$\mathcal{F}(t) \setminus \mathcal{A} = ([a_3(t), 0[\times \{t\}) \cup (]0, 5[\times \{t\}) \cup (]5, \infty[\times \{t\}) \,.$$

Thus, for $|t|$ sufficiently small, $f^{-1}(t)$ is a disjoint union of five line components. This means that the Betti numbers of $f^{-1}(t)$ do not depend on t, if $|t|$ is sufficiently small.

On the other hand, for $\varepsilon > 0$ sufficiently small, the restrictions $f : f^{-1}(] - \varepsilon, 0[) \to] - \varepsilon, 0[$ and $f : f^{-1}([0, \varepsilon[) \to [0, \varepsilon[$ are easily seen to be C^∞ trivial fibrations, while $f : f^{-1}(] - \varepsilon, \varepsilon[) \to] - \varepsilon, \varepsilon[$ is not a topological fibration.

Example 2.3.12 The polynomial $f(x, y) := x^2 y^2 + 2xy + (y^2 - 1)^2$ has the property that conditions (nV) and (nS) are satisfied at zero, but zero is an atypical value. Namely the set $\mathcal{K}$ contains the lines in $\mathcal{A}$ and the graph of the function $\varphi(y) := y^4 - 2y^2$. This function has a local maximum, for $y = 0$, and two local minimums, for $y = \pm 1$. The set $\mathcal{G}$ consists of the lines in $\mathcal{A}$ and the region of the plane situated above the graph of φ. For $t < 0$ with $|t|$ sufficiently small, the curve $f^{-1}(t)$ has two circle components. For $t \geq 0$ sufficiently small, the curve $f^{-1}(t)$ has two line components.

Example 2.3.13 The polynomial:

$$f(x, y) := x^2 y^3 (9 - y^2)^2 + 2xy(9 - y^2)(y^3 + y + 6) + 2(y^5 - 6y^3 + 6y^2 + 25y + 6)$$

has the property that the (nS) and (E) are satisfied at zero, but zero is an atypical value. The set $\mathcal{K}$ contains the lines in $\mathcal{A}$ and also the graph of the function $y \longmapsto \varphi(y) := \frac{(y^2 - 1)(y^2 - 4)(y^2 - 9)}{y}$. This graph has two connected components, separated by the vertical asymptote $\{y = 0\}$. For $|t|$ sufficiently small, the equation $\varphi(y) = t$ has six real solutions. There exists $a \in]1, 2[$ and $b \in]2, 3[$ such that the local maxima of φ are $-b$ and a, and the local minima of φ are $-a$ and b. The set $\mathcal{G}$ consists of the lines in $\mathcal{A}$ and the region of the plane situated between the two connected components of the graph of φ. For $|t| \neq 0$ sufficiently small, the curve $f^{-1}(t)$ has a circle component and four line components. The curve $f^{-1}(0)$ has only four line components.

Example 2.3.14 Let $f : \mathbb{R}^2 \to \mathbb{R}$ be defined by $f(x, y) := 2x^2 y^3 - 9xy^2 + 12y$. Then f is a trivial C^∞ fibration because the map $F : \mathbb{R}^2 \to \mathbb{R}^2$:

$$F(x, y) := \left(f(x, y), \frac{x}{g(x, y)} \right), \quad \text{where } g(x, y) := 2x^2 y^2 - 9xy + 12 ,$$

is a diffeomorphism of order two and $F^{-1} = F$.

The polynomial $h : \mathbb{R}^2 \to \mathbb{R}$, defined by $h(x, y) := f(x + y, y) = 2x^2 y^3 + 4xy^4 - 9xy^2 + 2y^5 - 9y^3 + 12y$, has line fibres with a large perturbation in the shape of an 'S' disappearing at infinity as the value of t tends to zero. This phenomenon implies that the polar curve $\Gamma(x_0, f)$ is not empty at the fibre

$f^{-1}(0)$. However, the fibre X_0 is not atypical. This contrasts with the complex setting, where we have seen that the nonemptiness of the polar curve $\Gamma(x_0, f)$ at $f^{-1}(0)$ guarantees that the value 0 is atypical. See also Exercise 2.9.

Exercises

2.1 Show that f is a $\mathcal{B}$-type polynomial if and only if dim Sing$f \leq 0$ and dim $\Sigma \leq 0$.

2.2 Show that f is a $\mathcal{F}$-type polynomial if and only if dim Sing$f \leq 0$ and dim $W \leq 0$.

2.3 Prove: $\overline{\text{Sing} f} \cap H^\infty \subset \Sigma$ and dim Sing$f \leq 1 + \dim \Sigma$.

2.4 Prove that Sing$\tau \cap (\mathbb{X}^\infty \setminus \mathbb{X}_{\text{sing}}) = \emptyset$. In particular, if f has no $\mathcal{B}$-singularities (i.e. $\Sigma = \emptyset$), then f has no $\mathcal{W}$-singularities (i.e. Sing$^\infty \tau = \emptyset$).

2.5 Show that Theorem 2.2.5 holds under the weaker assumption dim $\mathbb{X}_{t_0} \cap$ (Sing$^\infty f \cup$ Sing$f) \leq 0$, where $\mathbb{X}_{t_0}$ is a notation for the fibre $\tau^{-1}(t_0)$.

2.6 Consider the singular surface $X := \{x^2 + y^2 - z^2 = 0\} \cup \{z = 0\} \subset \mathbb{R}^3$ and let $f = y - 2z$. Show that the curve X_t, for $t \neq 0$, is the disjoint union of a line and an oval, and that X_0 is a line. Show that in the family $\{X_t\}_t$ an oval is vanishing at the origin when t tends to zero. Show that $\chi(X_t)$ is constant.

2.7 Show that Theorem 2.3.9 still holds if we allow singularities in the total space X but we require that the singular locus of X does not intersect the fibre X_{t_0}.

2.8 Show that if we drop the algebricity condition on $X \subset \mathbb{R}^n$ but suppose that $X \subset \mathbb{R}^n$ is closed, that the Betti numbers of X_t are finite and that the set of atypical values of f is discrete, then the equivalences of Theorem 2.3.9 still hold.

2.9 Let $f : \mathbb{R}^2 \to \mathbb{R}$ be a polynomial having 0 as a regular value. Show that $0 \in$ Atyp$f$ if and only if[7] we have the condition (nV) and for any connected component $Y_t \subset X_t$, there is an even number of intersection points $a(t)_j \in \Gamma(l, f) \cap Y_t$ with the generic polar curve $\Gamma(l, f)$ such that $|a(t)_j| \to \infty$ for $t \to 0$, where $t < 0$ or $t > 0$.

3

Local and global fibrations

3.1 Fibrations

In case of a nonconstant holomorphic germ $g : (\mathbb{C}^n, 0) \to (\mathbb{C}, 0)$, $n \geq 2$, we have the following well-known *Milnor fibration** in a small enough ball B_ε.

Theorem 3.1.1 [Mi2] *For any sufficiently small $\varepsilon > 0$ and any $0 < \delta \ll \varepsilon$, the folowing restriction of g:*

$$g_| : B_\varepsilon \cap g^{-1}(D_\delta^*) \to D_\delta^* \tag{3.1}$$

is a locally trivial C^∞ fibration. Its fibre is called Milnor fibre. *The isotopy type of the fibration does not depend on the choice of the radii ε and δ.*

Moreover, if g has an isolated singularity at 0, then the restriction of g to the sphere S_ε:

$$g_| : S_\varepsilon \cap g^{-1}(D_\delta) \to D_\delta \tag{3.2}$$

is a trivial fibration. Its fibre is called the link of g *and does not depend on the choices of ε and δ.*

In case of a nonconstant polynomial $f : \mathbb{C}^n \to \mathbb{C}$, $n \geq 2$, we would look for a large-scale fibration theorem. Actually, in this global situation we encounter elements that indicate a certain duality with respect to the local situation of the Milnor fibration, namely: instead of *small* balls we work with *big* balls B_M, and instead of their *interior* we consider their *exterior*, i.e. the complement $\complement B_M$. Let us fix some value $t_0 \in \mathbb{C}$ and denote by $D_\delta^*(t_0)$ the disk of radius δ centred at t_0.

Theorem 3.1.2 *Let $f : \mathbb{C}^n \to \mathbb{C}$, $n \geq 2$, be a nonconstant polynomial. For any large enough $M \gg 0$ there exists $\delta(M) > 0$ such that, for any $\delta \leq \delta(M)$,*

---

* For singular underlying spaces, see Theorems A1.1.2 and A1.1.3 in the Appendix.

34

the following restriction of f :

$$f_1 : \mathbb{C}B_M \cap f^{-1}(D_\delta^*(t_0)) \to D_\delta^*(t_0) \tag{3.3}$$

is a locally trivial C^∞ fibration. The isotopy type of the fibration does not depend on the choice of M and δ.

If moreover the fibre $f^{-1}(t_0)$ has isolated singularities, then the restriction to the sphere:

$$f_1 : \partial B_M \cap f^{-1}(D_\delta(t_0)) \to D_\delta(t_0) \tag{3.4}$$

is a trivial C^∞ fibration.

Proof Since the fibre $f^{-1}(t_0)$ might be singular, let us fix a complex semi-algebraic Whitney stratification on it. By the Tarski–Seidenberg principle, all the levels of the distance function ρ_E, except for finitely many, are transversal to the strata of $f^{-1}(t_0)$. This means that there exists a sufficiently large value $M_0 > 0$ such that the level $\rho_E = M$ is transversal to $f^{-1}(t_0)$, for any $M \geq M_0$.

Let us then fix such a value M. We claim that there exists a small enough $\delta > 0$, depending on M, such that the level $\rho_E = M$ is transversal to the (necessarily nonsingular) fibre $f^{-1}(t)$, for any $t \in D_\delta^*$. Indeed, by Theorem A1.1.7 the Whitney stratification of the fibre $f^{-1}(t_0)$ has the Thom (a_f)-regularity property. This means that the stratified transversality of $\{\rho_E = M\}$ to the fibre $f^{-1}(t_0)$ implies its transversality to the nearby fibres. Our claim then follows.

Up to now, we have shown how to define the radii M and δ. We continue by proving the fibration assertion. We need to compactify the picture, i.e. to work with the extension $\tau : \mathbb{X} \to \mathbb{C}$ of f. We consider the restriction of the Whitney stratification $\mathcal{W}$ on $\mathbb{X}$ to the subspace $\mathbb{X} \cap \tau^{-1}(D_\delta)$. The projection:

$$\tau_1 : \mathbb{C}B_M \cap \mathbb{X} \cap \tau^{-1}(D_\delta^*) \to D_\delta^* \tag{3.5}$$

is a stratified submersion. Indeed, the fibres of τ_1 are nonsingular and they are transversal to the sphere ∂B_M. They are transversal to the strata of the Whitney stratification that are included in $\mathbb{X}^\infty$, by the Tarki–Seidenberg principle and after eventually diminishing the radius δ. The only fibre which might be not transversal to the strata within $\mathbb{X}^\infty$ is $\tau^{-1}(t_0)$, but this fibre is avoided in our fibration.

In this singular setting, the Ehresmann's theorem, which is the key ingredient in the proof of Milnor's theorem 3.1.1, has to be replaced by the Thom–Mather first isotopy lemma, see §A1.1. As a matter of fact, in the proof of this lemma (for which we refer for instance to Verdier's paper [Ve]), we construct a special *rug* vector field (in French: *rugueux*), which is continuous, tangent to strata and

C^∞ on each stratum, and also tangent to the sphere ∂B_M and C^∞ on it. It is the flow produced by this vector field that defines a local trivialization of the projection (3.5). Since the flow is tangent to strata, this is moreover a stratified trivialization, i.e. respects the strata. This implies that we still have a locally trivial fibration if we take out some strata. So, after taking out the subspace $\mathbb{X}^\infty$, which is a collection of strata, we get that (3.3), is a locally trivial C^∞ fibration. Since the flow is also tangent to the boundary $\partial B_M \cap f^{-1}(D_\delta)$, it yields the sub-fibration (3.4). This one is moreover trivial since the disk D_δ is contractible. $\square$

Definition 3.1.3 The fibre $\partial B_M \cap f^{-1}(t_0)$ of the fibration (3.4) is called *the link at infinity of $f^{-1}(t_0)$*.

The first part of the proof of Theorem 3.1.2 shows that the link at infinity $\partial B_M \cap f^{-1}(t_0)$ is independent of the radius M, provided this is large enough. The same proof also shows that the link at infinity is the same for all ρ_E-regular fibres of f (up to C^∞ diffeomorphisms). We shall call it the *generic link at infinity of f*. This is a real submanifold of $\mathbb{C}^n$, of dimension $2n - 3$ and is a global analogous of the link (or the *boundary*) of the Milnor fibre of a holomorphic germ g.[1]

The first part of Theorem 3.1.2 is a general result, it holds at any fibre of a complex polynomial. We shall show below that we can prove other types of fibration theorems if we impose some natural conditions on the singularities of f. Our aim is twofold: global fibrations and local fibrations at points $y \in \mathbb{X}^\infty$.

A Milnor type fibration on large spheres. Milnor had proved in [Mi2] a second fibration theorem for holomorphic function germs: *For small enough $\varepsilon > 0$, the mapping to the circle:*

$$\phi = g/\|g\| : \partial B_\varepsilon \setminus g^{-1}(0) \to S^1 \tag{3.6}$$

is a locally trivial C^∞ fibration.[2]

In the global setting, Némethi and Zaharia proved a similar result for a particular class of polynomials, by following closely Milnor's proof in [Mi2].

Theorem 3.1.4 [NZ] *Assume that the polynomial $f : \mathbb{C}^n \to \mathbb{C}$ is ρ_E-regular at infinity, eventually except for the value zero. Then, for any large enough $M \gg 0$, the folowing map:*

$$\Phi := f/\|f\| : \partial B_M \setminus f^{-1}(0) \to S^1 \tag{3.7}$$

is a locally trivial C^∞ fibration. The isotopy type of the fibration does not depend on the choice of M.

General fibration theorem at infinity. In the fibration Theorem 3.1.2 we have used the complement of the ball as the neighbourhood of $\mathbb{X}^\infty$. Let us show that we can use other analytic functions instead of ρ_E.

Definition 3.1.5 Let $K \in \mathbb{K}^n$ be a compact set. We say that ϕ *defines* $\mathbb{X}^\infty$ if $\phi : (\mathbb{P}_k^n \setminus K) \times \mathbb{K} \to \mathbb{R}_{\geq 0}$ is a real analytic function and $\phi^{-1}(0) = \mathbb{X}^\infty$.

In the local setting, Looijenga [Loo] showed how to use any function defining zero in order to prove a fibration result like Theorem 3.1.1. We prove here the global counterpart, which is a generalization of the fibration Theorem 3.1.2.

Theorem 3.1.6 *Let ϕ be a function defining $\mathbb{X}^\infty$. Then, for any small enough $\varepsilon > 0$, there exists $\delta(\varepsilon) > 0$, such that, for any $\delta \leq \delta(\varepsilon)$, the restriction of f:*

$$f_| : \phi^{-1}(]0, \varepsilon]) \cap f^{-1}(D_\delta^*(t_0)) \to D_\delta^*(t_0) \tag{3.8}$$

is a locally trivial C^∞ fibration and its isotopy type does not depend on the choice of ε and δ.

If the fibre $f^{-1}(t_0)$ has isolated singularities, then the restriction:

$$f_| : \phi^{-1}(\varepsilon) \cap f^{-1}(D_\delta(t_0)) \to D_\delta(t_0) \tag{3.9}$$

is a trivial C^∞ fibration.

Proof The proof follows the one of Theorem 3.1.2 since there is a faithfull analogy between ϕ and the inverse of the distance function. The Tarski–Seidenberg principle applies to the analytic function ϕ and yields that all its levels close enough to zero are transversal to the strata of $f^{-1}(t_0)$. The rest of the proof is unchanged. $\square$

We have used in the proof of Proposition 1.2.12 the function $d^\infty(x, f(x)) = 1/\rho_E^2(x)$, which extends to an analytic function defining $\mathbb{X}^\infty$. Another example of a function defining $\mathbb{X}^\infty$ is the following one, which was and will be used several times.

Example 3.1.7 Let $h : \mathbb{P}^n \times \mathbb{K} \to \mathbb{K}$ be the function which is equal to x_0 in every chart $U_i \times \mathbb{K}$. A change of chart yields $x_0^{(i)} = x_0^{(j)}/x_i^{(j)}$, which shows that h is $\mathbb{K}$-analytic. This is actually a $\mathbb{K}$-analytic section of the line bundle on $\mathbb{P}^n \times \mathbb{K}$ associated to the divisor $H^\infty \times \mathbb{K}$. Then $\varphi = \|h\|^2$ is a function defining $\mathbb{X}^\infty$.

We may now supplement the statement of Theorem 3.1.6 by the following definition and proposition, which are needed in the proof of the determinacy scheme 1.1.

Definition 3.1.8 Let ϕ be a function defining $\mathbb{X}^\infty$. We say that f is ϕ-*controlled topologically trivial at* $t_0 \in \mathbb{C}$ if for any small enough $\varepsilon > 0$ there exists $\delta(\varepsilon) > 0$ such that for any $\delta \leq \delta(\varepsilon)$, the restriction of f:

$$f_| : \phi^{-1}(]0, \varepsilon]) \cap f^{-1}(D_\delta(t_0)) \to D_\delta(t_0) \tag{3.10}$$

is a trivial C^∞ fibration and its isotopy type does not depend on the choice of ε and δ.

We also say that f has *trivial* ϕ-*controlled relative homology at infinity at* $t_0 \in \mathbb{C}$ if:

$$H_\star(\phi^{-1}(]0, \varepsilon]) \cap f^{-1}(D_\delta(t_0)), \phi^{-1}(]0, \varepsilon]) \cap f^{-1}(t)) \equiv 0.$$

In the local version, we say that f has trivial ϕ-controlled *local* relative homology at infinity at $t_0 \in \mathbb{C}$ if:

$$H_\star(B_y \cap \phi^{-1}(]0, \varepsilon]) \cap f^{-1}(D_\delta(t_0)), B_y \cap \phi^{-1}(]0, \varepsilon]) \cap f^{-1}(t)) \equiv 0,$$

for any $y \in \mathbb{X}_{t_0}^\infty \cap \mathrm{Sing}^\infty f$ and any Milnor ball $B_y \subset \mathbb{X}$ centered at y, where ε and $\delta(\varepsilon)$ are much smaller with respect to the radius of B_y.

The notion of ϕ-controlled topological triviality depends on the function ϕ. We may prove for instance (see the notation φ in Example 3.1.7): φ-controlled $\Rightarrow \rho_E$-controlled, but the converse is not true, at least locally, see Example 1.3.4.

Proposition 3.1.9 If f has trivial φ-controlled local relative homology at infinity at t_0, then f is t-regular at t_0.

Proof By its Definition 1.2.10, the space $\mathrm{Sing}^\infty f$ is a closed analytic subspace of $\mathbb{X}^\infty$. By (2.6) we have $\mathrm{Sing}^\infty f \subset \mathrm{Sing}^\infty \tau$ and by Corollary 1.2.13, the latter is included in a finite number of fibres of τ.

Suppose, that $\mathrm{Sing}^\infty f \cap \mathbb{X}_{t_0}^\infty \neq \emptyset$. If y is an isolated point of this set, then by Theorem 2.1.7 we get that $\Gamma(x_0, \tau)_y^{(i)} \neq \emptyset$. Then the proof of (a)$\Rightarrow$(c) of Theorem 2.2.5 shows that the φ-controlled relative homology in a small Milnor ball B_y at y is not trivial, since there is a jump $\lambda_y \neq 0$ (see §3.3), so we get a contradiction. Let us point out that in this case we also get the global result, so we may drop the condition 'local' from the statement of Proposition 3.1.9 and from the implication (iv) of Figure 1.1.

If dim $\mathrm{Sing}^\infty f \cap \mathbb{X}_{t_0}^\infty > 0$, then we may endow this closed analytic set with a Whitney stratification $\mathcal{E}$ such that it is a local refinement of the canonical Thom–Whitney stratification at infinity, see §2.2. Let $\mathcal{E}_0$ be the open stratum (hence of maximal dimension). Take an affine plane $N \subset \mathbb{C}^n$ such that $\text{codim} N = \dim \mathcal{E}_0$ and the projective closure $\overline{N} \subset \mathbb{P}^n$ is transversal to $\mathcal{E}_0$ at y. Then $\overline{N} \cap \mathcal{E} = \{y\}$ and consequently the polynomial function $f_{|N}$ has an isolated $\mathcal{T}$-singularity at y. It follows, as explained above, that in the transversal slice $\mathbb{X} \cap \overline{N} \times \mathbb{C}$, the φ-controlled local fibration at y has nontrivial relative homology. But the Whitney conditions imply that we have a trivial transversal fibration structure along the stratum $\mathcal{E}_0$, hence the φ-controlled local fibration at y in $\mathbb{X}$ is a trivial product and therefore has the same relative homology. This contradiction shows that $\mathcal{E}_0 = \emptyset$, and this clearly implies $\mathcal{E} = \emptyset$, which is the desired contradiction.[3]

$\square$

Monodromy. By the definition of the atypical values, any polynomial function restricts to a locally trivial fibration:

$$f_{|} : \mathbb{C}^n \setminus f^{-1}(\text{Atyp} f) \to \mathbb{C} \setminus \text{Atyp} f. \qquad (3.11)$$

The fundamental group $\pi_1(\mathbb{C} \setminus \text{Atyp} f)$ acts therefore on the generic fibre $f^{-1}(t_0)$ up to isotopy. The image of this action is called the *geometric monodromy group*.

Definition 3.1.10 Let $\gamma : (S^1, 0) \to (\mathbb{C}, t_0)$ be a simple loop at t_0, i.e. such that γ is a homeomorphism on its image. We say γ is an *admissible loop* for f if $\text{Im} \gamma \cap \text{Atyp} f = \emptyset$.

More precisely, the restriction of f over any admissible loop based at t_0 is a locally trivial fibration and yields an automorphism $M_\gamma : f^{-1}(t_0) \to f^{-1}(t_0)$, called *geometric monodromy* associated to γ. This is independent on the base point, modulo conjugacy with an element in $\pi_1(\mathbb{C} \setminus \text{Atyp} f)$. In particular, M_γ induces an automorphism on each homology group $H_*(f^{-1}(t_0); \mathbb{Z})$, called *algebraic monodromy*. By Corollary 1.2.14, $\text{Atyp} f$ is a finite set and so the algebraic monodromy group Mon_{alg} is a finitely generated subgroup of $\text{Aut}(H_*(f^{-1}(t_0); \mathbb{Z}))$. A system of generators can be given as follows: choose a base point $t_0 \notin \text{Atyp} f$ and for each point $a \in \text{Atyp} f$ a differentiable admissible loop based at t_0, such that each such loop separates the point a from the other points in $\text{Atyp} f$. This system of generators is of course not canonical.

For any admissible loop γ, there is a locally trivial fibration:

$$f_{|} : f^{-1}(\text{Im} \gamma) \to \text{Im} \gamma \qquad (3.12)$$

called the *monodromy fibration over γ* of f. In particular, for large enough radius r, the disk $D_r \subset \mathbb{C}$ centered at 0 contains $\text{Atyp} f$ and for such a disk, the locally trivial fibration: $f_{|} : f^{-1}(\partial D_r) \to \partial D_r$ is called the *monodromy fibration at infinity*.

Lemma 3.1.11 For any polynomial function $f : \mathbb{C}^n \to \mathbb{C}$ and any admissible loop γ there exists $R_0 \in \mathbb{R}_+$ such that the fibration *(3.12)* is diffeomorphic to the following one:

$$f_| : f^{-1}(\operatorname{Im} \gamma) \cap B_R \to \operatorname{Im} \gamma, \qquad (3.13)$$

for all $R \geq R_0$, where $B_R \subset \mathbb{C}^n$ is the ball centered at 0, of radius R.

Proof One should use Corollary 1.2.13 in order to find suitable radii r_0 and R_0. The details are left to the reader.   □

We shall give more results about monodromy in §5.2, §8, and §9.4.

3.2 A global bouquet theorem

We recall that f is of $\mathcal{T}$-type if it has *isolated t-singularities*, more precisely if $\dim(\operatorname{Sing}^\infty f \cup \operatorname{Sing} f) \leq 0$ (cf. Definition 2.2.2). We also recall that we only consider nonconstant polynomials in $n \geq 2$ variables.

Theorem 3.2.1 *Let* $f : \mathbb{C}^n \to \mathbb{C}$ *be a polynomial of* $\mathcal{T}$*-type. Then the general fibre of f is homotopy equivalent to a bouquet of spheres of real dimension* $n - 1$.

Here, 'general fibre' means any fibre which has no t-singularities; we know from Corollary 1.2.13 that there are only finitely many fibres containing t-singularities. In the case where f has nonisolated t-singularities at infinity, Theorem 3.2.1 may fail; for instance, in Example 2.2.4(c) the general fibre is a circle, and not a bouquet of spheres of dimension 2 as it should have been if it had isolated t-singularities. In the local setting, we have the well-known Milnor fibre theorem: [Mi2] *The Milnor fibre of a holomorphic function* $g : (\mathbb{C}^n, 0) \to (\mathbb{C}, 0)$ *with isolated singularity is homotopy equivalent to a bouquet of spheres of dimension* $n - 1$. The number of spheres, which is also equal to the dimension of the $(n-1)$th homology group of the Milnor fibre, is called *the Milnor number of g*.

Proof of Theorem 3.2.1[4] Let $\varphi = \|h\|^2$ be the function defining $\mathbb{X}^\infty$, which was introduced in Example 3.1.7, where h denotes the holomorphic function on $\mathbb{X}$, which is equal to x_0 in each chart $U_i \times \mathbb{C}$. We adopt the following simplified notations:

$$F_V := f^{-1}(V)$$
$$F_V^{\text{int}} := f^{-1}(V) \cap (\mathbb{C}^n \setminus \varphi([0, \varepsilon[))$$
$$F_V^{\text{ext}} := f^{-1}(V) \cap \varphi([0, \varepsilon]),$$

where V is some subset of $\mathbb{C}^n$ and where the choice of ε will be clear from the context. Let:

$$\mathcal{B}_f := \tau(\text{Sing}^\infty f \cup \text{Sing} f)$$

denote the set of 'bad' values.

The proof is in two steps, and the most extensive is the first one, where we show that the reduced homology $\tilde{H}_\star(F_t, \mathbb{Z})$ of a general fibre is concentrated in dimension $n-1$. The second step collects the connectivity results proved in the first step, interprets them as attaching results, and concludes via Whitehead's theorem (cf. [Sp, 7.5.9]).

Step 1. Let $t \notin \mathcal{B}_f$, which means that the fibre F_t is nonsingular and has no t-singularities at infinity. For each $b \in \mathcal{B}_f$, let $D_\delta(b)$ be a closed disk centred at b. We choose some small enough common radius δ, such that every disk $D_\delta(b)$ meets $\mathcal{B}_f$ only at the centre b. It is then clear that any smaller δ will also have this property. Let us fix a point $t_b \in \partial D_\delta(b)$. We have the following general result, which we shall exploit in our case via the inclusion (1.7), namely Atyp$f \subset \mathcal{B}_f$:

Lemma 3.2.2 Let $f : \mathbb{C}^n \to \mathbb{C}$ be any polynomial (non-constant, $n \geq 2$). Then the $\mathbb{Z}$-homology of the pair $(\mathbb{C}^n, F_t)$ has the following direct sum splitting in all dimensions:

$$H_\star(\mathbb{C}^n, F_t) \simeq \oplus_{b \in \text{Atyp}f} H_\star(F_{D_\delta(b)}, F_{t_b}). \qquad (3.14)$$

Proof Let $\{\gamma_b\}_{b \in \text{Atyp}f}$ be a system of nonintersecting simple paths in $\mathbb{C} \setminus \cup_{b \in \text{Atyp}f} D_\delta(b)$ such that γ_b joins the point $t \in \mathbb{C}$ to the point $t_b \in \partial D_\delta$, see Figure 3.1.

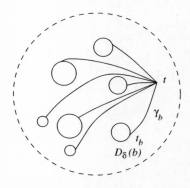

Figure 3.1. System of simple paths and disks

By the Definition 1.1.1 of Atyp$f$, the restriction:

$$f_| : \mathbb{C}^n \setminus f^{-1}(\cup_{b\in\mathrm{Atyp}f}D_\delta(b)) \to \mathbb{C} \setminus \cup_{b\in\mathrm{Atyp}f}D_\delta(b) \qquad (3.15)$$

is a locally trivial fibration. Using this trivialization and the fact that $\mathbb{C}$ contracts to $\bigcup_{b\in\mathrm{Atyp}f}(D_\delta(b) \cup \gamma_b)$, we get that, by a deformation retraction, $\mathbb{C}^n$ shrinks to $f^{-1}(\bigcup_{b\in\mathrm{Atyp}f}(D_\delta(b) \cup \gamma_b))$. We then get (3.14) by an excision argument.
$\square$

Decomposition of the relative homology $H_\star(F_{D_\delta(b)}, F_{t_b})$. Let us come back to the function φ defining $\mathbb{X}^\infty$. We take a small enough $\varepsilon > 0$ and an eventually smaller disk $D := D_\delta(b)$ centred at b such that the general fibration theorem 3.1.6 holds at the value b. The simpler notation t stands for t_b. Since the fibre F_b has isolated singularities only, by using the trivial fibration (3.4) and an excision we get the direct sum decomposition, see also Figure 3.2:

$$H_\star(F_D, F_t) \simeq H_\star(F_D^{\mathrm{ext}}, F_t^{\mathrm{ext}}) \oplus H_\star(F_D^{\mathrm{int}}, F_t^{\mathrm{int}}). \qquad (3.16)$$

In $\mathbb{C}^n \setminus \varphi([0, \varepsilon[)$ we may take small Milnor balls B_α at the isolated singularities $\alpha \in \mathrm{Sing}f \cap F_b$. Provided that the disk D is small enough, we have for each α a local Milnor fibration (cf. Theorem 3.1.1):

$$f_| : B_\alpha \cap f^{-1}(D^*) \to D^*$$

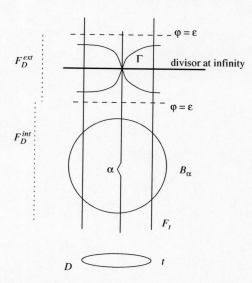

Figure 3.2. The 'interior' and 'exterior' parts

and a trivial fibration $\partial B_\alpha \cap f^{-1}(D) \to D$. Then, by applying an excision to $H_\star(F_D^{\text{int}}, F_t^{\text{int}})$, we decompose it as follows:

$$H_\star(F_D^{\text{int}}, F_t^{\text{int}}) \simeq \bigoplus_\alpha H_\star(B_\alpha \cap F_D, B_\alpha \cap F_t), \qquad (3.17)$$

where the direct sum is taken over the singular points α of F_b and where $H_\star(B_\alpha \cap F_D, B_\alpha \cap F_t)$ is the so-called *homological Milnor datum* at the point α. Since the singularities α are isolated hypersurface singularities, we may apply Milnor's bouquet theorem to conclude that the relative homology $H_\star(B_\alpha \cap F_D, B_\alpha \cap F_t)$ is concentrated in dimension $n = \dim F_D$.

Connectivity. In homotopy, the same Milnor's bouquet theorem implies that $B_\alpha \cap F_D$ is built out of $B_\alpha \cap F_t$ by attaching n-cells. It then follows that $F_D^{\text{int}} \overset{\text{ht}}{\simeq} F_t^{\text{int}} \cup (\cup_\alpha B_\alpha \cap F_t)$ is built from F_t^{int} by attaching n-cells.

The 'exterior' pair $(F_D^{\text{ext}}, F_t^{\text{ext}})$. Let us take a small open Milnor–Lê ball $B_j \subset \mathbb{P}^n \times \mathbb{C}$ at every singularity $a_j \in \overline{F_b} \cap \text{Sing}^\infty f$. The fibre F_b is t-regular at infinity at all points $\mathbb{X}^\infty \cap \overline{F_b} \setminus (\cup_j B_j)$, hence it is φ-regular at this compact set, since φ defines $\mathbb{X}^\infty$. If we denote by $\text{Sing}(\varphi, f)$ the singular locus of the map $(\varphi_{|\mathbb{C}^n}, f)$, then we have the following equality of germs:

$$\overline{\text{Sing}(\varphi, f)} = \Gamma(x_0, \tau)_y^{(i)}$$

in any chart $U_i \times \mathbb{C}$ and at any point $y \in \mathbb{X}^\infty$. Indeed, the inclusion 'left into right' is obvious and the equality follows since f is a complex analytic function. Therefore the real analytic set $\overline{\text{Sing}(\varphi, f)}$ intersects $\mathbb{X}^\infty$ exactly at the points a_j. It follows that for any small enough ε there is $\delta(\varepsilon) > 0$ such that for all disks $D := D_\delta$ with $\delta < \delta(\varepsilon)$ we have:

$$\overline{\text{Sing}(\varphi, f)} \cap \varphi^{-1}(]0, \varepsilon[) \cap F_{\partial D} \subset (\cup_j B_j) \cap F_{\partial D}.$$

Take the Whitney stratification $\mathcal{W}$ restricted to the subset $\mathbb{X} \cap \varphi^{-1}([0, \varepsilon])$ and refine it to a real Whitney stratification such that $\text{Sing}(\varphi, f)$ is a union of strata. We may construct a stratified vector field on F_D^{ext}, the flow of which defines a deformation retract of this set to its boundary $F_{\partial D}^{\text{ext}} \cup (\varphi^{-1}(\varepsilon) \cap F_D)$. This yields the following homotopy equivalence of pairs:

$$(F_D^{\text{ext}}, F_t^{\text{ext}}) \simeq (F_{\partial D}^{\text{ext}} \cup (\varphi^{-1}(\varepsilon) \cap F_D), F_t^{\text{ext}}).$$

We decompose the circle ∂D into two arcs I and J, which overlap along two small disjoint arcs and such that $t \in I \cap J$. Then we may continue with the following homotopy equivalences:

$$\simeq (F_I^{\text{ext}} \cup F_J^{\text{ext}} \cup (\varphi^{-1}(\varepsilon) \cap F_D), F_t^{\text{ext}})$$
$$\simeq (F_I^{\text{ext}} \cup F_J^{\text{ext}} \cup (\varphi^{-1}(\varepsilon) \cap F_D), F_J^{\text{ext}} \cup (\varphi^{-1}(\varepsilon) \cap F_D)),$$

since F_J^{ext} retracts to F_t^{ext} and $\varphi^{-1}(\varepsilon) \cap F_D$ retracts to $\varphi^{-1}(\varepsilon) \cap F_t$, both deformation retractions being consequences of the fibration theorem 3.1.6. We apply to the last pair the *homotopy excision*, i.e. Blakers–Massey theorem [BM] in the version of [BGr, Corollary 16.27], as in the proof of the forthcoming Proposition 10.1.6, and we get the pair:

$$(F_I^{\text{ext}}, F_{\partial I}^{\text{ext}} \cup (\varphi^{-1}(\varepsilon) \cap F_I)). \tag{3.18}$$

This excision is an isomorphism in homology. For homotopy groups, this is an isomorphism for $q < n-1$ and an epimorphism for $q = n-1$ if we show that the above pair (3.18) is $(n-1)$-connected. Let us observe that this pair is homotopy equivalent to the product $(F_t^{\text{ext}}, \varphi^{-1}(\varepsilon) \cap F_t) \times (I, \partial I)$, since $F_I^{\text{ext}} \overset{\text{ht}}{\simeq} F_t^{\text{ext}} \times I$. In homology we apply the Küneth formula and, resuming the above chain of isomorphisms, we get:

$$H_\star(F_D^{\text{ext}}, F_t^{\text{ext}}) \simeq H_{\star-1}(F_t^{\text{ext}}, \varphi^{-1}(\varepsilon) \cap F_t). \tag{3.19}$$

To show the desired $(n-1)$-connectivity of the pair (3.18) it is then sufficient to prove that the pair $(F_t^{\text{ext}}, \varphi^{-1}(\varepsilon) \cap F_t)$ is $(n-2)$-connected.

Connectivity. We may assume that B_j is included in some chart $U_{i(j)} \times \mathbb{C}$, where the $U_{i(j)}$ are not necessarily distinct, and that x_0 denotes the corresponding coordinate at infinity in each chart. Let us consider the holomorphic function h which defines the controll function φ of Example 3.1.7 and which is equal to x_0 in the coordinate charts. We have that the restriction:

$$h_| : F_t^{\text{ext}} \to D_\varepsilon^* \tag{3.20}$$

is a *proper fibration with finitely many isolated singularities* at the points $\{a_{j,k}\}_k := F_t \cap \Gamma(x_0, \tau)_{a_j}^{(j)}$. Let $\delta_{j,k} \subset D_\varepsilon^*$ be small closed disks around the images by $h_|$ of the points $a_{j,k}$ (if two $a_{j,k}$ have the same image, then the two $\delta_{j,k}$ coincide by definition) and let $\gamma_{j,k}$ be radial paths from $\delta_{j,k}$ to the boundary ∂D_ε^*. Then the pointed disk D_ε^* deforms by a retraction to $\partial D_\varepsilon^* \cup (\cup_{j,k} \delta_{j,k} \cup \gamma_{j,k})$. By using the fibration (3.20) and by applying the complex Morse theory to the holomorphic function h (see Chapter 10 for details on the Lefschetz method of slicing by pencils), we get that F_t^{ext} is obtained from $h^{-1}(\partial D_\varepsilon^*) \cap F_t$ by attaching $h^{-1}(\delta_{j,k} \cup \gamma_{j,k})$, which means the attaching of $(n-1)$-cells corresponding to the singularities $a_{j,k}$ of $h_{|F_t}$, for each k. This shows that the pair $(F_t^{\text{ext}}, \varphi^{-1}(\varepsilon) \cap F_t)$ is $(n-2)$-connected, hence the pair $(F_t^{\text{ext}}, F_{\partial t}^{\text{ext}} \cup (\varphi^{-1}(\varepsilon) \cap F_t))$ is $(n-1)$-connected. It then follows by the homotopy excision argument that the pair $(F_D^{\text{ext}}, F_t^{\text{ext}})$ is $(n-1)$-connected.

In homology, by using the fibration (3.20) we get the excision:

$$H_\star(F_t^{\text{ext}}, \varphi^{-1}(\varepsilon) \cap F_t) \simeq \oplus_{j,k} H_\star(h_|^{-1}(\delta_{j,k}), h_|^{-1}(s_{j,k})), \qquad (3.21)$$

where $s_{j,k} := \delta_{j,k} \cap \gamma_{j,k}$. Moreover the homology $H_\star(h_|^{-1}(\delta_{j,k}), h_|^{-1}(s_{j,k}))$ is concentrated in dimension $n-1 = \dim_{\mathbb{C}} F_t^{\text{ext}}$ and $\dim H_{n-1}(h_|^{-1}(\delta_{j,k}), h_|^{-1}(s_{j,k}))$ is the intersection multiplicity of F_t with $\Gamma(x_0, \tau)_{a_j}^{(j)}$ at a_j, denoted by $\text{mult}_{a_{j,k}}(F_t, \Gamma(x_0, \tau)_{a_{j,k}}^{(j)})$.

By collecting the results (3.21), (3.19), (3.17), and finally (3.16), we may conclude that $H_\star(F_D, F_t)$ is concentrated in dimension n. Via Lemma 3.2.2, this yields that $\tilde{H}_\star(F_t)$ is concentrated in dimension $n-1$ and thus concludes our proof of Step 1.

Step 2. The space F_D is obtained from F_t by attaching to it F_D^{ext} and F_D^{int}. We have proved above that the attaching of F_D^{int} amounts to attaching n-cells. Next, we have proved that the pair $(F_D^{\text{ext}}, F_t \cap F_D^{\text{ext}})$ is $(n-1)$-connected. Then, by Switzer's result [Sw, Prop. 6.13], the attaching of F_D^{ext} to F_t means attaching of cells of dimension $\geq n$.

It then follows that F_D is (up to homotopy) the result of attaching a finite number of cells of dimension $\geq n$ to F_t.

Finally, following the decomposition (3.15) and Figure 3.1, the whole space $\mathbb{C}^n = F_{\mathbb{C}}$ is obtained by attaching a finite number of cells of dimensions $\geq n$ to a general fibre F_t. Since F_t has the homotopy type of a $(n-1)$-dimensional CW-complex, we get that it is $(n-2)$-connected. It then has the same homotopy groups up to the level $n-1$ as a bouquet $\vee_\gamma S^{n-1}$. In case $n = 2$, by the classification of surfaces of surfaces we deduce that F_t is homotopy equivalent to such a bouquet of circles. When $n > 2$, by Step 1, there is a continuous map $\vee_\gamma S^{n-1} \hookrightarrow F_t$ which induces an isomorphism of homology groups at all levels. By Whitehead's theorem (cf. [Sp, 7.5.9]) this induces an isomorphism of homotopy groups, hence we get a *weak homotopy equivalence*. For CW-complexes, weak homotopy equivalence coincides with homotopy equivalence (cf. [Sp, 7.6.24]), and this finishes our proof. $\qquad\square$

3.3 Computing the number of vanishing cycles

Lemma 3.2.2 shows that for any complex polynomial f (as usual, nonconstant and in $n \geq 2$ variables) we have a direct sum decomposition of the *vanishing cycles* of the general fibre, in all dimensions:

$$\tilde{H}_{*-1}(F_t; \mathbb{Z}) \simeq \oplus_{b \in \text{Atyp} f} H_*(F_{D_\delta(b)}, F_{t_b}; \mathbb{Z}).$$

This is of course not canonical, as depending on the choices of paths in the proof of Lemma 3.2.2. In case of a $\mathcal{T}$-*type polynomial* f, we derive from this and from Theorem 3.2.1 a direct sum decomposition of the vanishing cycles, which are concentrated in dimension $n - 1$. Each term $H_n(F_{D_\delta(b)}, F_{t_b})$ is a free $\mathbb{Z}$-module, and the decompositions (3.16) and (3.17) show that we have:

$$\dim H_n(F_{D_\delta(b)}, F_{t_b}) = \mu_{F_b} + \lambda_{F_b}, \qquad (3.22)$$

where μ_{F_b} denotes the sum of Milnor numbers of f at the singularities a_i of the fibre F_b. The nonnegative integer λ_{F_b} represents the number of *vanishing cycles which are concentrated at the singularities at infinity of F_b*. Indeed, according to (3.19) and (3.21), this number splits as follows:

$$\lambda_{F_b} = \sum_{a_j \in \mathbb{X}_b \cap \text{Sing}^\infty f} \lambda_{a_j}, \qquad (3.23)$$

where

$$\lambda_{a_j} := \sum_k \dim H_{n-1}(h_|^{-1}(\delta_{j,k}), h_|^{-1}(s_{j,k})).$$

The proof of Theorem 3.2.1 yields that each term of this sum equals the corresponding intersection number $\text{mult}_{a_{j,k}}(F_t, \Gamma(x_0, \tau)_{a_{j,k}}^{(j)})$. By the intersection theory, the sum of these intersection numbers is precisely the intersection number $\text{mult}_{a_j}(\mathbb{X}_b, \Gamma(x_0, \tau)_{a_j}^{(j)})$.

We therefore get the following interpretation of the local number of vanishing cycles of a $\mathcal{T}$-type polynomial.

Corollary 3.3.1 (Polar number interpretation)
Let f be a $\mathcal{T}$-type polynomial. The number λ_{a_j} of vanishing cycles which are concentrated at the singularity at infinity $a_j \in \mathbb{X}_b \cap \text{Sing}^\infty f$ is equal to the intersection number $\text{mult}_{a_j}(\mathbb{X}_b, \Gamma(x_0, \tau)_{a_j}^{(j)})$. In particular, the latter does not depend on the chart U_j.

Definition 3.3.2 Whenever $\dim \text{Sing} f \leq 0$, we denote by μ_f *the total Milnor number*[5] of f, i.e. the sum $\sum_{c \in f(\text{Sing} f)} \mu_{F_c} = \sum_{a \in \text{Sing} f} \mu_a$ of the Milnor numbers at all the singularities of f.

Whenever dim Sing$^\infty f \leq 0$, we denote by λ_f the sum $\sum_{b \in \text{Atyp}f} \lambda_{F_b}$ and call it *the total number of vanishing cycles at infinity*.

Corollary 3.3.3 Let $f : \mathbb{C}^n \to \mathbb{C}$ be a polynomial of $\mathcal{T}$-type and let $t \notin \text{Atyp}f$. Then:

(a) dim $H_{n-1}(F_t) = \mu_f + \lambda_f$. In particular, λ_f is invariant under diffeomorphisms of $\mathbb{C}^n$.

(b) For any $b \in \text{Atyp}f$ and $t \notin \text{Atyp}f$, we have:

$$\chi(F_b) - \chi(F_t) = (-1)^n (\mu_{F_b} + \lambda_{F_b}).\tag{3.24}$$

(c) $\text{Atyp}f = \mathcal{B}_f$.

Proof (a). We first prove this for $t \in \mathcal{B}_f := \tau(\text{Sing}^\infty f \cup \text{Sing}f)$; we leave this task as an exercise. Then use (c) proved below to finish the argument.

(b). Let us remark that, despite the fact that F_b is not a deformation retract of $F_{D_\delta(b)}$, we have an equality at the level of Euler characteristics:

$$\chi(F_{D_\delta(b)}) = \chi(F_b).\tag{3.25}$$

Then our claim follows from (3.22).

(c). The inclusion $\text{Atyp}f \subset \mathcal{B}_f$ is true in whole generality, cf. (1.7). If $b \in \mathcal{B}_f$, then there are some isolated t-singularities $a_j \in X_b^\infty$ and so $\lambda_{a_j} > 0$, by Corollary 3.3.1 and since $\Gamma(x_0, \tau)_{a_j}^{(j)} \neq \emptyset$ (cf. Theorem 2.2.5). By using (b), this implies that f cannot be a topologically trivial fibration above any disk centered at b. $\qquad\square$

In the more general case of a polynomial with isolated singularities in the affine space, the t-singularities at infinity may be nonisolated and we do not have the numbers λ_p anymore. However, we can still give a meaning to the number λ_{F_a} by taking the relation (3.24) as its definition:

Definition 3.3.4 Let $f : \mathbb{C}^n \to \mathbb{C}$ be a polynomial with isolated singularities. The following number:

$$\lambda_{F_b} := (-1)^n (\chi(F_t) - \chi(F_b)) - \mu_{F_b}$$

may be called *Euler–Milnor number at infinity of the fibre F_b*.

If F_b is t-regular at infinity, then $\lambda_{F_a} = 0$. The converse is not true in general (see e.g. Example 1.3.5).

Number of vanishing cycles in the $\mathcal{W}$-class. Let f be a $\mathcal{W}$-type polynomial. At some $\mathcal{W}$-singularity $a_j \in \mathbb{X}_b \cap \operatorname{Sing} \tau$, we have a well-defined Milnor–Lê fibration of the function τ (see Theorem A1.1.2). Since the stratified singularity of τ is isolated and since the underlying space $(\mathbb{X}, a_j)$ is a hypersurface germ, we know from Lê's [Lê5, Theorem 5.1] that the Milnor fibre of τ at a_j is homotopy equivalent to a bouquet of spheres of dimension $n - 1$, and therefore the Milnor–Lê number $\mu_{a_j}(\tau)$ is well defined.

Proposition 3.3.5 (Milnor–Lê number interpretation)
At any $\mathcal{W}$-singularity $a_j \in \mathbb{X}_b \cap \operatorname{Sing} \tau$ of a $\mathcal{W}$-type polynomial, we have:

$$\lambda_{a_j} = \mu_{a_j}(\tau). \tag{3.26}$$

Proof Recall from §2.1 the definition $\Gamma(x_0, \tau)_{a_j}^{(i)} := \operatorname{closure}(\operatorname{Sing} \Phi_{|\mathbb{C}^n \cap (U_i \times \mathbb{C})} \setminus \operatorname{Sing} \tau_{|\mathbb{C}^n \cap (U_i \times \mathbb{C})}) \subset \mathbb{X}$, where $\Phi := (x_0, \tau) : (\mathbb{X} \cap (U_i \times \mathbb{C}), a_j) \to (\mathbb{C}^2, (0, b))$.

The stratified singular set of the function x_0 is $\operatorname{Sing} \mathbb{X} \subset \mathbb{X}^\infty \cap (U_i \times \mathbb{C})$ and the function τ is transversal to the strata of $\mathbb{X}^\infty$ except at finitely many points, the $\mathcal{W}$-singularities. Following [Lê2], there is a fundamental system of privileged open polydisks in $U_i \times \mathbb{C}$, centred at a_j, of the form $(P_\alpha \times D_\alpha)_{\alpha \in L}$ and a corresponding fundamental system $(D'_\alpha \times D_\alpha)_{\alpha \in L}$ of 2-disks at $(0, b)$ in $\mathbb{C}^2$, such that Φ induces, for any $\alpha \in L$, a map

$$\Phi_\alpha : \mathbb{X} \cap (P_\alpha \times D_\alpha) \cap x_0^{-1}(D'_\alpha) \to D'_\alpha \times D_\alpha,$$

which is a locally trivial fibration over $D'_\alpha \times D_\alpha \setminus \Phi_\alpha(\operatorname{Sing} \Phi)$. The Milnor fibre of the germ $\tau_| : (\mathbb{X}, a_j) \to (\mathbb{C}, b)$ is homotopy equivalent to $\mathbb{X}_t \cap (P_\alpha \times D_\alpha) \cap x_0^{-1}(D'_\alpha)$, for some $t \in D_\alpha^*$. By Lê's result [Lê5, Theorem 5.1] recalled above, this Milnor fibre is homotopy equivalent to a bouquet of $\mu_{a_j}(\tau)$ spheres of real dimension $n - 1$. See Figure 3.3.

We have a finite number of isolated singularities of the function:

$$x_0 : \mathbb{X}_t \cap (P_\alpha \times D_\alpha) \cap x_0^{-1}(D'_\alpha \setminus \{0\}) \to (D'_\alpha \setminus \{0\}) \times \{t\}, \tag{3.27}$$

which are precisely the intersections of the polar curve $\Gamma(x_0, \tau)_{a_j}^{(i)}$ with $\mathbb{X}_t \cap (P_\alpha \times D_\alpha)$. All of them project to $(D'_\alpha \setminus D''_\alpha) \times \{t\}$, where $D''_\alpha \subset D'_\alpha$ is a small enough disk centred at 0. We know that the sum of the Milnor numbers at these points of the restriction of x_0 to $\mathbb{X}_t$ is equal to λ_{a_j}, by Corollary 3.3.1.

We claim that the space $\mathbb{X}_t \cap (P_\alpha \times D_\alpha) \cap x_0^{-1}(D''_\alpha)$ is contractible, if D''_α is small enough. The stratified singularities of $x_{0|\mathbb{X}_t}$ are all on $\{x_0 = 0\}$, and

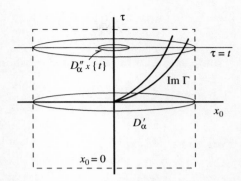

Figure 3.3. The polydisk neighbourhood and the small disk $D''_\alpha \times \{t\}$

therefore, by a result of Durfee [Du1], $\mathbb{X}_t \cap (P_\alpha \times D_\alpha) \cap x_0^{-1}(D''_\alpha)$ is a semi-algebraic tubular neighbourhood of the compact set $\mathbb{X}_t^\infty \cap (P_\alpha \times D_\alpha)$, defined by the rug function $|x_0|^2$. Consequently, these two sets are homotopy equivalent. In turn, the set $\mathbb{X}_t^\infty \cap (P_\alpha \times D_\alpha)$ coincides (up to homotopy equivalence) with the Milnor–Lê fibre at a_j of the restriction of τ to the space $\mathbb{X}^\infty$. The projection $\tau_| : \mathbb{X}^\infty \to \mathbb{C}$ is a stratified trivial fibration since $\mathbb{X}^\infty = \{f_d = 0\} \times \mathbb{C}$ is a product and may be endowed with a product Whitney stratification. This implies that the Milnor–Lê fibre* of the function $\tau_|$ is contractible, and our claim is proved.

To conclude the proof, we use complex Morse theory to show that $\mathbb{X}_t \cap (P_\alpha \times D_\alpha) \cap x_0^{-1}(D'_\alpha)$ is built, up to homotopy type, from the contractible space $\mathbb{X}_t \cap (P_\alpha \times D_\alpha) \cap x_0^{-1}(D''_\alpha)$ by attaching to it a number of λ_{a_j} cells of dimension $n-1$ corresponding to the singularities of the restriction (3.27). This shows that $\mathbb{X}_t \cap (P_\alpha \times D_\alpha) \cap x_0^{-1}(D'_\alpha)$ is homotopy equivalent to a bouquet of λ_{a_j} spheres of dimension $n-1$. $\qquad\square$

Number of vanishing cycles in the $\mathcal{B}$-class. As we have shown, $\mathcal{B}$-type polynomials are special $\mathcal{W}$-type polynomials. Inside the $\mathcal{B}$-class, we may obtain an interpretation of λ_{a_j}, which is not visible in the setting that we have used in the proof of Proposition 3.3.5. We have dim $\Sigma \leq 0$, and therefore $\mathbb{X}_{\text{sing}} = \Sigma \times \mathbb{C}$ is a finite union of lines $\cup_j \{p_j\} \times \mathbb{C}$, where $p_j \in \Sigma \subset H^\infty$ is by definition an isolated singular point of the hypersurface $\mathbb{X}_t := \tau^{-1}(t)$, for all $t \in \mathbb{C}$. Let $\mu_{p_j}(\mathbb{X}_t)$ denote its Milnor number. This number is constant, except for a finite number of values of t, where it may increase (Exercise 3.1). Let us denote by

---

* See Theorem A1.1.2 and §A1.1.

$\mu_{p_j,gen}$ the value of $\mu_{p_j}(\mathbb{X}_t)$ for generic t. At a special value $t = b$, we may then have: $\mu_{p_j}(\mathbb{X}_b) \geq \mu_{p_j,gen}$.

Proposition 3.3.6 (Jump of Milnor numbers interpretation)
At any singularity $a_j = (p_j, b) \in \mathbb{X}_b \cap \mathrm{Sing}\,\tau$ of a $\mathcal{B}$-type polynomial, we have:

$$\lambda_{a_j} = \Delta\mu_{p_j}(\mathbb{X}_b), \tag{3.28}$$

where $\Delta\mu_{p_j}(\mathbb{X}_b) := \mu_{p_j}(\mathbb{X}_b) - \mu_{p_j,gen}$ is the jump of Milnor numbers at a_j.

Proof We parallel a part of the proof of Proposition 3.3.5. Instead of the map Φ we consider here the map $\Psi := (\mathcal{F}, \tau) : U_i \times \mathbb{C} \to \mathbb{C}^2$, where $\mathcal{F}(x_0, x, t) := \tilde{f}(x_0, x) - tx_0^d$, $d = \deg f$ and τ is the projection to the t-coordinate. In the fixed chart $U_i \times \mathbb{C}$ we have $\{\mathcal{F} = 0\} = \mathbb{X}$ and $\mathrm{Sing}\,\mathcal{F} = \mathrm{Sing}\,\mathbb{X} = \Sigma \times \mathbb{C}$. We may consider the germ of $\mathbb{X}$ at a_j as the total space of a family of hypersurface germs $(\mathbb{X}_t, p_j)$ with isolated singularities, along the line germ $\{p_j\} \times (\mathbb{C}, b)$. Then the function $\mathcal{F}$ is, for each fixed t, a smoothing of $(\mathbb{X}_t, p_j)$.

The polar locus:

$$\Gamma(\mathcal{F}, \tau)_{a_j}^{(i)} := \mathrm{closure}(\mathrm{Sing}\,(\mathcal{F}, \tau)_{|U_i \times \mathbb{C}} \setminus (\Sigma \times \mathbb{C})) \tag{3.29}$$

is either the germ of a curve or it is empty. This is due to the fact that the projection τ is transversal to the nonsingular part at infinity $\mathbb{X}^\infty \setminus (\Sigma \times \mathbb{C})$, and transversal to the singular set $\Sigma \times \mathbb{C}$ except at the points a_j (see also Exercise 3.3). We may then apply a reasoning analogous to the one used before for the map Φ and show the following (in the same notations):

(i) $\tau^{-1}(t) \cap (P_\alpha \times D_\alpha) \cap \mathcal{F}^{-1}(D_\alpha'')$ is a semi-algebraic tubular neighbourhood of the compact set $\tau^{-1}(t) \cap (P_\alpha \times D_\alpha) \cap \mathcal{F}^{-1}(0) = \mathbb{X}_t \cap (P_\alpha \times D_\alpha)$ and therefore homotopy equivalent to it.

(ii) the space $\tau^{-1}(t) \cap (P_\alpha \times D_\alpha) \cap \mathcal{F}^{-1}(D_\alpha')$ is built from $\tau^{-1}(t) \cap (P_\alpha \times D_\alpha) \cap \mathcal{F}^{-1}(D_\alpha'')$ by attaching a number of n-cells. This number is equal to the local polar intersection multiplicity:

$$\gamma_{a_j} := \mathrm{mult}_{a_j}(\tau^{-1}(t), \Gamma(\mathcal{F}, \tau)_{a_j}^{(i)}).$$

We now claim that we have the equality:

$$\gamma_{a_j} = \lambda_{a_j}.$$

Indeed, the space $\tau^{-1}(t) \cap (P_\alpha \times D_\alpha) \cap \mathcal{F}^{-1}(D'_\alpha)$ is contractible since it is the Milnor fibre at a_j of the linear function τ on the nonsingular space $U_i \times \mathbb{C}$. It then follows from (i) and (ii) that $\mathbb{X}_t \cap (P_\alpha \times D_\alpha)$ is homotopy equivalent to a bouquet of γ_{a_j} spheres of dimension $n - 1$. On the other hand, since P_α is itself a polydisk of type $D_\alpha^\# \times P'_\alpha$, we have $\mathbb{X}_t \cap (P_\alpha \times D_\alpha) \overset{\text{ht}}{\simeq} \mathbb{X}_t \cap (P_\alpha \times D_\alpha) \cap x_0^{-1}(D_\alpha^\#)$ and in the proof of Proposition 3.3.5 we have shown that the latter is homotopy equivalent to a bouquet of λ_{a_j} spheres of dimension $n - 1$. Our claim is proved.

Finally, let us show that the polar intersection multiplicity γ_{a_j} equals the jump of Milnor numbers $\Delta\mu_{p_j}(\mathbb{X}_b)$. This is actually a well-known result due to Teissier [Te3], to which as we shall refer in some other sections, see §7.3, (7.11). The sketch of its proof goes as follows: Let $s \in \partial D'_\alpha$. The space $\tau^{-1}(t) \cap (P_\alpha \times D_\alpha) \cap \mathcal{F}^{-1}(D'_\alpha)$, which as we have shown is contractible, is built from the regular fibre $\tau^{-1}(t) \cap (P_\alpha \times D_\alpha) \cap \mathcal{F}^{-1}(s)$ of the restriction $\mathcal{F}_|$ to the former space, by attaching n-cells. There are γ_{a_j} cells corresponding to the singularities of $\mathcal{F}_|$ over $D'_\alpha \setminus \{0\}$ and $\mu_{p_j,gen}$ cells from the singularities of $\mathbb{X}_t \cap (P_\alpha \times D_\alpha)$. On the other hand, the space $\tau^{-1}(b) \cap (P_\alpha \times D_\alpha) \cap \mathcal{F}^{-1}(D'_\alpha)$, which is also contractible (by the local conical structure of analytic sets [BV]), is built from the regular fibre $\tau^{-1}(b) \cap (P_\alpha \times D_\alpha) \cap \mathcal{F}^{-1}(s) \overset{\text{ht}}{\simeq} \tau^{-1}(t) \cap (P_\alpha \times D_\alpha) \cap \mathcal{F}^{-1}(s)$ of the restriction $\mathcal{F}_|$, by attaching exactly $\mu_{p_j}(\mathbb{X}_b)$ cells of dimension n. This shows the equality $\mu_{p_j}(\mathbb{X}_b) = \gamma_{a_j} + \mu_{p_j,gen}$ and ends our proof. $\square$

Exercises

3.1 Show that in a holomorphic one-parameter family of hypersurface germs (like $(\mathbb{X}_t, p_j)$ for running parameter t), the Milnor number is constant except at some discrete values of the parameter, where it may increase.

3.2 Prove that $F_t \overset{\text{ht}}{\simeq} \vee S^{n-1}$ for the general fibre F_t of any complex polynomial in n variables with isolated singularities which is ρ-regular at all points $y \in \mathbb{X}^\infty$. (*Indication.* Use the definition of ρ-regularity and the proof of Theorem 3.2.1.)

3.3 Prove that the 'interior' polar locus $\Gamma(x_0, \tau)_a^{(i)}$ is isomorphic to the 'exterior' polar locus $\Gamma(\mathcal{F}, \tau)_a^{(i)}$. Deduce the following equivalence, for a $\mathcal{T}$-type polynomial f: $\text{mult}_a(\mathbb{X}_b, \Gamma(\mathcal{F}, \tau)_a^{(i)}) = 0 \Leftrightarrow f$ is locally trivial at a.

3.4 Let f be a $\mathcal{T}$-type polynomial and let $a \in \mathbb{X}_b \cap \text{Sing}^\infty f$. Prove the equalities:

$$\text{mult}_a(\mathbb{X}_b, \Gamma(x_0, \tau)_a^{(i)}) = \text{mult}_a(\mathbb{X}_b, \Gamma(\mathcal{F}, \tau)_a^{(i)}) = \lambda_a.$$

3.5 Let f be a $\mathcal{B}$-type polynomial and let $(p, b) \in \mathrm{Sing}^{\infty} f$. Show that, if $\mu_p(\mathbb{X}_t)$ is independent of t at (p, b), then f is locally trivial at (p, b). How does this constancy compare with the μ^*-constant condition at the point (p, b)? See also Parusiński's paper [Pa3] on the local μ-constant condition, and the forthcoming section §5 on the numerical control over the global triviality in a family of polynomials.

4

Families of complex polynomials

By *deformation* of the polynomial function f we mean a family of polynomial functions $f_s(x) = P(x, s)$ such that $f_0 = f$, where $P : \mathbb{C}^n \times \mathbb{C}^k \to \mathbb{C}$ and the parameter s varies in a small neighbourhood of $0 \in \mathbb{C}$. We shall assume that our family depends *holomorphically* on the parameter $s \in \mathbb{C}^k$ (unless otherwise stated). There is a well-defined general fibre G_s of the polynomial function f_s, since we have seen that the set of atypical values Atyp f_s is a finite set. When specializing f_s to f_0, the topology of the general fibre may change, and the number of atypical values may vary. First we prove some general results on the topology, then we focus on the transformation of the set of singularities Sing$f \cup$Sing$^\infty f$ in the neighbourhood of infinity, and finally we give a criterion for the topological triviality of a family.

4.1 Deformations to general hypersurfaces

Definition 4.1.1 We say that an affine hypersurface $Y \subset \mathbb{C}^n$ of degree d is *general* if its projective closure is nonsingular and transverse to the hyperplane at infinity $H^\infty = \mathbb{P}^n \setminus \mathbb{C}^n$.

Any hypersurface $Y = X_0 \subset \mathbb{C}^n$ (say of degree $d \geq 2$) can be deformed in a constant degree family $\{X_s\}_{s \in \delta}$ such that X_s is general for $s \neq 0$. Indeed, let $Y := \{f = 0\}$ and let $f_s = (1 - s)f + s(g_d - 1)$, where $g_d = x_1^d + \cdots + x_n^d$. The deformation f_s is linear in s. Then $X_s := \{f_s = 0\}$ defines a family $\{X_s\}_{s \in \delta}$, which has the desired property for a small enough disk δ centred at 0.

Computation of Euler characteristics. It is well-known that the Euler characteristic $\chi(n, d)$ of a general affine hypersurface of degree d is equal to $1 + (-1)^{n-1}(d - 1)^n$ (Exercise 4.1). For a family of hypersurfaces such that X_s is general for any $s \neq 0$ close enough to zero (as it is the case for holomorphic families), we may compute the jump of the Euler characteristic by involving the compactifications $\bar{X}_s \subset \mathbb{P}^n$ and their restrictions to the hyperplane at infinity $\bar{X}_s \cap H^\infty$, as follows. We first have the following formula, since χ is a *constructible function*:

$$\chi(X_0) - \chi(X_s) = \chi(\bar{X}_0) - \chi(\bar{X}_s) - \chi(\bar{X}_0 \cap H^\infty) + \chi(\bar{X}_s \cap H^\infty). \quad (4.1)$$

If X_0 has isolated singularities only, then some vanishing cycles of X_s vanish, as s tends to zero, at each singular point $q \in \operatorname{Sing} X_0$. Since the germ at $s = 0$ of the family X_s is a smoothing of the hypersurface germ (X_0, q), these vanishing cycles are of pure dimension $n - 1$ and their number is equal to the Milnor number $\mu_q(X_0)$ of the isolated hypersurface singularity germ (X_0, q). If moreover the projective closure $\bar{X}_0$ has isolated singularities, then we have the same property at singular points $q \in (\operatorname{Sing} \bar{X}_0) \cap H^\infty$. The jump of Euler characteristic is then:

$$\chi(\bar{X}_0) - \chi(\bar{X}_s) = (-1)^n \sum_{q \in \operatorname{Sing} \bar{X}_0} \mu_q(\bar{X}_0).$$

We may apply the same analysis to the hypersurface $\bar{X}_0 \cap H^\infty \subset \mathbb{P}^{n-1}$ of dimension $n - 2$, provided that this has isolated singularities. Note also that such a singular point may be either a singular point of $\bar{X}_0$, or a point where $\bar{X}_0$ is nonsingular but is tangent to H^∞ at p. We get:

$$\chi(\bar{X}_0 \cap H^\infty) - \chi(\bar{X}_s \cap H^\infty) = (-1)^{n-1} \sum_{p \in \operatorname{Sing}(\bar{X}_0 \cap H^\infty)} \mu_p(\bar{X}_0 \cap H^\infty).$$

Formula (4.1) then yields the following jump formula[1] for the Euler characteristic:

Lemma 4.1.2 Let $X_0 \subset \mathbb{C}^n$ be an affine hypersurface such that $\bar{X}_0$ and $\bar{X}_0 \cap H^\infty$ have isolated singularities only. Let $\{X_s\}_{s \in \delta}$ be a deformation of X_0 such that X_s is general for $s \neq 0$. Then:

$$\chi(\bar{X}_0) - \chi(\bar{X}_s)$$
$$= (-1)^n \left(\sum_{q \in \operatorname{Sing} \bar{X}_0} \mu_q(\bar{X}_0) + \sum_{p \in \operatorname{Sing}(\bar{X}_0 \cap H^\infty)} \mu_p(\bar{X}_0 \cap H^\infty) \right).$$

We may derive from it a formula for the Euler characteristic of the fibres F_t of a $\mathcal{F}$-type polynomial. In this setting we note that, in contrast to $\mu_p(\mathbb{X}_t)$, which may jump at a finite number of values of t, the number $\mu_p(\mathbb{X}_t^\infty)$ is independent of t and therefore we shall use the notation μ_p^∞ for it.

Corollary 4.1.3 Let $f : \mathbb{C}^n \to \mathbb{C}$ be a polynomial of $\mathcal{F}$-type. Then, for any $t \in \mathbb{C}$, we have:

$$\chi(F_t) = 1 + (-1)^{n-1}(d-1)^n$$
$$+ (-1)^n \sum_{q \in \text{Sing } F_t} \mu_q(F_t) + (-1)^n \sum_{p \in W} (\mu_p(\mathbb{X}_t) + \mu_p^\infty). \qquad (4.2)$$

Number of vanishing cycles. Formula (4.2) holds for any value of t. But if $t \notin \text{Atyp} f$, then F_t is the general fibre G of the $\mathcal{F}$-type polynomial f, and since we know from Theorem 3.2.1 that G is in this case a bouquet of $(n-1)$-spheres up to homotopy type, we get a formula for the $(n-1)$th Betti number:

$$b_{n-1}(G) = (d-1)^n - \sum_{p \in W} (\mu_p(\mathbb{X}_t) + \mu_p^\infty). \qquad (4.3)$$

We also have a formula for $b_{n-1}(G)$ from Corollary 3.3.3, namely: $b_{n-1}(G) = \mu_f + \lambda_f$. Plugging in the interpretation of λ_{a_i} as the jump of Milnor numbers, cf. Proposition 3.3.6, we get the second presentation of the number of vanishing cycles, which applies to $\mathcal{B}$-type polynomials:

$$b_{n-1}(G) = \sum_{b \in \text{Sing} f} \mu_{F_b} + \sum_{b \in \text{Atyp} f} \sum_{p \in \Sigma} \Delta \mu_p(\mathbb{X}_b). \qquad (4.4)$$

Generic-at-infinity polynomials. If we restrict the genericity condition to a neighbourhood of infinity, then we obtain the following class of polynomials, which we shall call the $\mathcal{G}$-class:

Definition 4.1.4 A polynomial $f : \mathbb{C}^n \to \mathbb{C}$ is called *generic-at-infinity* (or, of $\mathcal{G}$-type) if, for any $t \in \mathbb{C}$, the compactified fibre $\overline{F_t} \subset \mathbb{P}^n$ is transversal to the hyperplane at infinity H^∞.

In particular, the fibres of a $\mathcal{G}$-type polynomial may have at most isolated singularities (easy exercise). Comparing with Definition 2.2.2, we get the obvious inclusion:

$$\mathcal{G}\text{-class} \subset \mathcal{F}\text{-class}.$$

We have the following numerical characterization of the $\mathcal{G}$-class:

Proposition 4.1.5 A polynomial f of degree d is generic-at-infinity if and only if its general fibre G satisfies $b_{n-1}(G) = (d-1)^n$.

Proof If f is of $\mathcal{G}$-type, then f has only isolated singularities and $\Sigma = W = \emptyset$. The general fibre G of f is a general hypersurface. Since all the Milnor numbers $\mu_p(\mathbb{X}_t)$ and μ_p^∞ are zero, for all $(p,t) \in \mathbb{X}^\infty$, formula (4.3) shows that $b_{n-1}(G) = (d-1)^n$.

Reciprocally, let $b_{n-1}(G) = (d-1)^n$ and suppose that our polynomial f was not generic-at-infinity. We may assume that f has a nontrivial part of degree $d - 1$; if not, then we deform f by adding some homogeneous polynomial of degree $d - 1$ times a parameter. Next, we claim that we can deform f in a one-parameter family f_s such that, for all $s \neq 0$, the compactified fibres of f_s are nonsingular in the neighbourhood of the hyperplane at infinity, and tangent to this hyperplane at a single point. This can be done by deforming only the degree d homogeneous part of f, as follows. It is classical from G. Boole, cf. Cayley [Ca], that the discriminant of the space $\check{\mathbb{P}}^{N-1}$ of all the degree d hypersurfaces in $\mathbb{P}^{n-1}$ is a hypersurface $\mathcal{D} \subset \check{\mathbb{P}}^{N-1}$, where $N := \binom{n-1+d}{d}$. In turn, $\mathcal{D}$ contains a Zariski-open set, such that any of its points represents a hypersurface, which has just a single singular point of Morse type. For our specific needs, we may choose such a deformation, which, in addition, has the property that the single Morse point of $\{(f_d)_s = 0\}$ avoids the proper intersection $\{(f_{d-1})_s = 0, (f_d)_s = 0\}$ for $s \neq 0$.

Altogether, this shows the existence of a one-parameter polynomial family f_s such that $f_0 = f$, and that, for $s \neq 0$ small enough, W_s is one point and $\Sigma_s = \emptyset$. This means that f_0 deforms into polynomials f_s of $\mathcal{F}$-type. The formula (4.3) tells us that the generic fibre G_s of f_s verifies $b_{n-1}(G_s) = (d-1)^{n-1} - 1$. The semi-continuity result proved below, Proposition 4.2.1, shows that $b_{n-1}(G_0) \leq b_{n-1}(G_s)$. This yields the strict inequality $b_{n-1}(G) < (d-1)^{n-1}$, which contradicts our hypothesis. $\qquad\square$

4.2 Semi-continuity of numbers

General semi-continuity of Betti number. In case of holomorphic function germs $g_s : (\mathbb{C}^n, 0) \to (\mathbb{C}, 0)$ it is well known that the $(n-1)$th Betti number of the Milnor fibre is upper semi-continuous, i.e. does not decrease under the specialization $g_s \to g_0$. In case of a polynomial $f_s : \mathbb{C}^n \to \mathbb{C}$, the role of the Milnor fibre is played by the general fibre G_s of f_s. This is a *Stein manifold* of dimension $n - 1$ and, by Hamm's result [Hm2], has the homotopy type of a CW-complex of dimension $\leq n - 1$ (see also §6.1), which is moreover finite, since G_s is algebraic. It follows that the $(n-1)$th homology group with integer coefficients is free. Let us first prove a general specialization result, which reveals an opposite behaviour[2] with respect to the one in the local case. The exposition is based on [ST7].

Proposition 4.2.1 [ST7] Let $P : \mathbb{C}^n \times \mathbb{C}^k \to \mathbb{C}$ be any holomorphic deformation of a polynomial $f_0 := P(\cdot, 0) : \mathbb{C}^n \to \mathbb{C}$. Then the general fibre G_0 of f_0 can be naturally embedded into the general fibre G_s of f_s, for $s \neq 0$ close enough to zero.

The embedding $G_0 \subset G_s$ induces an inclusion $H_{n-1}(G_0) \hookrightarrow H_{n-1}(G_s)$ which is compatible with the intersection form. In particular $b_{n-1}(G_s) \geq b_{n-1}(G_0)$.

Proof It is of course enough to consider a one-parameter family of hypersurfaces $\{f_s^{-1}(t)\}_{s \in L} \subset \mathbb{C}^n$, for fixed t, where L denotes the germ of some line through 0. We denote by X_t the total space over L. By choosing t generic enough, we may assume that $f_s^{-1}(t)$ is a generic fibre of f_s, for s in a small enough neighbourhood L_ε of 0 in L (Exercise: prove this claim). Let $\sigma : X_t \to L_\varepsilon$ denote the projection. Now X_t is the total space of a family of nonsingular hypersurfaces. Since $\sigma^{-1}(0)$ is an affine hypersurface, by taking a large enough radius R, we get $\partial \bar{B}_{R'} \pitchfork \sigma^{-1}(0)$, for all $R' \geq R$. Moreover, the sphere $\partial \bar{B}_R$ is transversal to all nearby fibres $\sigma^{-1}(s)$, for small enough s. It follows that the projection σ from the pair of spaces $(X_t \cap (B_R \times \mathbb{C}), X_t \cap (\partial \bar{B}_R \times \mathbb{C}))$ to L_ε is a proper submersion and hence, by Ehresmann's theorem, it is a trivial fibration. This shows that $B_R \cap \sigma^{-1}(0) \overset{\text{diff}}{\simeq} B_R \cap \sigma^{-1}(s)$. It also follows that $B_R \cap \sigma^{-1}(0)$ is diffeomorphic to the general fibre G_0 and, obviously, the space $B_R \cap \sigma^{-1}(s)$ is included in the general fibre G_s. This proves our first claim.

The affine hypersurfaces $\sigma^{-1}(s)$ are finite CW-complexes of dimension $\leq n - 1$, by Hamm's result [Hm2]. By using Andreotti's and Frankel's classical argument for the distance function [AF] we show that the nonsingular hypersurface $\sigma^{-1}(s)$ is obtained from $B_R \cap \sigma^{-1}(s)$ by attaching cells of index at most $n - 1$. This implies that the nth relative homology $H_n(G_s, G_0)$ is trivial, so the second claim follows. Finally, the compatibility of the inclusion with the intersection form is a standard fact. $\qquad\square$

Semi-continuity at infinity. In the remainder we restrict to constant degree families, $\deg f_s = d$, where f_s runs within certain classes of polynomials: $\mathcal{F}$-class $\subset \mathcal{B}$-class $\subset \mathcal{W}$-class $\subset \mathcal{T}$-class, cf. Definition 2.2.2, Proposition 2.2.3. These classes have the property that the reduced homology of the general fibre G_s of f_s is concentrated in dimension $n - 1$ and is localizable at finitely many points (see also Proposition 8.1.3 for details), in the affine space (the μ-*singularities*) or in the part at infinity of the projective compactification of some fibre of f_s (the λ-*singularities*, see §3.2). We have also seen (Corollary 3.3.3) that the total number of vanishing cycles is related to the Euler characteristic

$\chi(G_s)$ by the formula $\chi(G_s) = 1 + (-1)^{n-1}(\mu(s) + \lambda(s))$. We are using here the simplified notations $\mu(s)$ and $\lambda(s)$ for $\mu(f_s)$ and $\lambda(f_s)$ respectively.

Our aim is to study the behaviour of the μ- and λ-singularities of f_s under the specialization $s \to 0$. It is easy to see that, for singularities which tend to a μ-singular point, the total number of local vanishing cycles is constant, in other words: we have *local conservation of* μ. In the global setting, the new phenomenon which occurs and which we investigate here is that some μ-singularities may change into λ-singularities, at points at infinity. In the case of a constant degree family $\{f_s(x)\}_s$ of $\mathcal{T}$-type polynomials we then have:

(a) $\mu(s)$ is lower semi-continuous under the specialization $s \to 0$.
(b) $\mu(s) + \lambda(s) \geq \mu(0) + \lambda(0)$ (by Proposition 4.2.1).

Whenever $\mu(s)$ decreases, we say that there is *loss of μ at infinity*. This may happen only when one of the two following phenomena occurs:

 (i) the modulus of some critical point tends to infinity and the corresponding critical value is bounded (see Examples 4.2.2 and 4.2.4).
(ii) the modulus of some critical value tends to infinity (cf. [ST5, Examples 8.2–3]).

In contrast to $\mu(s)$, it turns out that $\lambda(s)$ is *not semi-continuous under specialization*; it can increase or decrease (Examples 4.2.2–4.2.4). Moreover, the λ-values may behave like the critical values in case (ii) above, see Example 4.3.6.

Example 4.2.2 $f_s = (xy)^3 + sxy + x$, see Figure 4.1(a).
This is a deformation inside the $\mathcal{F}$-class, with constant $\mu + \lambda$, where λ increases. For $s \neq 0$: $\lambda = 1 + 1$ and $\mu = 1$. For $s = 0$: $\lambda = 3$ and $\mu = 0$.

Example 4.2.3 $f_s = (xy)^4 + s(xy)^2 + x$, see Figure 4.1(b).
This deformation is within the $\mathcal{F}$-class, has constant $\mu = 0$, and $\lambda(0) = 2$ at one point and $\lambda(s) = 1 + 1$ at two points at infinity which differ by the value of t only, namely $([0 : 1], s, 0)$ and $([0 : 1], s, -s^2/4)$. The notations are explained below.

Example 4.2.4 $f_s = xy^4 + s(xy)^2 + y$, see Figure 4.1(c). This is an $\mathcal{F}$-class family where λ decreases at $s = 0$. For $s \neq 0$: $\lambda = 2$ and $\mu = 5$. For $s = 0$: $\lambda = 1$ and $\mu = 0$.

In Figure 4.1 we call *splitting* the change of a λ-singularity into μ- and λ-singularities, from $s = 0$ to $s \neq 0$.

In order to understand the behaviour of $\lambda(s)$ in some more detail, we stick to the $\mathcal{B}$-class. We first need to introduce more involved notations for families.

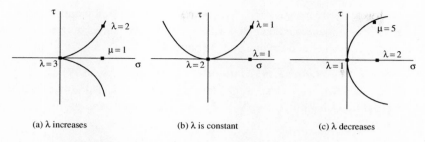

Figure 4.1. Mixed splitting in (a) and (c); pure λ-splitting in (b)

We remind that in the case of a single polynomial f, we have defined the space $\mathbb{X}$ and the proper projection $\tau : \mathbb{X} \to \mathbb{C}$.

Notations. We associate to the deformation $P(x, s) = f_s(x)$ of constant degree $= d$ the following analytic hypersurface:

$$\mathbb{Y} = \{([x : x_0], s, t) \in \mathbb{P}^n \times \mathbb{C}^k \times \mathbb{C} \mid \tilde{P}(x, x_0, s) - t x_0^d = 0\},$$

where $\tilde{P}$ denotes the homogenized of P by the variable x_0, considering that s varies in a small neighbourhood of $0 \in \mathbb{C}^k$. The projection $\tau : \mathbb{Y} \to \mathbb{C}$ to the t-coordinate extends the map P in the sense that $\mathbb{C}^n \times \mathbb{C}^k$ is embedded in $\mathbb{Y}$ (via the graph of P) and $\tau_{|\mathbb{C}^n \times \mathbb{C}^k} = P$. Let $\sigma : \mathbb{Y} \to \mathbb{C}^k$ denote the projection to the s-coordinates.

Furthermore, let $\mathbb{Y}_{s,*} := \mathbb{Y} \cap \sigma^{-1}(s)$, $\mathbb{Y}_{*,t} := \mathbb{Y} \cap \tau^{-1}(t)$ and $\mathbb{Y}_{s,t} := \mathbb{Y}_{s,*} \cap \tau^{-1}(t) = \mathbb{Y}_{*,t} \cap \sigma^{-1}(s)$. Note that $\mathbb{Y}_{s,t}$ is the closure in $\mathbb{P}^n$ of the affine hypersurface $f_s^{-1}(t) \subset \mathbb{C}^n$.

Let $\mathbb{Y}^\infty := \mathbb{Y} \cap \{x_0 = 0\} = \{P_d(x, s) = 0\} \times \mathbb{C}$ be the hyperplane at infinity of $\mathbb{Y}$, where P_d is the degree d homogeneous part of P in the variables $x \in \mathbb{C}^n$. It is important to remark that, for any fixed s, the space $\mathbb{Y}_{s,t}^\infty := \mathbb{Y}_{s,t} \cap \mathbb{Y}^\infty$ does not depend on t.

The singular locus $\text{Sing}\,\mathbb{Y} := \{x_0 = 0, \frac{\partial P_d}{\partial x}(x, s) = 0, P_{d-1}(x, s) = 0, \frac{\partial P_d}{\partial s}(x, s) = 0\} \times \mathbb{C}$ is included in $\mathbb{Y}^\infty$ and is a product-space by the t-coordinate. It depends only on the degrees d and $d - 1$ parts of P with respect to the variables x.

Let $\Sigma := \{x_0 = 0, \frac{\partial P_d}{\partial x}(x, s) = 0, P_{d-1}(x, s) = 0\} \subset \mathbb{P}^{n-1} \times \mathbb{C}^k$. The singular locus of $\mathbb{Y}_{s,*}$ is the analytic set $\Sigma_s \times \mathbb{C}$, where $\Sigma_s := \Sigma \cap \{\sigma = s\}$, and it is the union of the singularities at the hyperplane at infinity of the hypersurfaces $\mathbb{Y}_{s,t}$, for $t \in \mathbb{C}$.

We denote by $W_s := \{[x] \in \mathbb{P}^{n-1} \mid \frac{\partial P_d}{\partial x}(x, s) = 0\}$ the set of points at infinity where $\mathbb{Y}_{s,t}^{\infty}$ is singular, i.e. the points where $\mathbb{Y}_{s,t}$ is either singular or tangent to $\{x_0 = 0\}$; it does not depend on t and we have $\Sigma_s \subset W_s$.

Theorem 4.2.5 **(*Lower semi-continuity at λ-singularities*)** [ST7]
Let P be a constant degree one-parameter deformation inside the B-class. Then, locally at any λ-singularity $p \in \mathbb{Y}_{0,t}$ of f_0, we have the inequality:

$$\lambda_p(0) \leq \sum_i \lambda_{p_i}(s) + \sum_j \mu_{q_j}(s),$$

where p_i are the λ-singularities and q_j are the μ-singularities of f_s, which tend to the point p as $s \to 0$.

Proof We need to define a certain critical locus. First we endow $\mathbb{Y}$ with the coarsest Whitney stratification $\mathcal{W}$, without requiring that $\mathbb{Y}^{\infty}$ is a union of strata. Let $\Psi := (\sigma, \tau) : \mathbb{Y} \to \mathbb{C} \times \mathbb{C}$ be the projection. The *critical locus* Crit Ψ is the locus of points where the restriction of Ψ to some stratum of $\mathcal{W}$ is not a submersion; we denote by Disc $\Psi := \Psi(\text{Crit } \Psi)$ the *discriminant of* Ψ. It follows that Crit Ψ is a closed analytic set and that its affine part Crit $\Psi \cap (\mathbb{C}^n \times \mathbb{C} \times \mathbb{C})$ is the union, over $s \in \mathbb{C}$, of the affine critical loci of the polynomials f_s. Notice that both Crit Ψ and its affine part are, in general, not product spaces by the t-variable. In case of a constant degree one-parameter deformation in the B-class, the stratification $\mathcal{W}$ has a maximal stratum that contains the complement of the set $\Sigma \times \mathbb{C}$, which is of dimension ≤ 2. At any point of this complement, all the spaces $\mathbb{Y}$, $\mathbb{Y}_{s,*}$ and $\mathbb{Y}_{s,t}$ are nonsingular in the neighbourhood of infinity. In particular, Crit $\Psi \cap (\mathbb{Y}^{\infty} \setminus \Sigma \times \mathbb{C}) = \emptyset$. Since our deformation is within the B-class, it follows that the affine part Crit $\Psi \cap (\mathbb{C}^n \times \mathbb{C} \times \mathbb{C})$ is of dimension at most one. The map Ψ is submersive over some Zariski-open subset of any two-dimensional stratum included in $\Sigma \times \mathbb{C}$. Consequently the part at infinity of Crit Ψ has dimension < 2. We conclude altogether that dim Crit $\Psi \leq 1$.

In general, the functions σ and τ do not have isolated singularity with respect to our stratification $\mathcal{W}$. Nevertheless, except of finitely many values $\varepsilon \in \mathbb{C}$, the function $\sigma + \varepsilon \tau$ has an isolated singularity at p with respect to the stratification $\mathcal{W}$. Let us fix some ε close to zero and consider the local application $\Psi_{\varepsilon} = (\sigma + \varepsilon \tau, \tau): \mathcal{N} \to \mathbb{C}^2$ for some good neighbourhood $\mathcal{N} \subset \mathbb{Y}$ of $p \in \mathbb{Y}_{0,t}^{\infty}$.

The germ of the space $\mathcal{N} \cap (\sigma + \varepsilon \tau)^{-1}(0)$ is a germ of a complete intersection at p and the function $\tau_| : \mathcal{N} \cap (\sigma + \varepsilon \tau)^{-1}(0) \to \mathbb{C}$ has an isolated singularity at p. We may apply the stratified bouquet theorem in this setting (see Lê's theorems in the Appendix, §A1.1 and A1.1) and get that the Milnor–Lê fibre of $\tau_|$ is homotopy equivalent to a bouquet of spheres $\bigvee S^{n-1}$. It follows that the general fibre of Ψ_ε, which is $\mathcal{N} \cap \Psi_\varepsilon^{-1}(s, t)$, for some $(s, t) \notin \mathrm{Disc}\, \Psi$, is homotopy equivalent to the same bouquet $\bigvee S^{n-1}$; let ρ denote the number of S^{n-1} spheres in this bouquet.

On the other hand, the Milnor–Lê fibre at p of the function $\sigma + \varepsilon\tau$ is homotopy equivalent to a bouquet $\bigvee S^n$, by the same result *loc.cit.*; let ν denote the number of n-spheres.

In the remainder, we shall use the local Morse–Lefschetz theory (see also § 6.1 for the global affine theory) in order to count the vanishing cycles of dimension $n - 1$ in two ways (similarly to the counting performed in the proof of Proposition 3.3.5), namely as follows:

(I) along $(\sigma + \varepsilon\tau) = 0$,
(II) along $(\sigma + \varepsilon\tau) = u$, for $u \neq 0$ close enough to zero,

and we compare the results. Figure 4.2 is a guide to these computations; in the picture, the germ of the discriminant locus $\mathrm{Disc}\, \Psi$ at $\Psi(p)$ consists of the two axes and the germ of a curve.

(I). We start with the fibre $\mathcal{N} \cap \Psi_\varepsilon^{-1}(0, \delta)$, where δ is close enough to zero. To obtain $\mathcal{N} \cap (\sigma + \varepsilon\tau)^{-1}(0)$, which is contractible, one attaches to $\mathcal{N} \cap \Psi_\varepsilon^{-1}(0, \delta)$ a certain number of n cells corresponding to the vanishing cycles at infinity, as $t \to 0$, in the family of fibres $\Psi_\varepsilon^{-1}(0, t)$. This is exactly the number ρ defined above and it is the sum of two numbers, corresponding to the attaching, which is done in two steps, as we detail in the following. One is the number of cycles in $\mathcal{N} \cap \Psi_\varepsilon^{-1}(s, \delta)$, vanishing, as $s \to 0$, at points that tend to p when δ tends to zero; we denote this number by ξ. The other number is the number of cycles

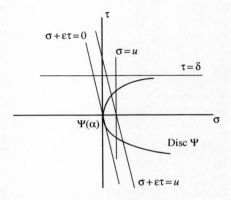

Figure 4.2. Two ways of counting vanishing cycles

in $\mathcal{N} \cap \Psi^{-1}(0, t)$, vanishing as $t \to 0$; this number is $\lambda_p(0)$, by definition. We thus have the equality: $\lambda_p(0) + \xi = \rho$.

(II). We start with the fibre $\mathcal{N} \cap \Psi_\varepsilon^{-1}(u, \delta)$, which is homeomorphic to $\mathcal{N} \cap \Psi_\varepsilon^{-1}(0, \delta)$ and to $\mathcal{N} \cap \Psi^{-1}(u, \delta)$. The Milnor fibre $\mathcal{N} \cap \{\sigma + \varepsilon\tau = u\}$ cuts the critical locus Crit Ψ at certain points p_k. The number of points, counted with multiplicities, is equal to the local intersection number $\text{mult}_p(\{\sigma + \varepsilon\tau = 0\}, \text{Crit } \Psi)$. We now use the pencil of (σ, τ)-fibres of the slice $\{\sigma + \varepsilon\tau = u\}$ and we get, by Morse–Lefschetz theory, that the space $\mathcal{N} \cap \{\sigma + \varepsilon\tau = u\}$ is homotopy equivalent to the fibre $\mathcal{N} \cap \Psi_\varepsilon^{-1}(u, \delta)$ to which we attach a number of n-cells corresponding to the vanishing cycles at Crit $\Psi \cap \{\sigma + \varepsilon\tau = u\} \cap \{\sigma = 0\}$, which is just the number ξ defined above, and to the vanishing cycles at points $\{\sigma + \varepsilon\tau = u\} \cap \overline{\text{Crit } \Psi \setminus \{\sigma = 0\}}$. For any ε small enough we have the inequality:

$$\text{mult}_p(\{\sigma + \varepsilon\tau = 0\}, \overline{\text{Crit } \Psi \setminus \{\sigma = 0\}}) \leq \text{mult}_p(\{\sigma = 0\}, \overline{\text{Crit } \Psi \setminus \{\sigma = 0\}}).$$

By the Morse–Lefschetz theory, in order to get $\mathcal{N} \cap \{\sigma = u\}$, we have to attach to $\mathcal{N} \cap \Psi^{-1}(u, \delta)$ a number of n-cells corresponding to the vanishing cycles at points p_i and q_j. By our hypothesis, this number is $\sum_i \lambda_{p_i}(u) + \sum_j \mu_{q_j}(u)$. We get the inequality: $\rho + \nu \leq \xi + \sum_i \lambda_{p_i}(u) + \sum_j \mu_{q_j}(u)$.

By collecting the results obtained at steps (I) and (II), we obtain:

$$\lambda_p(0) = \rho - \xi \leq \rho + \nu - \xi \leq \sum_i \lambda_{p_i}(u) + \sum_j \mu_{q_j}(u), \qquad (4.5)$$

which proves our claim.   □

4.3 Local conservation

Deformations with constant $\mu + \lambda$. The inequality (4.5) may become an equality under certain conditions; we then say that there is *local conservation* of the quantity of singularity.

Corollary 4.3.1 Let $P(x, s)$ be a constant degree deformation inside the $\mathcal{B}$-class such that $\mu(s) + \lambda(s)$ is constant for s in some neighbourhood of zero. Then:

(a) When $s \to 0$, there cannot be loss of μ or of λ such that the modulus of the corresponding atypical value tends to infinity.

(b) λ is upper semi-continuous, i.e. $\lambda(s) \leq \lambda(0)$.
(c) there is local conservation of $\mu + \lambda$ at any λ-singularity of f_0.

Proof (a). If there is loss of μ or of λ, then this loss must be compensated by the increase of λ at some singularity at infinity of f_0. But Theorem 4.2.5 shows that the local $\mu + \lambda$ cannot increase when $s \to 0$.
(b). This is clear since $\mu(s) + \lambda(s)$ is constant and $\mu(s)$ can only decrease when $s \to 0$.
(c). The global conservation of $\mu + \lambda$ together with the local semi-continuity (Theorem 4.2.5) imply local conservation. $\qquad\square$

Persistence of λ-singularities. We define below another class, between the $\mathcal{B}$-class and the $\mathcal{F}$-class, where we may study the exchanges $\mu \mapsto \lambda$ described in Theorem 4.2.5 in some more depth.

For a deformation $\{f_s\}_s$ inside the $\mathcal{B}$-class, the compactified fibres of f_s have only isolated singularities. Let us take some point $z(0) \in \Sigma_0$. Fix some value $t \notin \mathrm{Atyp} f_0$ and assume without dropping the generality that $t \notin \mathrm{Atyp} f_s$ for all small enough s. Then the Milnor number of $\mathbb{Y}_{0,t}$ at $(z(0), t)$ is larger or equal to the sum of the Milnor numbers of $\mathbb{Y}_{s,t}$ at the points $(z_i(s), t) \in \Sigma_s \times \{t\}$ such that $z_i(s) \to z(0)$.

(*) If the above local sum of Milnor numbers is conserved, then there is a unique singular point $z(s) \to z(0)$, by the *nonsplitting principle* (see page 68).

Definition 4.3.2 We say that a constant degree deformation $P(x, s)$ inside the $\mathcal{B}$-class has *constant generic singularity type at infinity* at some point $(z(0), t) \in \Sigma_0 \times \mathbb{C}$ when the condition (*) holds. If this holds for all points in $\Sigma_0 \times \mathbb{C}$, we say that P is a deformation with *constant generic singularity type at infinity*, briefly a *cgst-deformation*.

We prove below a result on λ-singularities and we refer to Exercises 4.5, 4.6 for more issues on *cgst-deformations*:

Theorem 4.3.3 [ST7] *Let $P(x, s)$ be a constant degree deformation, inside the $\mathcal{B}$-class, with constant generic singularity type at infinity. Then:*

(a) *the λ-singularities of f_0 are locally persistent in f_s.*
(b) *a λ-singularity of f_0 cannot split such that two or more λ-singularities belong to the same fibre of f_s.*

Part (a) means that a λ-singularity cannot completely disappear when deforming f_0, but of course it may split or its type may change, see Figure 4.1. Nevertheless a λ-singularity may split into several λs along a line $\{z(s)\} \times \mathbb{C}$, in disjoint fibres, see Example 4.2.3.

Proof (a). Let $(z, t_0) \in \Sigma_0 \times \mathbb{C}$ be a λ-singularity of f_0. Let us denote by $\mathcal{G}(y, s, t)$ the localization of the map $\tilde{P}(x, x_0, s) - t x_0^d$ at the point $(z, 0, t_0) \in \mathbb{Y}$, where y_0 is the coordinate defining the hyperplane at infinity of $\mathbb{P}^n$. The idea is to consider the two-parameter family of functions $\mathcal{G}_{s,t} \colon \mathbb{C}^n \to \mathbb{C}$, where $\mathcal{G}_{s,t}(y) = \mathcal{G}(y, s, t)$. Then $\mathcal{G}(y, s, t)$ is the germ of a deformation of the function $\mathcal{G}_{0,t_0}(y)$.

We consider the germ at $(z, 0, t_0)$ of the singular locus Γ of the map $(\mathcal{G}, \sigma, \tau) \colon \mathbb{C}^n \to \mathbb{C}^3$, which is the union of the singular loci of the functions $\mathcal{G}_{s,t}$, for varying s and t. We claim that Γ and the discriminant $D := \operatorname{Im}\Gamma \subset \mathbb{C}^3$ are both surfaces, in the sense that every irreducible component is a surface. Moreover, we claim that the projections $\Gamma \overset{(y_0, \sigma, \tau)}{\to} D$ and $D \overset{(s,t)}{\to} \mathbb{C}^2$ are finite (ramified) coverings.

These claims all follow from the following standard fact: the local Milnor number conserves in *deformations of functions*. In our case, the function germ $\mathcal{G}_{0,t_0}$ with Milnor number μ_0 deforms into a function $\mathcal{G}_{s,t}$ with finitely many isolated singularities, and the total Milnor number is conserved, for any couple (s, t) close to $(0, t_0)$.

Let us now remark that the germ of $\Sigma \times \mathbb{C}$ at $(z, 0, t_0)$ is a union of components of Γ and projects by (y_0, s, t) to the plane $D_0 := \{y_0 = 0\}$ of $\mathbb{C}^3$. The inclusion $D_0 \subset D$ cannot be an equality, since, by hypothesis, we have a jump $\lambda = \Delta\mu_{z,0}(t_0) > 0$ at the point of origin $(z, 0, t_0)$, and we may apply the above argument on the conservation of the total Milnor number. So there must exist some other components of D. Every such component is a surface in $\mathbb{C}^3$ and has to intersect the plane $D_0 \subset \mathbb{C}^3$ along a curve. Then, for every point (s', t') of such a curve, the sum of Milnor numbers of the function $\mathcal{G}_{s',t'}$ at points on the hypersurface $\{y_0 = 0\}$ that tend to the original point $(z, 0, t_0)$ when $s' \to 0$ is therefore strictly higher than the one computed for a generic point (s, t) of the plane D_0. Therefore our claim (a) will be proved if we show the following two facts:

(i) the singularities of $\mathcal{G}_{s',t'}$, which are on the hypersurface $\{y_0 = 0\}$ and which tend to the original point $(z, 0, t_0)$ when $s' \to 0$, are included in $\mathbb{Y} = \{\mathcal{G} = 0\}$;

(ii) there exists at least a component $D_i \subset D$ such that $D_i \cap D_0 \neq D \cap \{s = 0\}$.

To show (i), let $g_k(y, s)$ denote the degree k part of P after localizing $\tilde{P}$ at $(z, 0)$ and note that $\mathcal{G}(y, s, t) = g_d(y, s) + y_0(g_{d-1}(y, s) + \cdots) - t y_0^d$. Then observe that the set:

$$\Gamma \cap \{y_0 = 0\} = \left\{ \frac{\partial g_d}{\partial y} = 0, g_{d-1} = 0 \right\} \tag{4.6}$$

does not depend on the variable t and its slice by $\{\sigma = s, \tau = t\}$ consists of finitely many points. These points may fall into two types: (I) points on $\{g_d = 0\}$, and therefore on $\{\mathcal{G} = 0\}$, and (II) points not on $\{g_d = 0\}$. We show that type (II) points do not actually occur. This is a consequence of our hypothesis *cgst*-deformation, as follows. By choosing a generic $\hat{t}$ such that $\hat{t} \notin \mathrm{Atyp}f_s$ for all s, and by using the independence on t of the set (4.6), the *cgst* condition implies that, as $s \to 0$, type (II) points cannot collide with type (I) points along the slice $\{y_0 = 0, \sigma = s, \tau = \hat{t}\}$. If there were a collision, then there would exist a singularity in the slice $\{\mathcal{G} = 0, y_0 = 0, \sigma = 0, \tau = \hat{t}\}$ with the Milnor number higher than the generic singularity type at infinity. This proves the equality:

$$\Gamma \cap \{y_0 = 0\} = \Gamma \cap \{\mathcal{G} = y_0 = 0\}, \tag{4.7}$$

which finishes the proof of (i). Now observe that the equality (4.7) also proves (ii), by a similar reason: if there were a component D_i such that $D_i \cap D_0 = D \cap \{s = 0\}$, then there would exist a singularity in the slice $\{\mathcal{G} = 0, y_0 = 0, \sigma = 0, \tau = \hat{t}\}$ with the Milnor number higher than the generic singularity type at infinity. Notice that we have in fact proved more than (ii), namely:

(ii$'$). There is no component $D_i \neq D_0$ such that $D_i \cap D_0 = D \cap \{s = 0\}$.

This ends the proof of (a).

(b). If there were collision of some singularities, out of which two or more λ-singularities are in the same fibre, then there would be at least two different points $z_i \neq z_j$ of Σ_s such that (z_i, s, t) and (z_j, s, t) collide when $s \to 0$. This is exactly what is forbidden by the *cgst* assumption. $\qquad\square$

We may prove (see Exercise 4.4 for details) that the deformations in the $\mathcal{F}$-class such that $\mu + \lambda = $ constant are *cgst*-deformations. Therefore, we has the following consequence of Theorem 4.3.3(a):[3]

Corollary 4.3.4 Inside the $\mathcal{F}$-class, a λ-singularity of f_0 cannot be deformed into only μ-singularities by a constant degree deformation with constant $\mu + \lambda$.

Example 4.3.5 $f_s = x^4 + sz^4 + z^3 + y$.
This is a deformation inside the $\mathcal{B}$-class with constant $\mu + \lambda = 0$, which is not *cgst* at infinity.

The fibre $\mathbb{Y}_{s,t}$ is singular only at $p := ([0 : 1 : 0], 0)$ and the singularities of $\mathbb{Y}_{0,t}^{\infty}$ change from a single smooth line $\{x^4 = 0\}$ with a special point p on it into the isolated point p, which is a $\tilde{E}_7$ singularity of $\mathbb{Y}_{s,t}^{\infty}$. We use the notation $\oplus$ for the Thom–Sebastiani sum of two types of singularities in separate variables. We have:

$s = 0$: the generic type of $\mathbb{Y}_{0,t}$ at infinity is $A_3 \oplus E_7$ with Milnor number 21, and $\chi(\mathbb{Y}_{0,t}^{\infty}) = 2$.

$s \neq 0$: the generic type of $\mathbb{Y}_{s,t}$ at infinity is $A_3 \oplus E_6$ with Milnor number 18, and $\chi(\mathbb{Y}_{s,t}^\infty) = 5$.

The jumps of $+3$ and -3 compensate each other in the formula (4.8).

Example 4.3.6 $f_s = x^2 y + x + z^2 + s z^3$.

This is a *cgst*-deformation, but $\mu + \lambda$ is not constant. Moreover, f_s is $\mathcal{F}$-type for all $s \neq 0$, whereas f_0 is not $\mathcal{F}$-type (but still $\mathcal{B}$-type). The generic type at infinity is D_4, for all s, and there is a jump $D_4 \to D_5$ for $t = 0$ and all s. For $s \neq 0$, a second jump $D_4 \to D_5$ occurs for $t = c/s^2$, where c is some constant.

There are no affine critical points, i.e. $\mu(s) = 0$ for all s, but $\lambda(s) = 2$ if $s \neq 0$ and $\lambda(0) = 1$. We have $\mathrm{Atyp} f_s = \{0, c/s^2\}$ for all $s \neq 0$, and χ^∞ changes from 3 if $s = 0$ to 2 if $s \neq 0$. There is a persistent λ-singularity in the fibre over $t = 0$ and there is a branch of the critical locus $\mathrm{Crit}\,\Psi$, which is asymptotic to $t = \infty$.

Rigidity of singularities in the $\mathcal{F}$-class. Let $P(x, s) = f_s(x)$ define a family of $\mathcal{F}$-type polynomials, of constant degree d, where s varies in a small enough disk $\delta \subset \mathbb{C}$ centered at 0. We suppose that the coefficients of the polynomial $f_s(x)$ are polynomial functions of the parameter s, so that the space $\mathbb{Y}$ is algebraic.

Notations. We recall a few notations and set some more. Let $s \in \delta$ be fixed and let $p \in \Sigma_s$. We know that $\mu_p(\mathbb{Y}_{s,t}) > \mu_{p,\mathrm{gen}}(s)$ for a finite number of values of t and that the number of vanishing cycles of f_s at (p, t) is the jump $\lambda_{p,t}(s) := \mu_p(\mathbb{Y}_{s,t}) - \mu_{p,\mathrm{gen}}(s)$. Let us denote $\lambda_t(s) := \sum_{p \in \Sigma_s} \lambda_{p,t}(s)$.

We denote by $\mathrm{A}_{\mathrm{inf}} f_s := f_s(\mathrm{Sing}^\infty f_s)$ the set of 'critical values at infinity' (i.e. the images of points where the jump is positive), and by $\mathrm{A}_{\mathrm{aff}} f_s := f_s(\mathrm{Sing} f_s)$ the set of critical values produced by affine critical points. We have:

$$\lambda(s) = \sum_{t \in \mathrm{A}_{\mathrm{inf}} f_s} \lambda_t(s) \quad \text{and} \quad \mu(s) = \sum_{t \in \mathrm{A}_{\mathrm{aff}} f_s} \mu_{F_{s,t}},$$

where $F_{s,t} := f_s^{-1}(t)$ is our usual notation. We also know that $\mathrm{Atyp} f_s = \mathrm{A}_{\mathrm{aff}} f_s \cup \mathrm{A}_{\mathrm{inf}} f_s$ and that $f_s : f_s^{-1}(\mathbb{C} \setminus \mathrm{Atyp} f_s) \to \mathbb{C} \setminus \mathrm{Atyp} f_s$ is a locally trivial fibration. As usual, $\mathbb{Y}_{s,*}^\infty := \mathbb{Y}_{s,*} \cap \{x_0 = 0\}$.

Definition 4.3.7 We say that a finite set Ω_s of points in $\mathbb{C}^k$, for some k, depending on a real parameter $s \in [0, 1]$, is an *analytic braid*[4] if the number $\#\Omega_s$ is constant, the multi-valued function $s \mapsto \Omega_s$ is continuous, and $\Omega = \cup_{s \in [0,1]}(\Omega_s \times \{s\})$ is a real analytic sub-variety of $\mathbb{C}^k \times [0, 1]$.

Proposition 4.3.8 Let $\{f_s\}_{s\in[0,1]}$ be an analytic family of $\mathcal{F}$-type polynomials. Suppose that the integers $\mu(s) + \lambda(s)$, #Atypf_s and degf_s are independent of $s \in [0, 1]$. Then:

(a) Σ_s, $A_{\mathrm{aff}}f_s$, $A_{\mathrm{inf}}f_s$, and Atypf_s are analytic braids;
(b) for any continuous function $s \mapsto p(s) \in \Sigma_s$, we have $\mu_{p(s),\mathrm{gen}} = constant$;
(c) for any continuous function $s \mapsto c(s) \in A_{\mathrm{inf}}f_s$, we have $\lambda_{p(s),c(s)} = constant$;
(d) for any continuous function $s \mapsto c(s) \in A_{\mathrm{aff}}f_s$, we have $\mu_q(F_{s,c(s)}) = constant$ for any $q \in \mathrm{Sing}\, F_{s,c(s)}$.

Proof Let us remark that our hypotheses imply that the family $\{f_s\}_s$ is a *cgst*-deformation, see Exercise 4.4. This means that there is no collision of points $p(s) \in \Sigma_s$ as $s \to 0$ and that $\mu_{p(s),\mathrm{gen}}(s)$ is constant. Consequently, Σ_s is an algebraic braid. This proves (b) and a part of (a).

Next, let us observe that the multi-valued map $s \mapsto \mathrm{Atyp}f_s$ is continuous. This is a consequence of the conservation of the local μ in families of functions (generally valid) and of the conservation of the local $\mu + \lambda$ shown in Corollary 4.3.1. By the same reasons, there is no loss towards infinity of values from Atypf_s, as $s \to 0$. This shows that the values Atypf_s are bounded within some disk of radius independent of s. The constancy of #Atypf_s shows that there is no collision of atypical values when $s \to 0$.

Let us now observe that, since Atypf_0 does not split and since λ-singularities are persistent by Theorem 4.3.3(a), we might only have splitting of $\lambda_{p(0),c(0)}$ in a single fibre of f_s (into λ-singularities and eventually μ-singularities). By Theorem 4.3.3(b), there cannot be more than one λ-singularity. Since the local $\lambda + \mu$ is independent on s, by using formula (4.9) and the jump interpretation of λ, which we recalled in the above notations, we deduce that we are in the presence of a family of hypersurface germs $\mathbb{Y}_{s,t}$, where an isolated singularity splits such that the total Milnor number is constant. This contradicts the nonsplitting principle (see its statement below). We have thus proved the claim (c), and also that #$A_{\mathrm{inf}}f_s$ is constant and that the map $s \mapsto A_{\mathrm{inf}}f_s$ is continuous.

Since we have local conservation of μ in the family f_s at any μ-singularity of f_0, it follows that a μ-singularity cannot split into a single fibre, again due to the nonsplitting argument. However, there cannot be loss of μ towards infinity, since we know now that both $\mu + \lambda$ and λ are conserved locally. This proves (d), that #$A_{\mathrm{aff}}f_s$ is constant and that the map $s \mapsto A_{\mathrm{aff}}f_s$ is continuous.

The maps $s \mapsto A_{\mathrm{inf}}f_s$ and $s \mapsto A_{\mathrm{aff}}f_s$ are also analytic, finite maps and their images are therefore analytic. The same is true for $s \mapsto \mathrm{Atyp}f_s$. $\qquad\square$

The local nonsplitting principle. We recall here the statement of the result due to A'Campo, Lazzeri, and Lê [A'C1, Laz, Lê5]. Let $\{X_t\}_{t \in D_\varepsilon}$ be an analytic family of hypersurfaces in some ball $B_\delta \subset \mathbb{C}^n$, where $D_\varepsilon \subset \mathbb{C}$ is some disk centered at the origin, such that X_0 has an isolated singularity at $x_0 \in X_0 \cap B_\delta$. For small enough radii ε and δ, any hypersurface X_t has isolated singularities in B_δ, say at the points $\{x_{t,1}, \ldots, x_{t,k}\} \subset X_t \cap B_\delta$, for some $k \geq 1$. Let $\mu(X_t, x_{t,i})$ denote the Milnor number of X_t at the point $x_{t,i}$. Then the *local nonsplitting theorem* says the following: if $\sum_{i=1}^k \mu(X_t, x_{t,i}) = \mu(X_0, x_0)$, then, $k = 1$, i.e. the singularity x_0 of X_0 does not split in the deformation, and moreover its trajectory $\{x_{t,1}\}_{t \in D_\varepsilon}$ is a nonsingular curve.

Exercises

4.1 Prove that, if $Y \subset \mathbb{C}^n$ is a general hypersurface of degree d, then $\chi(Y) = 1 + (-1)^{n-1}(d-1)^n$.

4.2 Show that the Euler characteristic of the smooth hypersurface $V_{gen}^{n,d}$ of degree d in $\mathbb{P}^n$ is $\chi^{n,d} := \chi(V_{gen}^{n,d}) = n + 1 - \frac{1}{d}\{1 + (-1)^n(d-1)^{n+1}\}$.

4.3 Let f be a $\mathcal{B}$-type polynomial and G its general fibre. Prove the formula:

$$b_{n-1}(G) = (-1)^{n-1}(\chi^{n,d} - 1) - \sum_{p \in \Sigma} \mu_{p,\mathrm{gen}} - (-1)^{n-1}\chi^\infty, \qquad (4.8)$$

where $\chi^\infty := \chi(\{f_d = 0\})$.

4.4 Consider a deformation $\{f_s\}_s$ of constant degree d, inside the $\mathcal{F}$-class. Show that formula (4.8) takes the following form:

$$\mu(s) + \lambda(s) = (d-1)^n - \sum_{p \in \Sigma_s} \mu_{p,\mathrm{gen}}(s) - \sum_{p \in W_s} \mu_p^\infty(s), \qquad (4.9)$$

where $\mu_p^\infty(s)$ denotes the Milnor number, independent of t, of the singularity of $\mathbb{Y}_{s,t} \cap H^\infty$ at the point $(p, t) \in W_s \times \mathbb{C}$. Conclude that $\mu_{p,\mathrm{gen}}(s)$ and $\mu_p^\infty(s)$ are constant in $\mu + \lambda$ constant families, and, therefore, that such families are *cgst*-deformations.

4.5 Remark that, in deformations within the $\mathcal{B}$-class, there is no inclusion relation between the properties 'constant generic singularity type' and '$\mu(s) + \lambda(s)$ constant', see Examples 4.3.5, 4.3.6. Show that in the $\mathcal{F}$-class the two conditions are equivalent.

4.6 Let $\Delta\chi^\infty := (-1)^n(\chi^\infty(s) - \chi^\infty(0))$ in formula (4.8) applied to deformations inside the $\mathcal{B}$-class. Show the following:

(i) If $\Delta\chi^\infty < 0$, then the deformation is not *cgst*.

(ii) If $\Delta\chi^\infty > 0$, then the deformation is not $\mu + \lambda = \text{constant}$.

(iii) If $\Delta\chi^\infty = 0$, then we have the equivalence: $\mu + \lambda = \text{constant} \Leftrightarrow cgst$ at infinity.

(iv) in deformations inside the $\mathcal{F}$-class, $\Delta\chi^\infty \geq 0$.

5

Topology of a family and contact structures

5.1 Topological equivalence of polynomials

A well-known result by Lê and Ramanujam [LêRa] for a smooth family of germs of holomorphic functions with isolated singularity $g_s : (\mathbb{C}^n, 0) \to (\mathbb{C}, 0)$, says that the constancy of the local Milnor number implies that the hypersurface germs $g_0^{-1}(0)$ and $g_s^{-1}(0)$ have the same topological type whenever $n \neq 3$. Under the same conditions, Timourian [Tim] and King [Ki] showed moreover that the family of function germs is topologically trivial over a small enough disk δ. The techniques available by now for proving the Lê–Ramanujam–Timourian–King theorem do not work for local nonisolated singularities.

In the global affine setting, we may ask a similar question: *under what conditions can the topological equivalence of polynomial functions be controlled by numerical invariants?*

We say that two polynomial functions $f, g : \mathbb{C}^n \to \mathbb{C}$ are *topologically equivalent* if there exist homeomorphisms $\Phi : \mathbb{C}^n \to \mathbb{C}^n$ and $\Psi : \mathbb{C} \to \mathbb{C}$ such that $\Psi \circ f = g \circ \Phi$. The global setting creates new problems, since we have to deal at the same time with several singular points and atypical values. Moreover, the singularities at infinity present a new and essential difficulty since, even if they are isolated, they are of a different type than the critical points of holomorphic germs. Our exposition follows [BT].

Local triviality at infinity. When studying a polynomial family $f_s(x) = P(x, s)$ of polynomial functions, we have to deal with the triviality problem in the neighbourhood of infinity.[1]

The Whitney stratification $\mathcal{W}_s$ of the singular space $\mathbb{Y}_{s,*}$ has in our setting the following strata: $\mathbb{Y}_{s,*} \setminus \mathbb{Y}_{s,*}^\infty$, $\mathbb{Y}_{s,*}^\infty \setminus \operatorname{Sing} \mathbb{Y}_{s,*}$, the complement in $\operatorname{Sing} \mathbb{Y}_{s,*}$ of the singularities at Infinity and the finite set of singular points at infinity. We

70

also recall that the restriction $\tau_{|\mathbb{Y}_{s,*}}$ is transversal to all the strata of $\mathbb{Y}_{s,*}$ except at the singular points at infinity.

Let (p, c) be a singularity at infinity of f_0, i.e. such that $\lambda_{p,c}(0) > 0$ and let $g_s : \mathbb{Y}_{s,*} \to \mathbb{C}$ be the localization at $(p(s), c(s))$ of the map $\tau_{|\mathbb{Y}_{s,*}}$. We denote by $B_\epsilon \subset \mathbb{C}^n \times \mathbb{C}$ the closed $2n + 2$-ball of radius ϵ centred at (p, c), such that $B_\epsilon \cap \mathbb{Y}_{s,*}$ is a Milnor ball for g_0. We choose $0 < \eta \ll \epsilon$ such that we get a Milnor tube $T_0 = B_\epsilon \cap \mathbb{Y}_{s,*} \cap g_0^{-1}(D_\eta(c))$. Then, for all $t \in D_\eta(c)$, $g_0^{-1}(t)$ intersects transversally $S_\epsilon = \partial B_\epsilon$. We recall that $g_0 : T_0 \setminus g_0^{-1}(c) \to D_\eta(c) \setminus \{c\}$ is a locally trivial fibration and that $g_0 : T_0 \to D_\eta(c)$ is a trivial fibration if and only if $\lambda_{p,c}(0) = 0$.

By using Proposition 4.3.8(a) and an analytic change of coordinates, we may assume that $(p(s), c(s)) = (p, c)$ for all $s \in [0, u]$, for some small enough $u > 0$. We set $T_s = B_\epsilon \cap \mathbb{Y}_{s,*} \cap g_s^{-1}(D_\eta(c))$ and notice that B_ϵ does not necessarily define a Milnor ball for g_s whenever $s \neq 0$. For some $u > 0$, let $T = \bigcup_{s \in [0,u]} T_s \times \{s\}$, and let $G : T \to \mathbb{C} \times [0, u]$ be defined by $G(z, s) = (g_s(z), s)$.

Theorem 5.1.1 [BT] *Let* $f_s(x) = P(x, s)$ *be a one-parameter polynomial family of polynomial functions of* $\mathcal{F}$-*type and of constant degree, such that the numbers* $\mu(s) + \lambda(s)$ *and* #Atypf_s *are independent of* s. *If* $n \neq 3$, *then there exists* $u > 0$ *and a stratified homeomorphism* α *such that the following diagram commutes:*

$$\begin{array}{ccc} T & \xrightarrow{\ \alpha\ } & T_0 \times [0, u] \\ {\scriptstyle G}\downarrow & & \downarrow{\scriptstyle g_0 \times \mathrm{id}} \\ D_\eta(c) \times [0, u] & \xrightarrow[\ \mathrm{id}\ ]{} & D_\eta(c) \times [0, u], \end{array}$$

and such that α *sends the strata of tube* T_s *to the corresponding strata of* T_0, *for all* s.

The stratification of some tube T_s is the one induced by $\mathcal{W}_s$, and has therefore three strata: $\{T_s \setminus (\{p\} \times D_\eta(c)), \{p\} \times D_\eta(c) \setminus (p, c), (p, c)\}$.

We cannot apply directly Timourian's proof [Tim] to the family g_s because here the space germ $(\mathbb{Y}_{s,*}, (p, c))$ is singular. Nevertheless we can adapt it to our situation. We start from the fact that (p, c) is the only singularity of g_s in T_s, by Proposition 4.3.8. One may show how the assumptions of [Tim, Lemma 3] are fulfilled in our new setting. We refer to [BT] for the details of this proof, which, in particular, adapts Lê's and Ramanujam's arguments of [LêRa] to the new setting. These arguments include the h-cobordism theorem, which is responsible for the excepted case $n = 3$.

5.1.1 The topological equivalence theorem

Theorem 5.1.2 [BT] *Let* $\{f_s\}_{s\in[0,1]}$ *be a continuous family of complex polynomials of $\mathcal{F}$-type in $n \neq 3$ variables. If the Euler characteristic of the general fibre $\chi(G_s)$, the number of atypical values #Atypf$_s$ and the degree $\deg f_s$ are independent of $s \in [0,1]$, then the polynomials f_0 and f_1 are topologically equivalent.*

Example 5.1.3 Let $f_s(x, y, z, w) = x^2 y^2 + z^2 + w^2 + xy + (1 + s)x^2 + x$. For $s \in \mathbb{C} \setminus \{-2, -1\}$, we have Atypf$_s = \{0, -\frac{1}{4}, -\frac{1}{4}\frac{s+2}{s+1}\}$, $\mu(s) = 2$, and $\lambda(s) = 1$. It follows that $\chi(G_s) = -2$ and #Atypf$_s = 3$. Since f_s is not of $\mathcal{F}$-type, we have to consider $f_s'(x, y) := x^2 y^2 + xy + (1 + s)x^2 + x$. We then have $f_s = f_s' \oplus (z^2 + w^2)$. By applying Theorem 5.1.2 to the family f_s' and then coming back to f_s, we get that f_0 is topologically equivalent to f_s if and only if $s \in \mathbb{C} \setminus \{-2, -1\}$, since #Atypf$_{-1}$ = #Atypf$_{-2}$ = 2.

We recall that the $\mathcal{F}$-class of polynomials includes all polynomials of two variables with reduced fibres. These may have λ-singularities.

The constancy of $\chi(G_s)$ and of #Atypf$_s$ are *necessary* conditions for the topological triviality. The constancy of $\deg f_s$ is only a technical condition which enables us to work with the space $\mathbb{Y}$.

We first prove Theorem 5.1.2 in case the coefficients of the family $P(x, s)$ are polynomials in the variable s; the general case will follow by a constructibility argument.

The rough idea of the proof is to use the local triviality Theorem 5.1.1 and then patch together the pieces in order to obtain a global topological equivalence.[2] We outline below the steps of the proofs.

Start of the proof: trivialization in a large ball. Let $R_1 > 0$ such that for all $R \geq R_1$ and all $c \in A_{\text{inf}}(0)$ the intersection $f_0^{-1}(c) \cap S_R$ is transversal. We choose $0 < \eta \ll 1$ such that for all $c \in A_{\text{inf}}(0)$ and all $t \in D_\eta(c)$ the intersection $f_0^{-1}(t) \cap S_{R_1}$ is transversal. We set

$$K(0) = D \setminus \bigcup_{c \in A_{\text{inf}}(0)} \mathring{D}_\eta(c)$$

for a sufficiently large disk D of $\mathbb{C}$. There exists $R_2 \geq R_1$ such that for all $t \in K(0)$ and all $R \geq R_2$ the intersection $f_0^{-1}(t) \cap S_R$ is transversal (see Exercise 1.8).

By Proposition 4.3.8, Atypf$_s$ is an analytic braid, so we may assume that for a large enough D, Atypf$_s \subset \mathring{D}$ for all $s \in [0, u]$. Moreover, there exists a diffeomorphism $\phi : \mathbb{C} \times [0, u] \to \mathbb{C} \times [0, u]$ with $\phi(x, s) = (\phi_s(x), s)$ and such that $\phi_0 = \mathrm{id}$, that $\phi_s(\mathrm{Atyp} f_s) = \mathrm{Atyp} f_0$ and that ϕ_s is the identity on $\mathbb{C} \setminus \mathring{D}$, for all $s \in [0, u]$. We set $K(s) = \phi_s^{-1}(K(0))$.

We may choose u sufficiently small such that for all $s \in [0, u]$, for all $c \in A_{\inf}(0)$ and all $t \in \phi_s^{-1}(D_\eta(c))$ the intersection $f_s^{-1}(t) \cap S_{R_1}$ is transversal. We may also suppose that for all $s \in [0, u]$, for all $t \in K(s)$ the intersection $f_s^{-1}(t) \cap S_{R_2}$ is transversal. Notice that the intersection $f_u^{-1}(t) \cap S_R$ may not be transversal for all $R \geq R_2$ and $t \in K(s)$.

We denote:

$$B'(s) = \left(f_s^{-1}(D) \cap B_{R_1}\right) \cup \left(f_s^{-1}(K(s)) \cap B_{R_2}\right), \quad s \in [0, u].$$

By using Timourian's theorem at the affine singularities and by gluing the pieces by vector fields (for the details, see [BT]), we get the following trivialization:

$$
\begin{array}{ccc}
B' & \xrightarrow{\ \Omega^{aff}\ } & B'(0) \times [0, u] \\
{\scriptstyle F} \downarrow & & \downarrow {\scriptstyle f_0 \times \mathrm{id}} \\
D \times [0, u] & \xrightarrow[\ \phi\ ]{} & D \times [0, u],
\end{array}
$$

where $B' = \bigcup_{s \in [0, u]} B'(s) \times \{s\}$ and $F(x, s) = (f_s(x), s)$.

Gluing trivializations at infinity. It remains to deal with the part at infinity $f_s^{-1}(D) \setminus \mathring{B}'(s)$ according to the decomposition of D as the union of $K(s)$ and of the disks around each $c \in A_{\inf} f_s$. For each singular point $(p, c(0))$ at infinity we have a Milnor tube $T_{p,0}$ defined by a Milnor ball of radius $\epsilon(p, c(0))$ and a disk of radius η, small enough in order to be a common value for all such points.

Let g_s be the restriction to $\mathbb{Y}_{s,*}$ of the compactification τ of f_s, let $G :$ $\mathbb{Y} \longrightarrow \mathbb{C} \times [0, u]$ be defined by $G(x, s) = (g_s(x), s)$ and let $C'(s) = g_s^{-1}(\phi_s^{-1}(D_\eta(c(0)))) \setminus (\mathring{B}_{R_1} \cup \bigcup_{p(s)} \mathring{T}_{p(s)})$. Now g_s is transversal to the following manifolds: to $T_{p(s)} \cap \partial B_\epsilon$, for all $s \in [0, u]$, by the definition of a Milnor tube, and to $S_{R_1} \cap C'(s)$, by the definition of R_1. We shall call the union of these subspaces the *boundary of $C'(s)$*, denoted by $\delta C'(s)$. Recalling the definition of the Whitney stratification on $\mathbb{Y}_{s,*}$, and noticing that $C'(s) \cap \mathrm{Sing}\, \mathbb{Y}_{s,*} = \emptyset$, we get that g_s is transversal to the stratum $C'(s) \cap \mathbb{Y}_{s,*}^\infty$.

For $C' = \bigcup_{s \in [0, u]} C'(s) \times \{s\}$, we have that $C' \cap \mathrm{Sing}\, \mathbb{Y} = \emptyset$ and that the stratification $\{C' \setminus \mathbb{Y}^\infty, \mathbb{Y}^\infty\}$ verifies Whitney's conditions. Then by our assumptions and for small enough u, the function G has maximal rank on $\bigcup_{s \in [0, u]} \mathring{C}'(s) \times \{s\}$, on its boundary $\delta C' = \bigcup_{s \in [0, u]} \delta C'(s) \times \{s\}$ and on $C' \cap \mathbb{Y}^\infty_{s,*}$. By Thom–Mather's first isotopy theorem, G is a trivial fibration on $C' \setminus \mathbb{Y}^\infty$. More precisely, we may construct a vector field on C', which lifts the vector field $(\frac{\partial\phi_s}{\partial s}, 1)$ of $D \times [0, u]$ and which is tangent to the boundary $\delta C'$ and to $C' \cap \mathbb{Y}^\infty$. We may in addition impose the condition that it is tangent to the sub-variety $g_s^{-1}(\partial\phi_s^{-1}(D_\eta(c(0)))) \cap S_{R_2}$. We finally get a trivialization of C', respecting fibres and compatible with ϕ.

Since this vector field is constructed so as to coincide at the common boundaries with the vector field defined on each tube T_s of Theorem 5.1.1, and with the vector field on B' as defined above, this enables us to glue all the resulting trivializations over $[0, u]$. Namely, for

$$B''(s) := \left(f_s^{-1}(D) \cap B_{R_2}\right) \cup \left(f_s^{-1}(D \setminus \mathring{K}(s))\right) \text{ and } B'' := \bigcup_{s \in [0,u]} B''(s) \times \{s\}$$

we get a trivialization:

$$
\begin{array}{ccc}
B'' & \xrightarrow{\quad\Omega\quad} & B''(0) \times [0, u] \\
\scriptstyle F \downarrow & & \downarrow \scriptstyle f_0 \times \mathrm{id} \\
D \times [0, u] & \xrightarrow{\quad\phi\quad} & D \times [0, u].
\end{array}
$$

This diagram proves the topological equivalence of the maps $f_{0|} : B''(0) \longrightarrow D$ and $f_{u|} : B''(u) \longrightarrow D$.

Extending topological equivalences. By the transversality of $f_0^{-1}(K(0))$ to the sphere S_R, for all $R \geq R_2$, it follows that the map $f_0 : B''(0) \longrightarrow \mathring{D}$ is topologically equivalent to $f_0 : f_0^{-1}(\mathring{D}) \longrightarrow \mathring{D}$, which in turn is topologically equivalent to $f_0 : \mathbb{C}^n \longrightarrow \mathbb{C}$.

By applying to f_u the transversality result of Exercise 1.8 we get that there exists $R_3 \geq R_2$ such that $f_u^{-1}(t)$ intersects transversally S_R, for all $t \in K(u)$ and all $R \geq R_3$. Now, with arguments similar to the ones used in the proof of the classical Lê–Ramanujam theorem [LêRa], we may show as in [Ti3, Ti7] that the hypothesis of the constancy of $\mu + \lambda$ allows the application of the h-cobordism theorem on $B'''(u) \setminus \mathring{B}''(u)$, where $B'''(u) = \left(f_s^{-1}(D) \cap B_{R_3}\right) \cup \left(f_s^{-1}(D \setminus \mathring{K}(s))\right)$. Consequently, we get a topological equivalence between $f_u : B''(u) \longrightarrow D$ and $f_u : B'''(u) \longrightarrow D$. Finally, $f_u : B'''(u) \longrightarrow \mathring{D}$ is topologically equivalent to $f_u : f_u^{-1}(\mathring{D}) \longrightarrow \mathring{D}$ by the transversality evoked above, and this is in turn topologically equivalent to $f_u : \mathbb{C}^n \longrightarrow \mathbb{C}$.

This proves Theorem 5.1.2 under the hypothesis that the coefficients of the family $P(x, s)$ are polynomials in the parameter s. The extension to continuous coefficients is shown in the following.

On the continuity of the coefficients. Let $\mathcal{P}_{\leq d}$ be the vector space of all polynomials f in n complex variables of degree at most d. Let $\mathcal{P}_d(\mu + \lambda, \#\mathrm{Atyp}f)$ be the set of polynomials of degree d, with isolated singularities in the affine space and at infinity, with a fixed number of vanishing cycles $\mu + \lambda$ and with a fixed number of atypical values $\#\mathrm{Atyp}f$. We may prove (see [BT, Proposition 4.1]) that $\mathcal{P}_d(\mu + \lambda, \#\mathrm{Atyp}f)$ is a constructible set*. Since f_0 and f_1 are in the same connected component of $\mathcal{P}_d(\mu + \lambda, \#\mathrm{Atyp}f)$, we may connect f_0 to f_1 by a family g_s with $g_0 = f_0$ and $g_1 = f_1$ such that the coefficients of g_s are piecewise polynomial functions in the variable s. Since we have proved our result for each polynomial, we get that f_0 and f_1 are topologically equivalent and this ends the proof of Theorem 5.1.2. □

Remark 5.1.4 Theorem 5.1.2 has the following interpretation: to a connected component of $\mathcal{P}_d(\mu + \lambda, \#\mathrm{Atyp}f)$, one associates a unique topological type.** It follows that there is a finite number of topological types of complex polynomials of fixed degree and with isolated singularities in the affine space and at infinity. A more general fact, the finiteness of topological equivalence classes in $\mathcal{P}_{\leq d}$, has been conjectured by René Thom and proved by T. Fukuda [Fuk].

5.2 Variation of the monodromy in families

Let $f_s : \mathbb{C}^n \to \mathbb{C}, s \in [0, \varepsilon]$ for some $\varepsilon > 0$, be a family of polynomial functions depending smoothly on s. We shall discuss here the variation of the monodromy and refer to §2 for the basic definitions and notations.

In particular, Definition 3.1.10 tells that a simple loop $\gamma : S^1 \to \mathbb{C}$ is admissible for f if $\mathrm{Im}\,\gamma \cap \mathrm{Atyp}f = \emptyset$. For instance, we know from §3.1 that, for each fixed s, there is a large enough open disc $D_s \subset \mathbb{C}$, which is an admissible loop (and the monodromy above it is called the *monodromy at infinity*).

Now, if γ_s is an admissible loop for f_s, for any s, then we say that γ is *stably admissible*.

The 'Lê–Ramanujam problem' consists in finding the appropriate conditions under which the monodromy fibration is constant in a family. Lê and Ramanujam [LêRa] investigated the local setting, namely families of hypersurface germs with isolated singularity. In this case the Milnor fibres are homotopy equivalent to bouquets of spheres of dimension equal to the complex dimension of the fibres.

---

* A *constructible set* is a finite union of locally closed real algebraic sets.
** This does not exclude that two different connected components of $\mathcal{P}_d(\mu + \lambda, \#\mathrm{Atyp}f)$ may have the same topological type; for such an example see [Bo2].

In the affine global setting, let us call V-*class* the class of polynomials f : $\mathbb{C}^n \to \mathbb{C}$ such that their general fibre G is homotopy equivalent to a bouquet of spheres $\vee_\eta S^{n-1}$. By Theorem 3.2, the V-class includes the T-class. We shall call the family $\{f_s\}_s$ a V-*deformation of* $f = f_0$ if f_s is of V-type for all $s \in [0, \varepsilon]$. If, moreover, the number η_s of spheres in the bouquet $\vee_{\eta_s} S^{n-1}$ is independent on s, then we say that we have a η-*constant* V-*deformation*.

Theorem 5.2.1 *Let* $\{f_s\}_{s \in [0, \varepsilon]}$ *be a smooth* η-*constant* V-*deformation. Let* γ *be a stably admissible loop. Then the monodromy fibrations* $f_s^{-1}(\mathrm{Im}\,\gamma)$ *over* γ *are fibre homotopy equivalent. If* $n \neq 3$, *then these fibrations are diffeomorphic to each other.*

Proof The original method of proof due to Lê and Ramanujam [LêRa] actually applies, with minor modifications, to the monodromy fibrations (3.13), for the V-class of polynomials under the assumed conditions. Then we use Lemma 3.1.11 to conclude. We refer to [Ti3, Ti7] for the details. $\qquad\square$

Corollary 5.2.2 Let $P(x, s)$ be a constant degree deformation of f inside the B-class, where $0 \leq s \leq \varepsilon$. If $\mu + \lambda$ is constant and $n \neq 3$, then the monodromy fibrations over any stably admissible loop are isotopic in the family, for small enough ε.

In particular, the monodromy fibrations at infinity are isotopic in the family.

Proof The first claim follows immediately from Theorem 5.2.1. For the second claim we argue as follows. If s is fixed, then there is a large enough open disk $D_s \subset \mathbb{C}$ such that $\mathrm{Atyp} f_s \subset D_s$, as we have seen in §3.1. The radii of the disks D_s may vary with s, but, under our assumptions, Corollary 4.3.1(a) tells that there is no atypical value tending to infinity as $s \to 0$. This implies that there is a sufficiently large disk $D \subset \mathbb{C}$ such that $\mathrm{Atyp} f_s \subset D$ for all s close enough to zero. Therefore $\partial \bar{D}$ is a stably admissible loop and we may apply Theorem 5.2.1. $\qquad\square$

Example 5.2.3 Let $\{f_s\}_s$ be a family of polynomials such that f_s is ρ_s-regular at infinity, for some ρ_s as in Definition 1.2.1. Then f_s is of V-type and $\mathrm{Atyp} f_s = f_s(\mathrm{Sing} f_s)$. Let us assume that the deformation is η-constant. Since η_s is equal to the total Milnor number $\mu(f_s)$, it follows that there is no loss of μ as $s \to 0$ and therefore that the critical values of f_s are bounded inside some large enough disk D for all small enough s. So Theorem 5.2.1 applies to such a situation. See also Exercise 5.1.

Example 5.2.4 Here are some families of polynomial functions in two variables, with $\mu + \lambda = $ constant and $\lambda \neq 0$ taken from the classification lists due to Siersma and Smeltink [SiSm]:

$$\mathbf{A^2BC}_{5,\lambda} \ : \ x^4 - x^2 y^2 + Qxy + Kx^3 + Px^2 + A, \qquad Q \neq 0; \lambda = 1, 2, 3, 4;$$
$$\mu + \lambda = 5.$$
$$\mathbf{A^2B^2}_{4,\lambda} \ : \ x^2 y^2 + y^3 + Qxy + Ry^2 + By, \quad Q \neq 0; \lambda = 1, 2, 3; \ \mu + \lambda = 4.$$
$$\mathbf{A^2B^2+}_{3,\lambda} \ : \ x^2 y^2 + Qxy + Ry^2 + By, \qquad QR \neq 0; \lambda = 1, 2; \ \mu + \lambda = 3.$$

The parameters are A, B, K, P, Q, R and they depend on the variable s. Since $n = 2$ and since all listed polynomials have isolated critical points, they are $\mathcal{F}$-type and Corollary 5.2.2 applies. Actually there is a singularity at infinity which occurs at the point $p = ([0 : 1 : 0], \frac{Q^2}{4})$ in the first family, resp. at the point $p = ([0 : 1 : 0], -\frac{Q^2}{4})$ in the two last families. In each case, an atypical value is either a critical value, or the value $Q^2/4$, resp. $-Q^2/4$.

5.3 Contact structures at infinity

As we have seen in §3.1, Milnor's theorem 3.1.1 tells that the link $S_\varepsilon \cap g^{-1}(0)$ of an isolated singularity of a holomorphic function $g : (\mathbb{C}^n, 0) \to (\mathbb{C}, 0)$ is a closed oriented $(2n - 3)$-dimensional smooth manifold and does not depend, up to isotopy, on the radius ε of the sphere S_ε provided this is small enough. The couple $(S_\varepsilon, S_\varepsilon \cap g^{-1}(0))$, determines completely the topological type of the hypersurface germ $g^{-1}(0)$, as shown by Saeki [Sae]. Scherk [Sche] proved that the natural CR-structure on the link $S_\varepsilon \cap f^{-1}(0)$, defined by the maximal complex hyperplane distribution in its tangent bundle, determines its analytical type.

We explain here the *contact structure* on the link, which is an intermediate invariant. In case of a global polynomial function $f : \mathbb{C}^n \to \mathbb{C}$ we have a well-defined link at infinity of f, replacing small spheres by large ones. It will follow that there are well-defined contact boundaries attached to isolated singularities or to complex polynomials.

These situations are examples of the following type of problem: *find the contact manifolds which are 'fillable' in the sense of Mumford, i.e. such that they are of the form $(\partial V, \xi)$, where V is a complex space with isolated singularities and boundary ∂V, and ξ is the contact form.*

Such contact boundaries are, in our setting, analytic or algebraic invariants, but a priori not topological invariants, and we shall discuss their variation in topologically trivial families of hypersurfaces, following [Cau, CT1, CT2].

Families of contact manifolds. Let us recall some basic definition and refer to Blair's book [Bl] for basic facts on contact structures and to Eliashberg's paper [El2] for an overview of more recent results.

Definition 5.3.1 A *contact form* on a $(2n - 1)$-dimensional manifold M is a global 1-form α on M satisfying the nonintegrability condition $\alpha \wedge (d\alpha)^{n-1} > 0$. A *contact structure* on M is the hyperplane distribution defined by a contact form.

Any smooth level set of a strictly plurisubharmonic function on a complex manifold admits a natural contact structure, given by the maximal complex distribution in its tangent bundle: if φ denotes the function and J the complex structure, then $\alpha := J^* d\varphi$ is a contact form. One may consult Eliashberg's paper [El1] on contact structures and plurisubharmonic functions. We shall use the following general criterion for the contact isotopy in a family:

Theorem 5.3.2 [JGr] *Let* $\pi : M \to B$ *be a smooth fibre bundle such that its fibre is a closed, oriented, odd-dimensional manifold. Let* ξ_b *be a contact structure on each fibre* M_b *depending smoothly on the point* $b \in M$. *Then the contact manifolds* $(M_b, \xi_b)_{b \in B}$ *are all contact isotopic.*

Definition 5.3.3 Let M be a smooth closed oriented odd-dimensional manifold. An *almost contact structure* on M is a hyperplane distribution $\xi \subset TM$ endowed with a complex multiplication $J : \xi \to \xi$. An *almost contact manifold* is a manifold endowed with an almost contact structure.

Two almost contact manifolds (M_1, ξ_1, J_1) and (M_2, ξ_2, J_2) are called *almost contact homotopic* if they are isotopic submanifolds of a manifold W and if any isotopy between them carries (ξ_1, J_1) on an almost contact structure on M_2, which is homotopic to (ξ_2, J_2) in the space of almost contact structures on M_2.

Any real, co-oriented hypersurface M in an almost complex manifold (W, J) is naturally endowed with an almost contact structure: the complex hyperplane distribution (ξ_M, J_M) is just the maximal complex subbundle $TM \cap J\,TM$ of the tangent bundle to the hypersurface equipped with the complex structure induced by J.

Lemma 5.3.4 [Cau] *If* M_0 *and* M_1 *are two isotopic real hypersurfaces in an almost complex manifold* (W, J), *then they are almost contact homotopic.*

Proof We may identify almost contact structures with 2-forms of maximal rank when a Riemannian metric is fixed.

So let us fix a Riemannian metric on W. The almost complex structure J is associated to a nondegenerate 2-form $\omega \in \Omega^2(W, \mathbb{R})$, whose restriction to any hypersurface gives the natural almost contact structure. In particular, the almost contact structures induced by the inclusions $i_j : M_j \hookrightarrow W$ are given by the 2-forms $i_j^* \omega$, where $j = 0, 1$. If $\Phi_t : W \times [0, 1] \to W$ denotes the isotopy between M_0 and M_1, then the one-parameter family $((\Phi_t \circ i_0)^* \omega)_{t \in [0,1]}$

provides a homotopy between the almost contact structure on M_0 and the pull back by Φ_1 of that on M_1. $\qquad\square$

This result shows that, if we define the *formal homotopy class* of a hypersurface in an almost complex manifold as the homotopy class of its almost contact structure, this formal homotopy class is an isotopy invariant, providing the most primitive global invariant of contact structures (see [El2]).

The contact link. Let $g : (\mathbb{C}^n, 0) \to (\mathbb{C}, 0)$ be a holomorphic function germ with isolated singularity. We call *contact link* the link $S_\varepsilon \cap g^{-1}(0)$ equipped with the contact structure defined by the strictly plurisubharmonic squared distance function $z \mapsto \|z\|^2$. The following result, proved by Varchenko with help of Gray's theorem 5.3.2, shows that the contact link is a well-defined analytic invariant of the hypersurface germ $(g^{-1}(0), 0)$.

Proposition 5.3.5 [Var3] The contact link of an isolated hypersurface singularity does not depend on the radius of the small sphere S_ε defining it, up to contact isotopy. Moreover, it does not depend on the choice of analytic coordinates.

Global setting. While focussing on the behaviour at infinity of the fibres of polynomials, we have used in §1.2 the notion of rug function. We need here to adapt it to our problem.

Definition 5.3.6 A *pseudo-convex rug function at infinity* is a proper real polynomial map $\rho : \mathbb{C}^n \to \mathbb{R}_{\geq 0}$ which is strictly plurisubharmonic.

The squared distance function $d_{z_0} : z \mapsto \|z - z_0\|^2$, for any point z_0 in $\mathbb{C}^n$, is an example of pseudo-convex rug function at infinity. It is clear that the class of pseudo-convex rug functions at infinity is right-invariant under the action of the automorphism group Aut $\mathbb{C}^n$.

Proposition 5.3.7 [CT1] Let $V \subset \mathbb{C}^n$ be a complex affine algebraic set with at most isolated singularities. For any pseudo-convex rug function at infinity $\rho : \mathbb{C}^n \to \mathbb{R}_{\geq 0}$ there exists $R_\rho \in \mathbb{R}_+$ such that for any $R \geq R_\rho$ the intersection $\rho^{-1}(R) \cap V$ is a contact manifold. This is independent of the choice of ρ and $R \geq R_\rho$, up to contact isotopy. We call it *the contact link at infinity* of V.

Proof The pseudo-convex rug function at infinity ρ has a finite number of critical values on any nonsingular semi-algebraic subset of $\mathbb{C}^n$ (see [Mi2]). Choosing R_ρ greater than any critical value of ρ on V ensures that $\rho^{-1}(R) \cap V$ is a contact manifold, for any $R \geq R_\rho$. Its contact isotopy class does not depend on R, as a straightforward application of Theorem 5.3.2 (see Exercise 5.2).

Let $\rho_0, \rho_1 : \mathbb{C}^n \to \mathbb{R}_{\geq 0}$ be two pseudo-convex rug functions at infinity. We claim that there exist $a_0 > R_{\rho_0}$ and $a_1 > R_{\rho_1}$ such that $\rho_0^{-1}] - \infty, a_0] \subset$

$\rho_1^{-1}] - \infty, a_1)$ and that $d\rho_i(z)$ are positive multiple of each other on the closed set $V_{reg} \cap (\varphi_0^{-1}] - \infty, a_0] \setminus \varphi_1^{-1}] - \infty, a_1[)$. Then we use Exercise 5.3 to conclude.

Let us prove our claim. If not true, then there exists a sequence z_n of points tending to infinity in the set of points $z \in V_{reg}$ such that the $d\rho_0(z)$ and $d\rho_1(z)$ are not zero and are negative multiple of each other. Applying the *curve selection lemma* [Mi2] in this semi-algebraic setting, there exists an analytic curve γ : $[-\varepsilon, 0] \to \mathbb{P}^n$ such that $\gamma[-\varepsilon, 0[\subset V_{reg}, \gamma(0) \in H^\infty \cap \bar{V}_{reg}$, and the two analytic functions $\rho_0 \circ \gamma$ and $\rho_1 \circ \gamma$ having derivatives of opposite sign. This is clearly impossible. $\qquad\square$

Let $f : \mathbb{C}^n \to \mathbb{C}$ be a complex polynomial function. We have seen in Corollary 1.2.14 that the set Atyp$f$ of atypical values of f is finite. It also follows from Proposition 5.3.7 that any fibre $f^{-1}(t)$, which contains at most isolated singularities, has a well-defined contact link at infinity.

We want to compare the contact link of the fibres and show that there is a generic contact link. We first recall the ρ-*regularity* introduced in §1.2. and the fact that the set of values t of f such that $f^{-1}(t)$ is not ρ_E-regular at infinity is a finite set, Corollary 1.2.13. Then we say that the fibre $f^{-1}(t)$ is *regular-at-infinity* if there exists a pseudo-convex rug function at infinity ρ such that $f^{-1}(t)$ is ρ-regular at infinity. It then follows that the complement of the set Reg$^\infty f$ of regular-at-infinity values of f is a finite set.

It also follows from this definition that a regular-at-infinity fibre can have at most isolated singularities, see also Exercise 5.1.

Theorem 5.3.8 [CT1] *Let $f : \mathbb{C}^n \to \mathbb{C}$ be a complex polynomial function. The contact link of a regular-at-infinity fibre $f^{-1}(t)$ does not depend, up to contact isotopy, on the choice of the value $t \in$ Reg$^\infty f$.*

Proof We first assume that $t_1, t_2 \in$ Reg$^\infty f$ are both ρ_E-regular at infinity values. Then pick up a path Γ from t_1 to t_2 such that ImΓ consists of only values of f which are ρ_E-regular at infinity. This is possible by the above remark on the finiteness of non-ρ_E-regular values. For sufficiently large R, $f_{|\Gamma} : f^{-1}(\Gamma) \cap S_R \to \Gamma$ is a smooth fibration. We then apply Gray's Theorem 5.3.2 to show that the contact links at infinity of the fibres $f^{-1}(t_1)$ and $f^{-1}(t_2)$ are contact isotopic.

Let us consider the general case: say t_1 is a ρ_1-regular at infinity value and t_2 is a ρ_2-regular at infinity value. Then we can find disk neighbourhoods of t_1 and t_2 consisting of ρ_1-regular and ρ_2-regular values, respectively. We connect each of t_1 and t_2 by paths Γ_1 and Γ_2, respectively, to some point in its disk neighbourhood, which is moreover a ρ_E-regular value. Then apply the above reasoning to these paths, and once again to conclude. $\qquad\qquad\square$

This theorem gives the precise meaning to the notion of the *generic contact link* of a polynomial f, which we shall denote by $\mathcal{B}^\infty f$. It does not depend on changes of coordinates in the automorphism group Aut $\mathbb{C}^n$, since the class of pseudo-convex rug functions at infinity is right-invariant under this action.

μ-constant deformations. In the local setting we have the following invariance result, which uses the proof of Lê–Ramanujam's theorem [LêRa]:

Theorem 5.3.9 [Cau] *Let $g_s : (\mathbb{C}^n, 0) \to (\mathbb{C}, 0)$ be a smooth family of holomorphic function germs with isolated singularity. If the Milnor number $\mu(g_s)$ is constant for $s \in [0, 1]$ and if $n \neq 3$, then the contact boundaries of g_0 and g_1 are almost contact homotopic.*

The statement excludes the surface case, $n = 3$, due to the use of the h-cobordism theorem in [LêRa]. For normal surface germs, we may refer to [CNP], which uses the recent work of E. Giroux and others on contact 3-manifolds in order to show that the contact link is in fact a topological invariant.

Global setting. Let $f_s : \mathbb{C}^n \to \mathbb{C}$ family of polynomials as in §5.2, depending smoothly on the parameter $s \in [0, \varepsilon]$, for some real $\varepsilon > 0$. As we have seen in §5, even if we suppose that f_s has isolated singularities, the total Milnor number is not enough to control the topology of f_s. We have defined in §5.2 the class of $\mathcal{V}$-type polynomials, which includes the other classes defined previously ($\mathcal{T}$-type, etc), and the notions of $\mathcal{V}$-*deformation*, respectively η-*constant $\mathcal{V}$-deformation*.

It is clear that if a $\mathcal{V}$-deformation is topologically trivial, then it is η-constant. The converse is not true, as we have seen in Example 5.2.4.

Theorem 5.3.10 [CT1] *Let $\{f_s\}_{s\in[0,\varepsilon]}$ be a smooth η-constant $\mathcal{V}$-deformation of polynomial functions $f_s : \mathbb{C}^n \to \mathbb{C}$, for $n \neq 3$. Then the contact links at infinity $\mathcal{B}^\infty f_0$ and $\mathcal{B}^\infty f_\varepsilon$ are almost contact homotopic.*

Proof It appears that the arguments used in the local case by Lê–Ramanujam ([LêRa]) can be applied in case of η-constant $\mathcal{V}$-deformations (cf. [Ti3, Ti7]), as we have also done in Theorem 5.2.1. For a ρ_E-regular at infinity fibre $f_0^{-1}(t)$, for large enough R_0, for small enough s and a good choice of δ, we get the following contact isotopies:

$$\mathcal{B}^\infty f_0 \overset{\text{cont}}{\simeq} f_0^{-1}(t) \cap S_{R_0} \overset{\text{cont}}{\simeq} f_s^{-1}(t) \cap S_{R_0} \overset{\text{cont}}{\simeq} f_s^{-1}(t + \delta) \cap S_{R_0},$$

where we pass from t to a nearby value $t+\delta$ since $f_s^{-1}(t)$ might be not ρ_E-regular at infinity. For large enough R_s we also have:

$$\mathcal{B}^\infty f_s \overset{\text{cont}}{\simeq} f_s^{-1}(t+\delta) \cap S_{R_s}.$$

Under our η-constancy hypothesis, the cobordism $W := f_s^{-1}(t+\delta) \cap (B_{R_s} \setminus \overset{\circ}{B}_{R_0})$ between the two contact boundaries is a product h-cobordism. We may then use Lemma 5.3.4 to conclude.   □

Remark 5.3.11 In [El1], Eliashberg raises the following 'J-convex h-cobordism problem': suppose we have a strictly plurisubharmonic function f on a product $W = M \times [0, 1]$, such that the two boundary components $M \times \{i\}$, $i = 0, 1$, are level sets of f. Is it possible to find another strictly plurisubharmonic function g, which coincides with f on the boundary and without critical points? Were this true, the proofs of Theorem 5.3.9 and Theorem 5.3.10 would show the invariance of the contact link (at infinity).[3]

Exercises

5.1 Prove that if f is ρ-regular, then f has at most isolated singularities and is of $\mathcal{V}$-type. Show that a family of ρ_E-regular polynomial functions with constant total Milnor number fits in the setting of Example 5.2.3.[4]

5.2 Let M be a complex manifold and $\varphi : M \to \mathbb{R}$ a proper, strictly plurisubharmonic function. Show that if $[a, b] \subset \mathbb{R}$ contains no critical value of φ, then the levels $\varphi^{-1}(a)$ and $\varphi^{-1}(b)$ are isotopic contact manifolds.

5.3 Let M be a complex manifold and $\varphi_0, \varphi_1 : M \to \mathbb{R}$ two proper, strictly plurisubharmonic functions such that $\varphi_0^{-1}] - \infty, a_0] \subset \varphi_1^{-1}] - \infty, a_1)$ for some $a_0, a_1 \in \mathbb{R}$. Let $h : \varphi_0^{-1}] - \infty, a_0] \setminus \varphi_1^{-1}] - \infty, a_1[\to [0, 1]$ be the smooth function defined by

$$h(z) = \frac{\varphi_0(z) - a_0}{(\varphi_0(z) - a_0) + (a_1 - \varphi_1(z))}.$$

Under what conditions is this function a proper submersion? Whenever this is the case, prove that $\varphi_0^{-1}(a_0)$ and $\varphi_1^{-1}(a_1)$ are contact isotopic manifolds. (*Hint:* use that $\varphi_t := (1 - t)\varphi_0 + t\varphi_1$ is strictly plurisubharmonic for $t \in [0, 1]$, and Gray's theorem 5.3.2.)

PART II

The impact of global polar varieties

6

Polar invariants and topology of affine varieties

6.1 The Lefschetz slicing principle

By a classical result due to Karčjauskas [Ka], any complex affine algebraic space has the structure of a CW-complex of dimension $\leq n$. A more general result has been proved by Hamm [Hm2, Hm3] for Stein spaces and its proof is based on real Morse theory.

We adopt here another viewpoint: we construct a model of any closed affine algebraic variety $Y \subset \mathbb{C}^N$ as the result of scanning it by a Lefschetz pencil. The genericity of our model, in the sense that we shall explain in detail, is the central property which makes it useful in several problems that we shall consider in the current chapter.

For a projective hyperplane $H \subset \mathbb{P}^{N-1}$, we denote by $l_H : \mathbb{C}^N \to \mathbb{C}$ its corresponding affine linear form and view H as an element of the dual projective space $\check{\mathbb{P}}^{N-1}$. In our affine setting, what we first need is a pencil of hypersurfaces on Y without 'singularities at infinity':

Proposition 6.1.1 (Affine generic pencils)
Let $Y \subset \mathbb{C}^N$ be a closed affine algebraic variety and let $\mathcal{A}$ be a Whitney stratification of its projective closure $\bar{Y} \subset \mathbb{P}^N$ such that the part at infinity $\bar{Y} \setminus Y \subset H^\infty$ is a union of strata. Then, for any hyperplane $H \subset \mathbb{P}^{N-1}$ such that H is transversal in $H^\infty = \mathbb{P}^{N-1}$ to all the strata of $\bar{Y} \setminus Y$, the restriction $l_H : Y \to \mathbb{C}$ is a locally topologically trivial stratified fibration in the neighbourhood of $\bar{Y} \cap H^\infty$.

Proof Let:

$$\mathbb{Y} := \text{closure}\{(x,t) \in Y \times \mathbb{C} \mid l_H(x) = t\} \subset \mathbb{P}^N \times \mathbb{C}$$

be the hypersurface in $\bar{Y} \times \mathbb{C}$ defined by the equation $l_H(x) - tx_0 = 0$.

Let $\hat{\mathcal{A}}$ be the (roughest) canonical Whitney stratification of $\mathbb{Y}$ induced by the stratification $\mathcal{A}$ restricted to Y, and such that $\mathbb{Y}^\infty$ is a union of strata. The projection $\tau_{(l_H)} : \mathbb{Y} \to \mathbb{C}$ extends l_H since its restriction to $\mathbb{Y} \setminus \mathbb{Y}^\infty$ identifies to l_H. Its stratified singularities are the set $\mathrm{Sing}_{\mathcal{A}}(l_H|Y)$. This set consists of finitely many points: if not so, then there would exist a nonempty curve $\Gamma_0 \subset \mathrm{Sing}_{\mathcal{A}}(l_H|Y)$, contained in a single stratum $\mathcal{A}_\alpha$ of the stratification $\mathcal{A}$ of Y. We have that $l_H(\Gamma_0) = a$ for some $a \in \mathbb{C}$. At any point $x \in \Gamma_0$ we then have the nontransversality:

$$\overline{\langle l_H^{-1}(a) \rangle} \not\pitchfork_x \langle \widehat{T_x \mathcal{A}_\alpha} \rangle, \tag{6.1}$$

where $\overline{\langle l_H^{-1}(a) \rangle} \subset \mathbb{P}^N$ denotes the projective hyperplane which is the closure of $l_H^{-1}(a)$, and $\langle \widehat{T_x \mathcal{A}_\alpha} \rangle$ denotes the projective plane of dimension $\dim \mathcal{A}_\alpha$ tangent to $\mathcal{A}_\alpha$ at x. Let $y \in \bar{\Gamma}_0 \cap H^\infty$ and let $\mathcal{A}_0$ be the stratum on $\bar{Y} \setminus Y$ to which y belongs. Passing to the limit point $x \to y$ in (6.1) yields the nontransversality $\overline{\langle l_H^{-1}(a) \rangle} \not\pitchfork_y \langle \widehat{T_x \bar{\mathcal{A}}_0} \rangle$ due to the Whitney condition (a) satisfied by the couple of strata $(\mathcal{A}_\alpha, \mathcal{A}_0)$. Since $\overline{\langle l_H^{-1}(a) \rangle} \cap H^\infty = H$, this gives a contradiction to the assumed transversality of H to the stratum $\mathcal{A}_0$.

It is therefore sufficient to prove that $\tau_{(l_H)}$ is a stratified submersion on every stratum included in $\mathbb{Y}^\infty := \mathbb{Y} \cap (H^\infty \times \mathbb{C})$. Let us remark that $\mathbb{Y} = \mathbb{H} \cap (\bar{Y} \times \mathbb{C})$, where $\mathbb{H} = \{(x_0, x) \in \mathbb{P}^N, t \in \mathbb{C} \mid l_H(x) - tx_0 = 0\}$ is a nonsingular hyperplane of $\mathbb{P}^N \times \mathbb{C}$. Since the axis H of the pencil l_H is transversal to the strata of $\bar{Y}$ and since, for any t, $\mathbb{H} \cap (H^\infty \times \{t\}) = H$, it follows that, for any stratum $S \subset \bar{Y} \setminus Y$, the hyperplane $\mathbb{H}$ is transversal to the stratum $S \times \mathbb{C}$ of the product stratification $\mathcal{A} \times \mathbb{C}$ of $\bar{Y} \times \mathbb{C}$. We know that a transversal intersection of a Whitney stratified set yields a stratification which is also Whitney (see §A1.1). It then follows that the strata at infinity of the canonical stratification $\hat{\mathcal{A}}$ of $\mathbb{Y}$ consist of only strata that are products by $\mathbb{C}$. This implies that the projection $\tau_{(l_H)}$ is transversal to these strata. Our statement is proved. $\square$

Corollary 6.1.2 (Affine Lefschetz pencils)
There exists a Zariski-open subset $\Omega_{Y,\mathcal{A}} \subset \check{\mathbb{P}}^{N-1}$ such that, for any $H \in \Omega_{Y,\mathcal{A}}$, the affine hyperplane pencil $l_H : Y \to \mathbb{C}$ has a finite number of stratified Morse singularities* and is a locally trivial stratified fibration in the neighbourhood of $\bar{Y} \cap H^\infty$.

Proof From Proposition 6.1.1 we have the local triviality as soon as H is transversal to the strata at infinity of $\bar{Y}$. This is a Zariski-open property. We have also seen in the proof of the same proposition that in this case the pencil

---

* The notions of stratified singularity and stratified Morse singularity are explained in Appendix A1.2.

$l_H : Y \to \mathbb{C}$ has a finite number of stratified singularities. It is well known that under small deformations these singularities split into Morse singularities. Therefore the hyperplanes H, such that the pencil l_H has only stratified Morse singularities, are generic. In the Zariski topology, this genericity precisely means that H belongs to a Zariski-open subset of $\check{\mathbb{P}}^{N-1}$. By intersecting it with the former Zariski-open subset we obtain a Zariski-open subset $\Omega_{Y,\mathcal{A}}$, which has all the required properties. $\qquad\square$

Example 6.1.3 Consider the case: Y is nonsingular of dimension n. By Corollary 6.1.2 the singular points a_i of an affine Lefschetz pencil l_H on Y are ordinary complex Morse points (see §A1.2) and by Proposition 6.1.1 the pencil is topologically trivial in the neighbourhood of infinity. Then the Lefschetz–Morse–Smale theory tells the following: *Y is built out of a general slice $Y \cap \{l_H = u\}$ by attaching a number of cells of pure dimension n*. Here 'general' means that the slice $\{l_H = u\}$ does not contain any of the points a_i. The pure dimension is explained by the fact that a complex Morse singularity of a holomorphic function germ on a nonsingular space of dimension n has index n.

The global polar numbers. We assume that the singular affine variety $Y \subset \mathbb{C}^N$ is of pure dimension $n > 0$ and we choose a Lefschetz pencil $l_H \in \Omega_{Y,\mathcal{A}}$ as in Corollary 6.1.2. The affine hyperplane pencil $l_H : Y \to \mathbb{C}$ has a finite number of stratified Morse singularities (see Definition A1.2.2) and we want to consider only those points on the highest-dimensional stratum Y_{reg}.

Let $\alpha_{Y,H}$ be the number of Morse points of the affine Lefschetz pencil l_H on the *regular part* Y_{reg}. The number $\alpha_{Y,H}$ depends *a priori* on H. Nevertheless there exists a Zariski-open subset $\Omega_{Y_{\mathrm{reg}}} \subset \Omega_{Y,\mathcal{A}}$ for which this number is locally constant with respect to the variation of $H \in \Omega_{Y_{\mathrm{reg}}}$, and therefore constant (since $\Omega_{Y_{\mathrm{reg}}}$ is connected). This can be proved as follows. Let $\hat{C}$ be the following algebraic variety:

$$\hat{C} := \mathrm{closure}\{(x, H, u) \mid l_H^{-1}(u) \,\pitchfork_x Y_{\mathrm{reg}}\} \subset Y \times \check{\mathbb{P}}^{N-1} \times \mathbb{C}$$

and its projection $\pi : \hat{C} \to \check{\mathbb{P}}^{N-1} \times \mathbb{C}$. Since $\dim \hat{C} = N - 1$ and $\dim \check{\mathbb{P}}^{N-1} \times \mathbb{C} = N$ and since π is an algebraic map, the discriminant of π is an algebraic variety of dimension $\le N - 1$. Therefore the complement $\mathcal{U}$ of this discriminant is Zariski-open in $\check{\mathbb{P}}^{N-1} \times \mathbb{C}$. The intersection $\mathcal{U} \cap (\Omega_{Y,\mathcal{A}} \times \mathbb{C})$ is also Zariski-open, and dense. Denoting $\mathcal{G} := \mathrm{closure}\{(x, H, u) \in Y \times \check{\mathbb{P}}^{N-1} \times \mathbb{C} \mid l_H(x) - u = 0\} \subset \bar{Y} \times \check{\mathbb{P}}^{N-1} \times \mathbb{C}$, it then follows that the map:

$$\gamma : \mathcal{G} \to \check{\mathbb{P}}^{N-1} \times \mathbb{C}$$

is a locally trivial fibre bundle over the Zariski-open subset $\mathcal{U} \cap (\Omega_{Y,\mathcal{A}} \times \mathbb{C})$. (The discussion concerning the stratification of $\mathcal{G}$ and the transversality of γ is similar to the last part of the proof of Proposition 6.1.1.) This implies that any two slices $l_H^{-1}(u) \cap Y$ are stratified homeomorphic, for (H, u) varying in the Zariski-open set $\mathcal{U} \cap (\Omega_{Y,\mathcal{A}} \times \mathbb{C})$. In particular the restrictions to the stratum Y_{reg} of any two such slices are homeomorphic.

We then consider, for such a point (H, u), the Lefschetz pencil l_H restricted to the stratum Y_{reg}. By the Lefschetz principle, applied as in Example 6.1.3, we get the following equality at the level of Euler characteristics:

$$\chi(Y_{\text{reg}}) = \chi(l_H^{-1}(u) \cap Y_{\text{reg}}) + (-1)^n \alpha_{Y,H}. \tag{6.2}$$

By the above discussion, $\chi(l_H^{-1}(u) \cap Y_{\text{reg}})$ is independent of $(H, u) \in \mathcal{U} \cap (\Omega_{Y,\mathcal{A}} \times \mathbb{C})$; it then follows that the number of Morse points $\alpha_{Y,H}$ is independent of the choice of H in the projection to $\check{\mathbb{P}}^{N-1}$ of the Zariski-open set $\mathcal{U} \cap (\Omega_{Y,\mathcal{A}} \times \mathbb{C})$, which is Zariski-open itself, and hence connected.

We may therefore denote this generic number $\alpha_{Y,H}$ by α_Y or, keeping trace of the dimension of the space, by $\alpha_Y^{(n)}$.

In the next step we replace Y by a general hyperplane section $Y \cap H_1$ of dimension $n - 1$ and defines as above the generic number $\alpha_{Y \cap H_1}$. It turns out that this does not depend on the choice of the affine hyperplane H_1, provided it is generic enough (see also Definition 6.3.9). Therefore we may denote $\alpha_{Y \cap H_1}$ by $\alpha_Y^{(n-1)}$, and keep in mind that this is by definition the number of Morse points of a general Lefschetz pencil on $Y_{\text{reg}} \cap H_1$.

This procedure continues by recursion and we finally get a sequence of nonnegative integers:

$$\alpha_Y^{(n)}, \alpha_Y^{(n-1)}, \cdots, \alpha_Y^{(1)}, \alpha_Y^{(0)}, \tag{6.3}$$

where the last one $\alpha_Y^{(0)}$ is by definition the *degree of* Y, i.e. the number of points of the intersection of Y_{reg} with a generic plane of codimension n in $\mathbb{C}^N$.

All these numbers are well-defined invariants of Y, by the connectivity of the Zariski-open sets of generic slices and of pencils which we use (see also §6.3). They depend however on the embedding of Y into $\mathbb{C}^N$. Let us remark that generic global polar numbers $\alpha_Y^{(i)}$ are well defined for any affine space Y, even if it is not purely dimensional, since the index i captures the information on the dimension of irreducible components. For a zero-dimensional connected space Y, it is natural to set $\alpha_Y^{(0)} = 1$.

These numbers[1] can also be interpreted as total multiplicities of global polar varieties. We refer the reader to §6.3 for the definitions and for comparison to the local setting. The multiplicities of the local polar varieties were defined and used by Lê–Teissier [LêTe], see also Piene [Pi1].

Euler characteristic. The first consequence of the construction of the sequence of generic polar numbers (6.3) is the following general CW-complex model: any nonsingular algebraic space $Y \subset \mathbb{C}^N$ of dimension n has the structure of a CW-complex of dimension $\leq n$, with $\alpha_Y^{(i)}$ cells in dimension i. Therefore we may express its Euler characteristic as follows:

$$\chi(Y) = \sum_{i=0}^{n} (-1)^i \alpha_Y^{(i)}. \tag{6.4}$$

The singular setting. Let now Y be a singular affine variety. As we have seen, a general Lefschetz pencil $l_H : Y \to \mathbb{C}$ has only isolated stratified singularities and is 'good' at infinity according to Corollary 6.1.2.

Definition 6.1.4 ([GM2], see also Definition A1.2.3) The *complex link* $CL(Y, y)$ of a space germ (Y, y) is the fibre in the local Milnor–Lê fibration defined by a general function germ at y (see Theorem A1.1.2). Up to homotopy type, this does not depend on the stratification, the choice of the representative of the space, or the general function.

'General function' means here a function g defined on an ambient nonsingular space, such that the covector $d_y g$ is noncharacteristic with respect to Y at y. Let us fix an algebraic Whitney (b)-regular stratification $\mathcal{A} = \{\mathcal{A}_i\}_{i \in \Lambda}$ on Y (see §A1.1).

Definition 6.1.5 ([GM2], see also Definition A1.2.3) The *complex link of the stratum* $\mathcal{A}_i$ of Y, denoted by $CL_Y(\mathcal{A}_i)$, is by definition the complex link of the germ $(\mathcal{N}_i, p_i)$, where $\mathcal{N}_i$ is a generic slice of Y at some $p_i \in \mathcal{A}_i$, by a plane in $\mathbb{C}^N$ of codimension equal to the dimension of $\mathcal{A}_i$.

It was also shown by Goresky and MacPherson that this does not depend on the choices of the point $p_i \in \mathcal{A}_i$ and of the generic slice, up to homotopy. We remark that the complex link of a point-stratum $\{y\}$ is precisely the complex link of the germ (Y, y).

Let $\text{Cone}(CL_Y(\mathcal{A}_i))$ denote the cone over the complex link and let $\text{NMD}(\mathcal{A}_i)$ denote the *normal Morse datum* at some point of $\mathcal{A}_i$, that is the pair of spaces $(\text{Cone}(CL_Y(\mathcal{A}_i)), CL_Y(\mathcal{A}_i))$ (see also §A1.2). After Goresky and MacPherson [GM2], the local normal Morse data are local invariants up to homotopy and do not depend on the various choices in cause. The complex link of a stratum $\mathcal{A}_0 \subset Y_{\text{reg}}$ is empty, and we set by definition $\chi(\text{NMD}(\mathcal{A}_0)) = 1$.

Theorem 6.1.6 *Let $Y \subset \mathbb{C}^N$ be a closed algebraic space and let $\mathcal{A} = \{\mathcal{A}_i\}_{i \in \Lambda}$ be some Whitney stratification of Y. Then:*

$$\chi(Y) = \sum_{i \in \Lambda} \chi(\mathrm{NMD}(\mathcal{A}_i)) \sum_{j=0}^{\dim \mathcal{A}_i} (-1)^j \, \alpha_{\bar{\mathcal{A}}_i}^{(j)}. \tag{6.5}$$

Proof Take a generic affine Lefschetz pencil $l_H : Y \to \mathbb{C}$. It has only finitely many stratified Morse singularities. By the definition of the generic global polar numbers and their geometric interpretation, the number of stratified Morse points on a stratum $\mathcal{A}_i$ is precisely $\alpha_{\bar{\mathcal{A}}_i}^{(\dim \mathcal{A}_i)}$.

According to the Lefschetz slicing method applied to singular spaces (cf. [GM2]), the space Y is obtained from a generic hyperplane slice $Y \cap \mathcal{H}$ of the pencil, to which are attached cones over the complex links of each singularity of the pencil. Goresky and MacPherson [GM2] have proved that the Milnor data of a stratified Morse function germ is the $(\dim \mathcal{A}_i)$-times suspension of $\mathrm{NMD}(\mathcal{A}_i)$. At the level of Euler characteristics, we then have:

$$\chi(Y) = \chi(Y \cap \mathcal{H}) + \sum_{i \in \Lambda} (-1)^{\dim \mathcal{A}_i} \alpha_{\bar{\mathcal{A}}_i}^{(\dim \mathcal{A}_i)} \chi(\mathrm{NMD}(\mathcal{A}_i)), \tag{6.6}$$

where the sign $(-1)^{\dim \mathcal{A}_i}$ is due to the repeated suspension of the normal Morse data. We then apply formula (6.6) to $Y \cap \mathcal{H}$ and to the successive generic slices in decreasing dimensions. In the final equality, we get $\sum_{j=0}^{\dim \mathcal{A}_i} (-1)^j \, \alpha_{\bar{\mathcal{A}}_i}^{(j)}$ as the coefficient of $\chi(\mathrm{NMD}(\mathcal{A}_i))$, for each $i \in \Lambda$. This ends our proof. $\square$

In Theorem 6.2.4, we shall interpret the sum $\sum_{j=0}^{\dim \mathcal{A}_i} (-1)^j \, \alpha_{\bar{\mathcal{A}}_i}^{(j)}$ as a global Euler obstruction.

The case of locally complete intersections. Let $Y \subset \mathbb{C}^N$ be a closed affine variety which is a local complete intersection of dimension n (hence pure dimensional*), with arbitrary singularities. This implies that the complex link of any stratum $\mathcal{A}_i$ is homotopy equivalent to a bouquet of spheres of dimension equal to $\mathrm{codim}_Y \mathcal{A}_i - 1$, by Lê's theorem [Lê5]. We may then write the formula (6.6) in the following form:

$$\chi(Y) = \chi(Y \cap \mathcal{H}) + (-1)^n (\alpha_Y^{(n)} + \beta_Y^{(n)}), \tag{6.7}$$

---

* A good reference for local complete intersections is [Loo].

where $\beta_Y^{(d)}$ collects the contributions from all the lower-dimensional strata in the sum (6.6). More precisely, under our assumption we have:

$$\beta_Y^{(d)} := \sum_{\mathcal{A}_i \not\subset Y_{\text{reg}}} \alpha_{\bar{\mathcal{A}}_i}^{(\dim \mathcal{A}_i)} \, b_{d - \dim \mathcal{A}_i - 1}(\text{CL}_Y(\mathcal{A}_i)),$$

where $b_{d - \dim \mathcal{A}_i - 1}(\text{CL}_Y(\mathcal{A}_i))$ denotes the Betti number of the complex link $\text{CL}_Y(\mathcal{A}_i)$. According to their definitions, $\alpha_Y^{(d)}$ and $\beta_Y^{(d)}$ are both nonnegative integers. Their sum represents the number of d-cells which have to be attached to $Y \cap \mathcal{H}$ in order to obtain Y.

6.2 Global Euler obstruction

The *local Euler obstruction* was introduced by MacPherson in [MP] as an invariant of complex analytic varieties that plays an essential role for studying the Chern classes of singular varieties. Roughly speaking, the local Euler obstruction $\text{Eu}_X(x)$ of an analytic space X at some point x is the obstruction for extending a continuous stratified radial vector field around x to a nonzero section of the Nash bundle over the Nash blow up $\tilde{X}$ of X, see [MP, Dub1, BrSc, LêTe].

The polar invariants turn out to play a key role in the study of the local Euler obstruction. They enter in the local Euler obstruction formula of Lê–Teissier [LêTe, (5.1.2)], see (6.2). Some other types of formulas have been proved during the time, see for instance [Dub1, BLS, Sch1, BMPS, Ti14]. Their relations were not clear until recently; we refer to Schürmann's book [Sch1] and [ScTi].

We shall define the *global Euler obstruction* $\text{Eu}(Y)$ for an affine singular variety $Y \subset \mathbb{C}^N$ of pure dimension n, and discuss all the ingredients in this global setting, following [STV1]. Our main aim is to explain how the generic global polar numbers enter in the description of the global Euler obstruction. We shall see that actually the local setting can be recovered by the following presentation, which is in terms of vector fields. We refer to [Sch1, ScTi] and to Appendix A1.2 for a dual viewpoint on the (local) Euler obstruction, and for its relations to the Chern–MacPherson cycles.

Definition of the global Euler obstruction. Let $\tilde{Y}$ denote the Nash blow-up of Y, that is:

$$\tilde{Y} := \text{closure}\{(x, T_x Y_{\text{reg}}) \mid x \in Y_{\text{reg}}\} \subset Y \times G(n, N),$$

where $G(n, N)$ is the Grassmannian of complex n-planes in $\mathbb{C}^N$. Let $\nu : \widetilde{Y} \to Y$ denote the algebraic natural projection. Let $\widetilde{T}$ denote the Nash bundle over $\widetilde{Y}$, i.e. the restriction over $\widetilde{Y}$ of the bundle $\mathbb{C}^N \times U(n, N) \to \mathbb{C}^N \times G(n, N)$, where $U(n, N)$ is the tautological bundle over $G(n, N)$. We consider a real vector field **v** on some open subset $V \subset Y$, which is continuous and tangent to the strata of $\mathcal{A}$; we call such a vector field a *stratified vector field*. Then **v** has a well-defined canonical lifting $\widetilde{\mathbf{v}}$ to $\nu^{-1}(V)$ as a section of the real bundle underlying the Nash bundle $\widetilde{T} \to \widetilde{Y}$.*

Definition 6.2.1 We say that a stratified vector field **v** is *radial-at-infinity* if it is defined on the restriction to Y of the complement of a ball B_M of a sufficiently large radius M, centered at the origin of $\mathbb{C}^N$, and it is transversal to the sphere S_R, pointing outwards, for any radius $R > M$. In particular, **v** is without zeros on $Y \setminus B_M$.

Remark 6.2.2 The existence of a 'sufficiently large' radius M follows essentially from Milnor's finiteness result [Mi2, Cor. 2.8] applied to the strata of the stratification $\mathcal{A}$ of Y (see Definition 3.1.3 for the notion of *link at infinity*). From this and from Thom's isotopy lemma, we derive that $Y \cap S_R$ is stratified homeomorphic to $Y \cap S_{R'}$ for $R, R' \geq M$, and that $Y \cap \overset{\circ}{B}_R$ is stratified homeomorphic to Y for $R \geq M$.

We explain, roughly, the construction of a radial-at-infinity vector field on $Y \cap (\mathbb{C}^N \setminus B_R)$. This is based on the general construction due to M.-H. Schwartz [Sc] and is similar to the one used in [BV] for proving the local conical structure of Whitney-stratified analytic sets.

Let grad ρ denote the gradient vector field of the usual distance function ρ on $\mathbb{C}^N$ and consider, at each point $y \in Y$, the orthogonal projection (with respect to the standard Euclidean structure) of grad $\rho(p)$ to the tangent space $T_y \mathcal{A}_i$, where $\mathcal{A}_i$ is the stratum to which y belongs. This yields a vector field **w** on Y which is tangent to the strata, smooth on each stratum, without zeros, transversal to the spheres $S_{R'}$, for $R' \geq R$, and pointing outwards. On each stratum $\mathcal{A}_i$, this is just the vector field $\mathbf{w}_i$ defined as the gradient of the restriction of ρ to $\mathcal{A}_i$. The vector field **w** might not be continuous on $Y \cap (\mathbb{C}^N \setminus B_R)$, but the Whitney condition (a) implies that we can glue the vector fields $\mathbf{w}_i$ together by a partition of unity to get a stratified (hence continuous) vector field on Y, away from the large ball B_R, which is radial-at-infinity.

Definition 6.2.3 Let $\widetilde{\mathbf{v}}$ be the lifting to a section of the Nash bundle $\widetilde{T}$ of a radial-at-infinity, stratified vector field **v** over $Y \setminus B_R$. We call *global Euler*

---

* A good reference for such constructions is [BrSc].

obstruction Y, and denote it by $\mathrm{Eu}(Y)$, the obstruction for extending $\tilde{\mathbf{v}}$ as a nowhere zero section of $\widetilde{T}$ within $\nu^{-1}(Y \cap B_R)$.

More precisely, the obstruction to extend $\tilde{\mathbf{v}}$ as a nowhere zero section of $\widetilde{T}$ within $\nu^{-1}(Y \cap B_R)$ is a relative cohomology class $o(\tilde{\mathbf{v}}) \in H^{2d}(\nu^{-1}(Y \cap B_R), \nu^{-1}(Y \cap S_R)) \simeq H_c^{2d}(\nu^{-1}(Y \cap B_R)) \simeq H_c^{2d}(\widetilde{Y})$. The second isomorphism comes from Remark 6.2.2 and it is valid for $R \geq M$. Thus $o(\tilde{\mathbf{v}})$ is the localisation of the top Chern–Mather class $c^d(\widetilde{T}) \in H^{2d}(\widetilde{Y})$ of the Nash bundle $\widetilde{T}$ (which is oriented as a complex vector bundle).

The global Euler obstruction of Y is the evaluation of $o(\tilde{\mathbf{v}})$ on the homology orientation class $[\widetilde{Y}]$ in the Borel–Moore homology $H_{2d}^{BM}(\widetilde{Y})$. Then $\nu_*(o(\tilde{\mathbf{v}}) \cap [\widetilde{Y}]) \in H_0(\widetilde{Y}) \simeq H_0(Y)$ is mapped, by the natural "forgetful map" (forgetting the compact support), to the top-degree Chern–Mather class $c_M^d(Y) = \nu_*(c^d(\widetilde{Y}) \cap [\widetilde{Y}]) \in H_0^{BM}(Y)$.

Thus $\mathrm{Eu}(Y)$ is an integer and does not depend on the radius of the sphere defining the link at infinity of Y. Since two radial-at-infinity vector fields are homotopic as stratified vector fields, it does not depend on the choice of $\mathbf{v}$ either. Elementary obstruction theory tells us that $o(\tilde{\mathbf{v}})$ is also independent of the way we extend the section $\tilde{\mathbf{v}}$ to $\nu^{-1}(Y \cap B_R)$, see [Str]. Moreover, as shown in [MP] or [Dub1], this is also independent on the blow-up ν, so one can work with any blow-up that extends the tangent bundle over Y_{reg}.

Some properties of the Euler obstruction. As is well known (see for instance [Dub1, BrSc, LêTe]), the local Euler obstruction Eu_X is a *constructible function* on an analytic space X, constant on each stratum of the fixed Whitney stratification $\mathcal{W}$. Thus, given a stratum $\mathcal{A}_i \subset Y$ of $\mathcal{W}$, we denote by $\mathrm{Eu}_Y(\mathcal{A}_i)$ the local Euler obstruction at some point of $\mathcal{A}_i$. The global Euler obstruction defined above has the following properties.

(a) $\mathrm{Eu}(Y)$ is a localization of the top-degree Chern–Mather class $c_M^d(Y)$.

(b) $\mathrm{Eu}(Y) = \sum_{\mathcal{A}_i \subset Y} \chi(\mathcal{A}_i) \mathrm{Eu}_Y(\mathcal{A}_i)$.

The property (a) has already been discussed. The property (b) tells us that $\mathrm{Eu}(Y)$ is a Euler–Poincaré characteristic weighted by the constructible function Eu_Y. In particular, this shows that $\mathrm{Eu}(Y)$ is an *intrinsic invariant* of Y, not depending on the affine embedding of Y. The equality (b) follows immediately by Dubson's [Dub2, Theorem 1] applied to our setting, and will be proved in §6.3 in larger generality.

We are now ready to state a global affine counterpart of a Lê–Teissier formula (6.10). The presentation follows [STV1]:

Theorem 6.2.4 [STV1] *If $Y \subset \mathbb{C}^N$ is a closed algebraic subset of pure dimension n, then its global Euler obstruction satisfies the following equality:*

$$\mathrm{Eu}(Y) = \sum_{i=0}^{n} (-1)^i \alpha_Y^{(i)}. \tag{6.8}$$

It turns out that $\mathrm{Eu}(Y)$ depends only on the regular part of Y. By comparing (6.8) with formula (6.4), we get the equality $\mathrm{Eu}(Y) = \chi(Y)$ for a nonsingular Y; this also follows from the definition of $\mathrm{Eu}(Y)$. In general, $\mathrm{Eu}(Y)$ is neither the Euler characteristic of Y nor that of Y_{reg}.

Proof of Theorem 6.2.4 Let us fix an affine Lefschetz pencil for Y, defined by a linear function l_H, where $H \in \Omega_{Y,\mathcal{A}}$ (cf. Corollary 6.1.2). Its restriction to Y is a stratified submersion away from a finite set of points $\Sigma = \{y_1, \ldots, y_k\}$, which are stratified Morse singularities. Let us then take a large enough disk $D \subset \mathbb{C}$ such that the discriminant $B := l_H(\Sigma)$, which is a finite set, is contained in D. A generic slice $H_t := l^{-1}(t)$, i.e. for $t \in D \setminus B$, is transversal to the strata of $\mathcal{A}$ and induces a Whitney stratification of $Y \cap H_t$, to which we refer in the following. We apply Remark 6.2.2 to Y and then to $Y \cap H_t$. This yields a radius M such that the sphere S_R is stratified transversal to Y and to $Y \cap H_t$, for all $R \geq M$. Let us fix some radius $R \geq M$.

The critical set Σ is contained in $Y \cap B_R \cap l_H^{-1}(D)$. We shall construct a special continuous stratified vector field **v** on $Y \cap l_H^{-1}(D)$, which is radial-at-infinity on $Y \cap (l_H^{-1}(D) \setminus B_R)$ and points outwards the tube $l_H^{-1}(\partial D)$. We start with a radial-at-infinity vector field **v** on $Y \cap (l_H^{-1}(D) \setminus B_R)$, which in addition is radial-at-infinity with respect to the strata of the fixed slice $Y \cap H_t$. This can be done as explained after Remark 6.2.2. Next we extend this to a continuous vector field on $Y \cap H_t \cap B_R$, such that the extension has isolated zeros (which we can always do by the general theory).

Then we may extend* this vector field to a continuous stratified vector field without zeros within a tubular neighbourhood of $Y \cap H_t \cap B_R$, by using the lift by l_H of a vector field on the disk D, which is radial from the point t. We may further extend and deform this without zeros outside the tubular neighbourhood such that it becomes the vector field $\mathrm{grad}_Y \, l_H$ in the neighbourhood of any point $y_i \in \Sigma$. See Figure 6.1.

The vector field $\mathrm{grad}_Y \, l_H$ is defined as follows: take first the complex conjugate of the gradient of l_H and project it to the tangent spaces of the strata of Y into a stratified vector field on Y. This may be not continuous, but we can make it continuous by 'tempering' it, as we have indicated after Remark 6.2.2.

---

* This has been done in the local setting in [BLS, BMPS].

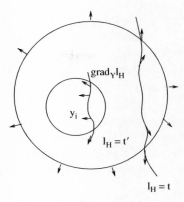

Figure 6.1. Construction of the vector field

We then get a continuous stratified vector field, well defined up to stratified homotopy; this is by definition our $\mathrm{grad}_Y\, l_H$.

Note that the only zeros of $\mathrm{grad}_Y\, l_H$ occur precisely at the points $y_i \in \Sigma$. If $\nu : \widetilde{Y} \to Y$ is the Nash blow-up of Y and $\widetilde{T}$ is the Nash bundle over $\widetilde{Y}$, then $\mathrm{grad}_Y\, l$ lifts canonically to a never zero section $\widetilde{\mathrm{grad}_Y\, l_H}$ of $\widetilde{T}$ restricted to $\nu^{-1}(Y \cap S_\varepsilon)$, where S_ε is a small Milnor sphere around some fixed y_i. In our problem, we are interested to compute the obstruction to extend $\widetilde{\mathrm{grad}_Y\, l_H}$ without zeros throughout $\nu^{-1}(Y \cap B_\varepsilon)$. This is a particular case of the following more general problem treated in [BMPS]: given a holomorphic function germ $f : (X, x_0) \to \mathbb{C}$, compute the obstruction to extending $\widetilde{\mathrm{grad}_X\, f}$ without zeros from $\nu^{-1}(X \cap S_\varepsilon)$ throughout $\nu^{-1}(X \cap B_\varepsilon)$. This is called *local Euler obstruction of f*, or "defect of f", and is denoted by $\mathrm{Eu}_f(X, x_0)$.* In case of Morse singular points, we may directly compute this obstruction as follows:

Lemma 6.2.5 Let l be a holomorphic function germ on (Y, y) with a stratified Morse singularity at the point y on the stratum $\mathcal{A}_0$. Then the local Euler obstruction of l is zero if $\dim \mathcal{A}_0 < \dim Y$, and is $(-1)^{\dim_\mathbb{C} Y}$ if $\mathcal{A}_0 \subset Y_{\mathrm{reg}}$.

Proof Take a small enough ball B_ε in $Y \subset \mathbb{C}^N$, centered at y and of radius $\varepsilon > 0$. Let v be the gradient vector field $\mathrm{grad}_Y\, l$ restricted to the link $Y \cap \partial B_\varepsilon$ and consider the tautological lift $\tilde{v}$ to the Nash blow-up $\widetilde{Y}$. By hypothesis the covector dl does not vanish on any limit of tangent spaces at points in the regular stratum Y_{reg}. By the definition of the space $\widetilde{Y}$ and the definition of stratified complex Morse points (see §A1.2, Definition A1.2.2), if $y \notin Y_{\mathrm{reg}}$, then the

---

* One may also consult [STV2] for the comparison between the Milnor number of f, the local Euler obstruction of f, and several versions of indices of vector fields.

section $\tilde{v}$ of $\widetilde{T}$ can be extended over $\nu^{-1}(Y \cap B_\varepsilon)$ without zeros as follows: view $\widetilde{T}$ as a subset of $(\mathbb{C}^N \times G(n, N)) \times \mathbb{C}^N$. Then at each point $(x, V) \in \widetilde{Y}$, we take the sum of $\tilde{v}$ with the tautological lift of the vector $\gamma(\|x\|) \cdot \mathrm{proj}_V(\overline{\mathrm{grad}_Y \, l})$, where proj_V denotes the orthogonal projection to V and $\gamma : [0, \varepsilon] \to [0, 1]$ is some continuous function such that $\gamma(\varepsilon) = 0$ and $\gamma(0) = 1$. This proves that $\mathrm{Eu}_l(Y, y) = 0$ for $y \notin Y_{\mathrm{reg}}$.

If $y \in Y_{\mathrm{reg}}$, the Nash bundle over some small neighbourhood U of y is the usual tangent bundle of $U \subset Y_{\mathrm{reg}}$ and then $\mathrm{Eu}_l(Y, y)$ is by definition the Poincaré–Hopf index of $\mathrm{grad}_Y \, l$ at y. As pointed out by Milnor [Mi2, Theorem 7.2] this index is $(-1)^{\dim_{\mathbb{C}} Y}$ for a complex Morse point. This concludes the proof of the lemma.   □

We continue the proof of Theorem 6.2.4. We have shown how we may extend the radial-at-infinity vector field $\mathbf{v}$ to $Y \cap B_R \cap l_H^{-1}(D)$. We may further extend it, and without zeros, to the exterior of the tube, $(Y \cap B_R) \setminus (Y \cap B_R \cap l_H^{-1}(D))$.

Altogether this construction shows that the global Euler obstruction $\mathrm{Eu}(Y)$ is precisely the obstruction to extend $\tilde{v}$ *inside* the lifted tube $\nu^{-1}(Y \cap B_R \cap l_H^{-1}(D))$. Moreover, it shows that this obstruction consists of the sum of two contributions:

(i) the obstruction to extend $\tilde{v}$ within the slice $\nu^{-1}(Y \cap H_t \cap B_R)$, as a lift of a stratified vector field with respect to the strata of $Y \cap H_t$. This is by definition $\mathrm{Eu}(Y \cap H_t)$, since the Whitney strata of $Y \cap H_t$ are precisely the intersections of the strata of Y with H_t, by the assumed transversality of H_t to the Whitney stratification $\mathcal{A}$ of Y.

(ii) the obstructions due to the isolated zeros of the gradient vector field $\mathrm{grad}_Y \, l_H$. By Lemma 6.2.5, the local obstruction at some stratified Morse point y_i is zero if $y_i \notin Y_{\mathrm{reg}}$ and it is $(-1)^n$ for $y_i \in Y_{\mathrm{reg}}$. So the sum of all such local obstructions is equal to $(-1)^n \alpha_Y^{(n)}$ since, as defined at §6.1, $\alpha_Y^{(n)}$ is the total number of Morse points on Y_{reg}. We therefore get:

$$\mathrm{Eu}(Y) = \mathrm{Eu}(Y \cap H_t) + (-1)^n \alpha_Y^{(n)}. \qquad (6.9)$$

Applying this formula to $Y \cap H_t$ in place of Y, we get $\mathrm{Eu}(Y \cap H_t) = \mathrm{Eu}(Y \cap H_t \cap H_{t'}) + (-1)^{n-1} \alpha_Y^{(n-1)}$. This goes on recursively. By collecting those formulas together, we get (6.8) and this finishes the proof of Theorem 6.2.4.   □

Comments on the local setting. We have seen that our formula (6.8), proved via (6.9), belongs to the family of Lefschetz type formulas. We may use the slicing by local Lefschetz pencils in a similar manner as in the global setting, and, by repeatedly applying the (completely similar) local version of the formula (6.9) to the Euler obstruction of the complex link of X at x_0, we obtain Lê–Teissier's formula for the local Euler obstruction [LêTe, Cor. 5.1.4]:

$$\mathrm{Eu}_X(x_0) = \sum_{i=0}^{n} (-1)^i m_{x_0}(\Gamma_i), \qquad (6.10)$$

where $m_{x_0}(\Gamma_i)$ denotes the multiplicity of the $(n - i)$-dimensional general polar curve germ Γ_i at x_0.

6.3 Affine polar varieties and MacPherson cycles

MacPherson [MP] has defined analogues of Chern classes* for singular algebraic or analytic spaces X, starting from the *Chern–Mather classes* on the Nash blow-up and the *local Euler obstruction* Eu_X. We have discussed in the preceding section the Euler obstruction and its global counterpart, from the point of view of its original definition [MP].

In the local analytic setting, Lê and Teissier [LêTe] showed that the Chern–MacPherson classes can be expressed in terms of the generic local polar varieties. Moreover, it was shown by Massey [Mas] that the polar varieties are also the key ingredients of 'geometric' representative cycles for the Chern–MacPherson homology classes.[2]

We shall deal here with the global affine algebraic setting, following [ScTi]. Let, therefore, Y be a complex algebraic proper subset of $\mathbb{C}^N$. It is well known that the Chern–MacPherson classes of Y may be represented by algebraic cycles. We show here how to produce a *global geometric* MacPherson cycle.

We refer to Appendix A1.2 for supplementary discussions on the notions we will use below and in particular on some ingredients of the stratified Morse theory. Let α be a constructible function with respect to some stratification S of a complex algebraic affine space $Y \subset \mathbb{C}^N$. The *kth MacPherson cycle* of α is defined as follows:

$$\Lambda_k(\alpha) = \sum_{S \in \mathcal{S}} (-1)^{\dim S} \eta(S, \alpha) P_k(\bar{S}), \qquad (6.11)$$

for $0 \le k \le n$, where $P_k(\bar{S})$ is the *kth affine global polar variety* of the algebraic closure $\bar{S} \subset \mathbb{C}^N$ of the stratum S, which will be explained below. The integer

---

* For the theory of Chern classes of complex analytic manifolds, one may consult Hirzebruch's book [Hirz].

$\eta(S, \alpha)$ denotes the normal Morse index, i.e. the Euler characteristic of the normal Morse data NMD(S) of S weighted by α (see Definition A1.2.4).

The following result explains the role of the global affine polar varieties in the expression of the dual MacPherson classes:

Theorem 6.3.1 [ScTi] *For any S-constructible function α, the fundamental class (or the cycle class) $[\Lambda_k(\alpha)]$ represents the kth dual MacPherson class $\check{c}_k^M(\alpha)$ in the Borel–Moore homology group $H_{2k}^{BM}(Y)$, or in the Chow group $CH_k(Y)$, respectively.*

We shall actually discuss a slightly more general result, Theorem 6.3.6. The proof relies on a reformulation of the *dual MacPherson–Chern class transformation* in terms of characteristic cycles of constructible functions. The former is defined as follows:

$$\check{c}_*^M := \check{c}_*^{Ma} \circ \check{\mathrm{Eu}}^{-1} : F(Y) \to H_*(Y), \qquad (6.12)$$

where $F(Y)$ denotes the group of constructible functions. The dual Euler transformation $\check{\mathrm{Eu}}$ associates to an irreducible subset $Z \subset Y$ the constructible function $\check{\mathrm{Eu}}_Z := (-1)^{\dim(Z)} \cdot \mathrm{Eu}_Z$, and extends to the group of cycles by linearity. Then $\check{\mathrm{Eu}}$ is an isomorphism of groups, since $\mathrm{Eu}_{Z|Z_{\mathrm{reg}}} \equiv 1$. The transformation $\check{c}_*^{Ma}$ is similarly defined by associating to an irreducible Z the total dual Chern–Mather class $\check{c}_*^{Ma}(Z)$ in $H_*(Y)$. We send to (6.14) for the definition of $\check{c}_*^{Ma}(Z)$.

Affine polar varieties and cycles. In the projective setting, the polar variety $P_k(Z)$ of a complex projective variety $Z \subset \mathbb{P}(\mathbb{C}^N)$ has been studied by Todd [To] for a nonsingular Z, and more recently by Piene [Pi1, Pi2] for singular Z. Teissier [Te3] studied the generic local polar varieties in the local analytic setting.

In all the singular versions, the definition of generic polar varieties is based on Kleiman's 'generic transversality' theorem [Kl]. This also holds in the global affine algebraic context, as we explain in the following. Unlike the results [LêTe, Pi1, Pi2, Te3], which are in terms of the Nash blow-up, we shall follow [Sch, ScTi] and use the language of conormal spaces, which relates most naturally to characteristic cycles.*

Kleiman's transversality theorem. Assume that M is a complex algebraic manifold of pure dimension N, with a trivial cotangent bundle T^*M

---

* For the relations between these two different viewpoints we send the reader to [HM, Me, Te3].

(e.g. $M := \mathbb{C}^N$), and let:

$$p : \mathbb{P}(T^*M) = M \times \mathbb{P}(\check{\mathbb{C}}^N) \to \mathbb{P}(\check{\mathbb{C}}^N)$$

be the projection to the last factor.

Let $X \subset M$ be a closed complex algebraic subset of pure dimension $n < N$, endowed with a Whitney stratification $\mathcal{S}$, let $T^*_X M = \overline{T^*_{X_{reg}} M} \subset T^*M$ be its conormal space. Let $\mathbb{P}(T^*_X M) \subset \mathbb{P}(T^*M)$ be its projectivization and note that $\mathbb{P}(T^*_X M)$ is pure $(N - 1)$-dimensional. Let $Z \subset X$ be some closed algebraic subset containing X_{sing}, such that $U := X \backslash Z$ is dense in X.

Let $G^i(\check{\mathbb{C}}^N)$ and $G^i(\mathbb{P}(\check{\mathbb{C}}^N))$ denote the Grassmannian of linear subspaces $\check{H}^i \subset \check{\mathbb{C}}^N$ (respectively, projective linear subspaces $\check{H}^i \subset \mathbb{P}(\check{\mathbb{C}}^N)$) of codimension i, where $0 \leq i \leq N - 1$.

In our setting, Kleiman's *generic transversality* result can be stated as follows:

Theorem 6.3.2 (Kleiman [Kl]) *For any $0 \leq i \leq N - 1$, there is a Zariski-open dense set $\Omega^i(X) \subset G^i(\check{\mathbb{C}}^N)$ such that the following properties hold for all $\check{H}^i \in \Omega^i(X)$:*

(a) *The intersection $\mathbb{P}(T^*_X M) \cap p^{-1}(\mathbb{P}(\check{H}^i))$ is of pure dimension $N - 1 - i$, or empty.*

(b) *$p^{-1}(\mathbb{P}(\check{H}^i))$ intersects $\mathbb{P}(T^*_U M)$ transversely and $\mathbb{P}(T^*_U M) \cap p^{-1}(\mathbb{P}(\check{H}^i))$ is dense in $\mathbb{P}(T^*_X M) \cap p^{-1}(\mathbb{P}(\check{H}^i))$.*

(c) *The dimension of $\mathbb{P}(T^*_{\overline{S'}} M) \cap \mathbb{P}(T^*_X M) \cap p^{-1}(\mathbb{P}(\check{H}^i))$ is strictly less than $N - 1 - i$, for any stratum $S' \in \mathcal{S}$ such that $S' \subset Z$.*

Corollary 6.3.3 The following intersection class in the Chow group:

$$[\mathbb{P}(T^*_X M)] \cap [M \times \mathbb{P}(\check{H}^i)] = \overline{[\mathbb{P}(T^*_U M) \cap p^{-1}(\mathbb{P}(\check{H}^i)))]} \in CH_{N-1-i}(\mathbb{P}(T^*M|_X))$$

equals $c^1(\mathcal{O}(1))^i \cap [\mathbb{P}(T^*_X M)]$ and is therefore independent of the choice of $\check{H}^i \in \Omega^i(X)$.

Proof The notation $\mathcal{O}(1)$ stays for the dual of the tautological line sub-bundle $\mathcal{O}(-1)$ on the projectivization

$$\pi : \mathbb{P}(T^*M|_X) \to X.$$

We have $c^*(T^*M|_X) = 1$ since $T^*M = M \times \check{\mathbb{C}}^N$ is trivial by hypothesis. We then get, for $\check{H}^i \in \Omega^i(X)$:

$$c^1(\mathcal{O}(1))^i \cap [\mathbb{P}(T^*_X M)] = [\mathbb{P}(T^*_X M)] \cap [M \times \mathbb{P}(\check{H}^i)]. \qquad \square$$

By using this corollary, the push-forward:

$$\pi_*([\mathbb{P}(T_X^*M)] \cap [M \times \mathbb{P}(\check{H}^i)]) \in CH_{N-1-i}(X) \qquad (6.13)$$

enters in the definition of the *Segre class* of the sub-cone $T_X^*M \subset T^*M$, as follows:

$$s_*(T_X^*M) := \pi_*(c^*(\mathcal{O}(-1))^{-1} \cap [\mathbb{P}(T_X^*M)]) = \sum_{i \geq 0} \pi_*(c^1(\mathcal{O}(1))^i \cap [\mathbb{P}(T_X^*M)]).$$

We refer to [Sch1, ScTi] for more details of this definition, as well as for the following interpretation of the *dual Chern–Mather class*:

$$\check{c}_*^{Ma}(X) := c^*(T^*M_{|X}) \cap s_*(T_X^*M) = s_*(T_X^*M), \qquad (6.14)$$

since $c^*(T^*M_{|X}) = 1$ in our setting.

Let us come back to the push-forward (6.13). If $N - 1 - i > n$, then this push-forward is equal to zero, by Theorem 6.3.2(a). Let $k := N - 1 - i$ and assume $0 \leq k \leq n$. Choose a basis $\omega = (\omega_1, \ldots, \omega_{k+1})$ of $\check{H}^i$, where we identify ω_j with the corresponding section of T^*M via the projection p. We have:

$$\pi\left(\mathbb{P}(T_U^*M) \cap p^{-1}(\mathbb{P}(\check{H}^i))\right) = \{x \in U \mid \text{rank } (\omega_1, \ldots, \omega_{k+1})_{|T_xU} \leq k\},$$

which is by definition the critical locus $\text{Crit}(\omega_1, \ldots, \omega_{k+1})_{|U}$.

Then by Theorem 6.3.2(b), we get:

$$\pi\left(\mathbb{P}(T_X^*M) \cap p^{-1}(\mathbb{P}(\check{H}^i))\right) = \overline{\text{Crit}(\omega_1, \ldots, \omega_{k+1})_{|U}} = \overline{\text{Crit}(\omega_1, \ldots, \omega_{k+1})_{|X_{\text{reg}}}}.$$

Since this is independent on the choice of the basis ω of $\check{H}^i$, we get the following definition.

Definition 6.3.4 We call the *affine global polar variety of X relative to $\check{H}^i$* the algebraic subset of X:

$$P_{N-1-i}(X, \check{H}^i) := \overline{\text{Crit}(\omega_1, \ldots, \omega_{N-i})_{|X_{\text{reg}}}}.$$

In case $M = \mathbb{C}^N$, we may use generic global affine coordinates x_j and trivialize $T^*\mathbb{C}^N$ by $(\omega_1, \ldots, \omega_N) := (dx_1, \ldots, dx_N)$.

Proposition 6.3.5 [ScTi] *For any $\check{H}^i \in \Omega^i(X)$, the global polar variety $P_{N-1-i}(X, \check{H}^i)$ is of pure dimension $N - 1 - i$ or it is empty, and represents the dual Chern–Mather class, i.e.:*

$$\check{c}_{N-1-i}^{Ma}(X) = [P_{N-1-i}(X, \check{H}^i)] \in CH_{N-1-i}(X). \qquad (6.15)$$

Proof By Theorem 6.3.2(a), dim $P_{N-1-i}(X,\check{H}^i) \leq N-1-i$. In case $i = N-1$ we have the equality $P_0(X,\check{H}^i) = \mathrm{Crit}\,(\omega_1)_{|U}$ since the later is at most a finite set of points, hence of dimension ≤ 0. Let us recall that $U := X \setminus Z$ is dense in X, where $Z \subset X$ is some closed algebraic subset containing X_{sing}.

In case $i < N - 1$ the map $\pi : \mathbb{P}(T_U^*M) \cap p^{-1}(\mathbb{P}(\check{H}^i)) \to \mathrm{Crit}\,(\omega_1,\ldots,\omega_{N-i})_{|U}$ is generically one-to-one, hence $P_{N-1-i}(X,\check{H}^i)$ is also purely $(N-1-i)$-dimensional or empty, by Theorem 6.3.2(a). From this we get:

$$\pi_*([\mathbb{P}(T_X^*M)] \cap [M \times \mathbb{P}(\check{H}^i)]) = [P_{N-1-i}(X,\check{H}^i)] \in CH_{N-1-i}(X). \quad (6.16)$$
$$\square$$

One may extend this result to constructible functions, as follows. Let us first recall the definition of the *characteristic cycle map* (see Appendix A1.2, Definition A1.2.5):

$$cc(\alpha) = \sum_{S \in \mathcal{S}} (-1)^{\dim S} \eta(S,\alpha) \cdot T_{\bar{S}}^*M$$

and the commutativity of diagram (1.8) in Appendix A1.2, from which we need the following piece:

$$F(X) \xleftarrow{\check{\mathrm{Eu}}_{\sim}} Z(X)$$
$$\underset{cc}{\searrow}\sim \quad cn \downarrow \sim$$
$$L(X,M)$$

where $Z(X)$ denotes the group of cycles and cn is the projection $Z \mapsto T_Z^*M$. By definition $L(X,M)$ is the group of all cycles generated by the conormal spaces T_Z^*M; these are *conic Lagrangian cycles*. We recall that $F(X)$ denotes the group of constructible functions on X. We then deduce:

$$\check{\mathrm{Eu}}^{-1}(\alpha) = cn^{-1} \circ cc(\alpha) =$$

$$cn^{-1}\left(\sum_{S \in \mathcal{S}} (-1)^{\dim S} \eta(S,\alpha) \cdot T_{\bar{S}}^*M\right) = \sum_{S \in \mathcal{S}} (-1)^{\dim S} \eta(S,\alpha)\bar{S}.$$

By applying the definition (6.12) and the identification (6.15) we get:

Theorem 6.3.6 *Let M be a complex algebraic manifold of pure dimension N with trivial cotangent bundle $T^*M = M \times \check{\mathbb{C}}^N$. Let $Y \subset M$ be a closed algebraic subset of dimension $n < N$, endowed with a complex algebraic*

Whitney stratification S. *Then for any* $\check{H}^{N-1-k} \in \cap_{S \in \mathcal{S}} \Omega^{N-1-k}(\bar{S})$, $0 \leq k \leq n$, *and for all* S-*constructible functions* α, *one has the equality*:

$$\check{c}_k^M(\alpha) = \sum_{S \in \mathcal{S}} (-1)^{\dim S} \eta(S, \alpha) \cdot [P_k(\bar{S}, \check{H}^{N-1-k})] \in CH_k(Y).$$

The definition of the Zariski-open set $\Omega^{N-1-k}(\bar{S})$ is given in Theorem 6.3.2, where we have to take $X := \bar{S}$, $Z := \bar{S} \setminus S$ and $U := S$. Let us point out that Theorem 6.3.6 is slightly more general than Theorem 6.3.1*. Indeed, this is a consequence of the fact that the polar classes $[P_k(\bar{S}, \check{H}^{N-1-k})]$ are independent of the choice of $\check{H}^{N-1-k} \in \cap_{S \in \mathcal{S}} \Omega^{N-1-k}(\bar{S})$ (cf. Proposition 6.3.5). Therefore the MacPherson cycle class $[\Lambda_k(\alpha)]$ is independent too.

Degrees of affine polar varieties. As before, let $X \subset \mathbb{C}^N$ be a closed complex algebraic subset of pure dimension $n < N$. Then let $\bar{X}$ be the closure of X in $\mathbb{P}^N \supset \mathbb{C}^N$, and let $H^\infty := \mathbb{P}(\mathbb{C}^N) \subset \mathbb{P}^N$ denote the hyperplane at infinity. We may endow $\bar{X}$ with an algebraic Whitney stratification $\hat{\mathcal{S}}$ such that $\bar{X} \cap H^\infty$ is a union of strata and that its restriction to X is the fixed Whitney stratification $\mathcal{S}$ on X. Let $\ker \check{H}^i \subset \mathbb{C}^N$ denote the dual of the linear subspace $\check{H}^i \subset \check{\mathbb{C}}^N$. The former has dimension i since the later has codimension i. We then denote by $\mathbb{P}(\ker \check{H}^i)$ the projectivization viewed as a subset of the hyperplane at infinity H^∞. Kleiman's transversality result implies the following:

Corollary 6.3.7 For any fixed i with $0 \leq i \leq N - 1$, there is a Zariski-open dense set $\hat{\Omega}^i(X) \subset \mathbf{G}^i(\check{\mathbb{C}}^N)$ such that, for any $\check{H}^i \in \hat{\Omega}^i(X)$, the following transversality property holds:

$$\mathbb{P}(\ker \check{H}^i) \pitchfork_{H^\infty} S, \ \forall S \in \hat{\mathcal{S}}, \ S \subset H^\infty. \tag{6.17}$$

The definition of the degrees of affine polar varieties relies on the following result, which we prove below, still following [ScTi]:[3]

Proposition 6.3.8 [ScTi] Let $\check{H}^{i+1} \subset \check{H}^i$ such that $\check{H}^{i+1} \in \hat{\Omega}^{i+1}(X)$ and $\check{H}^i \in \hat{\Omega}^i(X)$. Let $(\omega_1, \ldots, \omega_k)$ be some basis of $\check{H}^{i+1}$. Then the projection:

$$(\omega_1, \ldots, \omega_{N-1-i}) : P_{N-1-i}(X, \check{H}^i) \to \mathbb{C}^{N-1-i}$$

is a proper and finite map.

---

* The notation $P_k(\bar{S})$ used in Theorem 6.3.1 is a shorter notation for the polar variety $P_k(\bar{S}, \check{H}^{N-1-k})$.

Proof $P_0(X, \check{H}^{N-1})$ is algebraic and zero-dimensional, hence a finite set of points. We may then assume $i < N - 1$ in the rest of the proof. If the above map was not proper, then there is a sequence of points $x_k \in X_{\text{reg}} \cap \check{H}^i$ which tends to some point $y \in H^\infty$, with $\omega_j(x_k)$ bounded for all $1 \le j \le N - 1 - i$. Let $\hat{S}_\alpha \subset H^\infty$ be the stratum which contains y.

Let us denote by T_{x_k} the affine n-plane in $\mathbb{C}^N$ tangent to X_{reg} at x_k (which is obtained by translating the vector space $T_{x_k} X_{\text{reg}}$ such that its origin becomes the point x_k). Let then $\langle T_{x_k} \rangle$ denote the projective closure of T_{x_k} inside $\mathbb{P}^N$, i.e. $\langle T_{x_k} \rangle$ is the projective tangent space of X at x_k (this is a simple notion, which can be found for example in [Har, p.181]).

The assumed boundedness of $\omega_j(x_k)$ implies that $y \in \mathbb{P}(\ker \check{H}^{i+1})$. Taking eventually a sub-sequence, we may assume without loss of generality that $\omega_j(x_k)$ converges to some point $b_j \in \mathbb{C}$, for any $1 \le j \le N - 1 - i$, and that the limit of projective n-planes $\langle T_y \rangle := \lim_k \langle T_{x_k} \rangle$ exists in the Grassmannian $G^{N-n}(\mathbb{P}^N)$ of projective linear subspaces of codimension $N - n$. Let $b := (b_1, \dots, b_{N-1-i})$. Similarly, we denote by $\langle T_y \hat{S}_\alpha \rangle$ the projective plane in H^∞ which coincides with the tangent plane $T_y \hat{S}_\alpha$ at y, i.e. the projective tangent space of $\hat{S}_\alpha$ at y. We claim that the assumed transversality $\mathbb{P}(\ker \check{H}^{i+1}) \pitchfork_y \hat{S}_\alpha$ in H^∞ implies that $\omega' := (\omega_1, \dots, \omega_{N-1-i})$ is a submersion on $T_{x_k} X_{\text{reg}}$, for all x_k close enough to y. If this were not true, then we would have the following nontransversality in $\mathbb{P}^N$: $\overline{\omega'^{-1}(\omega'(x_k))} \not\pitchfork_{x_k} \langle T_{x_k} \rangle$, and, by passing to the limit, this implies:

$$(\overline{\omega'^{-1}(b)} \cap H^\infty) \not\pitchfork_y (\langle T_y \rangle \cap H^\infty). \tag{6.18}$$

We have $\overline{\omega'^{-1}(b)} \cap H^\infty = \mathbb{P}(\ker \check{H}^{i+1})$ and $\langle T_y \rangle \supset \langle T_y \hat{S}_\alpha \rangle$ due to the Whitney (a)-regularity. Therefore (6.19) contradicts the transversality $\hat{S}_\alpha \pitchfork \mathbb{P}(\ker \check{H}^{i+1})$.

Let now $\omega := (\omega_1, \dots, \omega_{N-i})$ be a basis of $\check{H}^i$ which extends the basis $\omega' = (\omega_1, \dots, \omega_{N-1-i})$ of $\check{H}^{i+1}$. Viewed as a linear map, ω is not surjective on the tangent spaces $T_{x_k} X_{\text{reg}}$, by our hypothesis. Due to the transversality of the intersection $\overline{\omega'^{-1}(\omega'(x_k))} \cap \langle T_{x_k} \rangle$ proved above, this nonsurjectivity is equivalent to the following nontransversality in $\mathbb{P}^N$:

$$\overline{\omega_{N-i}^{-1}(\omega_{N-i}(x_k))} \not\pitchfork_{x_k} \langle T_{x_k} \rangle \cap \overline{\omega'^{-1}(\omega'(x_k))},$$

which amounts to the inclusion:

$$\overline{\omega_{N-i}^{-1}(\omega_{N-i}(x_k))} \supset \langle T_{x_k} \rangle \cap \overline{\omega'^{-1}(\omega'(x_k))}$$

since the left-hand side is a hyperplane in $\mathbb{P}^N$. Slicing both sides by H^∞, we get the following inclusion in H^∞:

$$\mathbb{P}(\ker \omega_{N-i}) \supset (\langle T_{x_k} \rangle \cap H^\infty) \cap \mathbb{P}(\ker \check{H}^{i+1}).$$

Finally we pass to the limit and, observing that only the space $\langle T_{x_k} \rangle$ varies, we get:

$$\mathbb{P}(\ker \omega_{N-i}) \supset (\langle T_y \rangle \cap H^\infty) \cap \mathbb{P}(\ker \check{H}^{i+1}).$$

This implies $y \in \mathbb{P}(\ker \omega_{N-i})$ so that $y \in \mathbb{P}(\ker \check{H}^i)$ and contradicts the transversality $\hat{S}_\alpha \pitchfork \mathbb{P}(\ker \check{H}^i)$ in H^∞, since $\langle T_y \rangle \supset \langle T_y \hat{S}_\alpha \rangle$ by the Whitney (a)-regularity.

The properness of the map $(\omega_1, \ldots, \omega_{N-1-i}) : P_{N-1-i}(X, \check{H}^i) \to \mathbb{C}^{N-1-i}$ is now proved. This map has also finite fibres, because $P_{N-1-i}(X, \check{H}^i)$ is an affine algebraic variety (or it is empty). $\square$

Let $\check{H}^s \in \Omega^s(X) \cap \hat{\Omega}^s(X)$ for $s = N - 1 - k, N - k$, where $1 \le k \le n$. Then $P_k(X, \check{H}^{N-1-k})$ is purely k-dimensional or empty. The degree of the polar variety $P_k(X, \check{H}^{N-1-k})$, denoted by $\gamma_k(X, \check{H}^{N-1-k})$, is the degree of the proper finite map:

$$(\omega_1, \ldots, \omega_k) : P_k(X, \check{H}^{N-1-k}) \to \mathbb{C}^k,$$

i.e. the number of points in a generic fibre. We set $\gamma_0(X, \check{H}^{N-1}) := \#P_0(X, \check{H}^{N-1})$, which makes sense since $P_0(X, \check{H}^{N-1})$ is a finite set of points.

Here is the sketch of a direct argument that the number $\gamma_k(X, \check{H}^{N-1-k})$ is independent of the choice of $\check{H}^{N-1-k}$ in some Zariski-open dense subset of $\Omega^{N-1-k}(X) \cap \hat{\Omega}^{N-1-k}(X)$. By repeated slicing, it is sufficient* to prove this for the degree γ_0.

There exists a Zariski-open subset Ω of $G^{N-1}(\check{\mathbb{C}}^N)$ such that all the pencils l_H with $\check{H} \in \Omega$ have homeomorphic general fibres. This was proved in §6.1, the proof involves conormals and discriminants and uses the algebricity of the discriminants. Then one applies the Lefschetz principle at the level of Euler characteristics, more precisely one uses the relation (6.2) proved in §6.1, and gets:

$$\chi(X_{\text{reg}}) = \chi(l_H^{-1}(c) \cap X_{\text{reg}}) + (-1)^n \#\text{Morse points of the pencil } l_H \text{ on } X_{\text{reg}},$$

for some generic value c of the pencil l_H. Since the two Euler characteristics of this formula are independent of $\check{H} \in \Omega$, it follows that the number of Morse points is independent too, and this number is just $\gamma_0(X, \check{H})$ (see also the next

---

* One may check [ScTi] for the details of this reduction.

paragraph). Let us remark that $\gamma_0(X, \check{H})$ is precisely $\alpha_{X,H} = \alpha_X^{(n)}$, the polar number defined in §6.1.

Definition 6.3.9 We say that the number $\gamma_k(X) := \gamma_k(X, \check{H}^{N-1-k})$ is the *degree of the polar variety* $P_k(X, \check{H}^{N-1-k})$. It is independent on $\check{H}^{N-1-k}$ varying in some Zariski-open subset of $G^{N-1-k}(\check{\mathbb{C}}^N)$.

All the Zariski-open subsets which occurred up to now depend on the chosen Whitney stratification S of X. However the generic polar cycle classes $\check{c}_k^{Ma}(X) = [P_k(X, \check{H}^{N-1-k})]$ and the degrees $\gamma_k(X)$ do not.

The degree $\gamma_k(X)$ can also be described as follows:

$$\gamma_k(X) = (\omega_1, \ldots, \omega_k)_*([P_k(X, \omega)]) \in CH_k(\mathbb{C}^k) \simeq \mathbb{Z}.$$

The geometric interpretation of the degrees. $\gamma_k(X)$ is the number $\alpha_X^{(n-k)}$ of Morse points of a Lefschetz pencil on a generic k-times repeated hyperplane slice of X_{reg}, which was defined in §6.1. Degrees of global affine polar varieties appear implicitly in the definition of the asymptotical equisingularity of families of hypersurfaces (cf. [Ti6], see §7.2) and in the proof of the Lefschetz type formula for the global Euler obstruction, Theorem 6.2.4.

An index formula. We go on with some fixed algebraic Whitney stratification S of $X \subset \mathbb{C}^N$ and some extension of it to a stratification $\hat{S}$ of $\bar{X}$. For instance, we may choose as S the canonical (roughest) Whitney stratification provided by Teissier's theorem on the Whitney equisigularity [Te3]. Here we only need $n = \dim X < N$, but X need not to be pure dimensional.

As it has been discussed before, the polar classes $[P_k(\bar{S})]$ and their degrees $\gamma_k(\bar{S})$ are well defined, for any stratum $S \in S$. In particular, we have $P_{\dim S}(\bar{S}) = \bar{S}$ and $\gamma_{\dim S}(\bar{S}) = \deg S$. By definition $P_k(\bar{S}) = \emptyset$ and $\gamma_k(\bar{S}) = 0$ for $k > \dim S$.

We may define the *generalized degree* $\gamma_k(\alpha)$ of a S-constructible function α, as follows:

$$\gamma_k(\alpha) := \sum_{S \in S} (-1)^{\dim S} \cdot \eta(S, \alpha)\gamma_k(\bar{S}).$$

Let us recall that the *Euler characteristic weighted by* α is:

$$\chi(X, \alpha) := \sum_{S \in S} \alpha(S) \cdot \chi(H_c^*(S)),$$

where the compact support cohomology $H_c^*(S)$ is finite-dimensional since the strata are locally closed algebraic sets. We have the following index theorem:

Theorem 6.3.10 [ScTi] *For generic coordinates* $\{x_i\}_{i=1}^N$ *and for all S-constructible functions α and $0 \le k \le n$ we have:*

$$
\begin{aligned}
(-1)^k \gamma_k(\alpha) &= \chi(X \cap \{x_1 = t_1, \ldots, x_k = t_k\}, \alpha) \\
&\quad - \chi(X \cap \{x_1 = t_1, \ldots, x_{k+1} = t_{k+1}\}, \alpha).
\end{aligned}
\tag{6.19}
$$

It follows that $\gamma_k(\alpha)$ is independent of the choice of the coordinates $\{x_i\}$ and values $\{t_i\}$, provided all these are generic enough, since this is the case for the weighted Euler characteristic of the slices occurring in formula (6.20).

Proof of Theorem 6.3.10 We consider the generic Lefschetz pencil defined by the generic linear function x_{k+1} on the stratified transversal slice $X \cap \{x_1 = t_1, \ldots, x_k = t_k\}$ and we apply the Euler characteristic $\chi(\cdot, \alpha)$. By the geometric interpretation of the degree, the number of singular points of the pencil x_{k+1} on every stratum $S' := S \cap \{x_1 = t_1, \ldots, x_k = t_k\}$ is equal to $\gamma_k(\bar{S})$, for dim $S \ge k$. We get by the Lefschetz slicing principle: $\chi(X \cap \{x_1 = t_1, \ldots, x_k = t_k\}, \alpha) - \chi(X \cap \{x_1 = t_1, \ldots, x_{k+1} = t_{k+1}\}, \alpha) = \sum_{S \in \mathcal{S}} \gamma_k(\bar{S}) \chi(\mathrm{LMD}(x_{k+1|S'}, \alpha))$. The notation $\mathrm{LMD}(x_{k+1|S'})$ stands for the local Morse datum of x_{k+1} at some critical points on the stratum S' of the transversal slice $X \cap \{x_1 = t_1, \ldots, x_k = t_k\}$ and is independent on the choice of such a critical point. Notice that the terms of the sum corresponding to strata S of dimension $< k$ are zero since $\gamma_k(\bar{S}) = 0$ in this case.

We may use the decomposition of the local Morse datum $\mathrm{LMD}(x_{k+1|S'})$ as a product of normal and tangential Morse data [GM2, Theorem 3.7]. Then, by applying the Euler characteristic, we get:

$$
\chi(\mathrm{LMD}(x_{k+1|S'}, \alpha)) = (-1)^{\dim S'} \cdot \chi(\mathrm{NMD}(S', \alpha)),
$$

where $\dim S' = \dim S - k$. Normal Morse data do not change by taking generic slices so one has the equality $\chi(\mathrm{NMD}(S', \alpha)) = \chi(\mathrm{NMD}(S), \alpha))$. Since $\chi(\mathrm{NMD}(S), \alpha))$ is by definition the normal index $\eta(S, \alpha)$, formula (6.20) follows. $\square$

An immediate consequence is that a closed complex algebraic subset $X \subset \mathbb{C}^N$ of pure dimension $n < N$ satisfies the equality:

$$
\chi(X, \alpha) = \sum_{k=0}^{n} (-1)^k \gamma_k(\alpha).
\tag{6.20}
$$

For $\alpha = 1_X$, this calculates the global Euler characteristic $\chi(X)$.[4]

It is well known that the local Euler obstruction is a constructible function with respect to any Whitney stratification [Dub1, BrSc, LiTi, Sch]; in case

$\alpha = \mathrm{Eu}_X$, formula (6.21) specializes to the formula (6.8) for the global Euler obstruction, since one has:

$$\eta(S, \mathrm{Eu}_X) = \begin{cases} 1, & \dim S = \dim X \\ 0, & \dim S < \dim X. \end{cases} \tag{6.21}$$

As a final remark, the above study of the global affine algebraic setting applies to the local analytic setting too. One may recover in this way some of Massey's results [Mas], see [ScTi] for a few more details.

Exercises

6.1 *Isolated hypersurface singularities.* Let $Y \subset \mathbb{C}^{n+1}$ be a hypersurface with isolated singularities. Show that the Milnor number of the complex link $\mathrm{CL}_Y(\{q\})$ at some isolated singularity q is equal to the sectional Milnor–Teissier number $\mu_q^{\langle n-1 \rangle}(Y)$, see §7.3 for the definition. Prove the following formula:

$$\chi(Y) = \sum_{i=0}^{n} (-1)^i \alpha_Y^{(i)} + (-1)^n \sum_{q \in \mathrm{Sing}\, Y} \mu_q^{\langle n-1 \rangle}(Y). \tag{6.22}$$

Note that the sum $\sum_{i=0}^{n-1} (-1)^i \alpha_Y^{(i)}$ is precisely the Euler characteristic $\chi(Y \cap \mathcal{H})$ of the intersection with a generic hyperplane.

6.2 We recall that $\mathrm{Eu}_Y(\mathcal{A}_i)$ denotes the local Euler obstruction of some point of a stratum $\mathcal{A}_i$. In the above notations, prove the formulas:

$$\mathrm{Eu}(Y) = \sum_{\mathcal{A}_i \in \mathcal{A}} [\chi(H_t \cap \mathcal{A}_i) \cdot \mathrm{Eu}_Y(\mathcal{A}_i)] + (-1)^n \alpha_Y^{(n)}. \tag{6.23}$$

$$\mathrm{Eu}(Y) = \sum_{i \in \Lambda} \chi(\mathcal{A}_i)\, \mathrm{Eu}_Y(\mathcal{A}_i). \tag{6.24}$$

$$\chi(Y) = \sum_{i \in \Lambda} \mathrm{Eu}(\mathcal{A}_i)\, \chi(\mathrm{NMD}(\mathcal{A}_i)). \tag{6.25}$$

7

Relative polar curves and families of affine hypersurfaces

7.1 Local and global relative polar curves

Local polar curves stand as a key device, relating algebro-geometric properties to topological properties. They have been systematically used in the study of complex analytic spaces and of holomorphic functions in the 1970s by Kleiman, Lê, Hamm, Teissier, Merle, e.g. in [Kl, HmL1, Lê5, LêTe, Te2, Te3, Me]. Many other authors developed this study and found more and more applications of polar curves.[1]

In the preceding section we have discussed higher-dimensional polar loci of varieties. Polar curves appeared in §2 and §3. Here we focus on *relative polar curves* with respect to a given function on the space.

In the local setting, the *local polar curve theorem*, proved by Hamm and Lê in [HmL1, §2], states that, given a holomorphic function germ $f : (\mathbb{C}^{n+1}, 0) \to \mathbb{C}$, there exists a Zariski-open subset of linear forms $l : (\mathbb{C}^{n+1}, 0) \to \mathbb{C}$ such that the singular locus of the map germ (l, f) is a curve or it is empty. Lê D.T. also proved the following *equisingularity* result [Lê1]: if Sing$f$ is of dimension one, then the transversal singularity is μ-constant if and only if the generic local polar curve is empty. If this is the case, then Sing$f$ is a nonsingular curve germ. By a theorem due to Lê and Saito [LêSa], the absence of polar curve is equivalent to the Thom (a_f)-condition along the line Sing$f$.

The global affine setting. We have already defined in §2.1 the polar locus of a polynomial function. Here we give a more general definition, in the setting of singular underlying spaces, which involves the singular locus of a $\mathbb{K}$-analytic map on a singular space (Definition 2.2.1). As before, $\mathbb{K}$ stays for $\mathbb{R}$ or $\mathbb{C}$.

Definition 7.1.1 (**Polar locus**) Let $\mathcal{X} \subset \mathbb{K}^N$ be a $\mathbb{K}$-analytic set endowed with a $\mathbb{K}$-analytic, locally finite stratification $\mathcal{S} = \{S_i\}_{i \in I}$, which satisfies Whitney

(a) property. Given two $\mathbb{K}$-analytic functions $h, g : \mathcal{X} \to \mathbb{C}$, we say that the set:

$$\Gamma_S(h, g) := \text{closure}\{\text{Sing}_S(h, g) \setminus (\text{Sing}_S h \cup \text{Sing}_S g)\}$$

is the *polar locus* of (h, g) with respect to S.

If the function h is the restriction of a linear form $l : \mathbb{K}^N \to \mathbb{K}$, we say that $\Gamma_S(l, g)$ is the *polar locus of g relative to l*. As before, we denote by l_H the linear form associated to the hyperplane $H \in \check{\mathbb{P}}^{N-1}$.

We state a Bertini–Sard type result in our global setting, which reminds us of the spirit of Kleiman's transversality theorem 6.3.2. It extends to the global setting the above recalled local polar curve theorem [HmL1]. This will be the departure point for several streams of research, which we shall present hereafter.

Theorem 7.1.2 (Polar curve theorem) *Let $f : X \to \mathbb{K}$ be an algebraic function on an algebraic subset $X \subset \mathbb{K}^N$. There is a Zariski-open subset Ω_f of the dual projective space $\check{\mathbb{P}}^{N-1}$ such that, for any $H \in \Omega_f$, the polar locus $\Gamma_S(l_H, f)$ is a curve or it is empty.*

Proof We refer to the notations in §1.2. Since X may be singular, let $S = \{S_i\}_{i \in I}$ be a locally finite family which stratifies X with the Whitney (a) condition. By classical results due to Łojasiewicz [Łoj1] and Hironaka [Hiro] (see also [GWPL] and [GM2]), a stratification with complex algebraic, resp. real algebraic, strata exists and has finitely many strata.

We prove that the restriction of the polar locus $\Gamma_S(l_H, f)$ to any positive-dimensional stratum S_i is either a curve or empty. For some stratum S_i of dimension ≥ 1, we consider the projectivised relative conormal:

$$\mathbb{P}T^*_{f|S_i} := \text{closure}\{(x, \xi) \in S_i \times \check{\mathbb{P}}^{N-1}) \mid \xi(T_x(f^{-1}(f(x)) \cap S_i))$$

$$= 0\} \subset \bar{S}_i \times \check{\mathbb{P}}^{N-1},$$

where $\bar{S}_i$ denotes the closure of S_i in X; it is an algebraic set by our assumptions on the stratification S. We may then remark that $\mathbb{P}T^*_{f|S_i}$ is a $\mathbb{K}$-algebraic set of dimension N. Let us denote by C_i its closure in $\mathbb{P}^N \times \check{\mathbb{P}}^{N-1}$. For any $i \in I$, let $\pi : C_i \to \mathbb{P}^N$ be the projection to the first factor and $\gamma : C_i \to \check{\mathbb{P}}^{N-1}$ the one to the second factor. Then, for any $H \in \check{\mathbb{P}}^{N-1}$, the polar locus $\Gamma_S(l_H, f) \cap S_i$ is included in $\pi(\gamma^{-1}(l_H))$.

Since C_i is compact algebraic, the algebraic map γ is proper and we may apply Verdier's [Ve, Théorème 3.3]. The argument runs, briefly, as follows: the image of the algebraic subset of the points of C_i, where γ is not a stratified submersion (for some fixed algebraic stratification of C_i), is a strict algebraic subset of $\mathbb{P}^{N-1}$. Its complement is then a Zariski-open subset $\Omega_i \subset \check{\mathbb{P}}^{N-1}$. It has the property that, for any $H \in \Omega_i$, H is a regular value of the restriction $\gamma_{|(C_i)_{reg}}$ and, since $\dim(C_i)_{reg} = N$, that $\dim \gamma^{-1}(H) = 1$ or $\gamma^{-1}(H) = \emptyset$.

The restriction $\pi_{|\gamma^{-1}(H)}$ is obviously one-to-one, by the definition of $\mathbb{P}T^*_{f|S_i}$. We may then define $\Omega_f := \cap_{i \in I} \Omega_i$. □

Since we need more 'good' properties of the polar curve, there is the following supplement to Theorem 7.1.2:

Corollary 7.1.3
(a) There exists a Zariski-open subset $\Omega'_f \subset \Omega_f$ such that, for any $H \in \Omega'_f$, the polar curve $\Gamma_S(l_H, f)$ is reduced or it is empty.
(b) Let $c \in \mathbb{K}$ be a fixed value. There exists a Zariski-open subset $\Omega'_{f,c} \subset \Omega'_f$ such that, if $H \in \Omega'_{f,c}$, then no component of $\Gamma_S(l_H, f)$ is included into $X_c := f^{-1}(c)$ and the restriction $l_{H|} : X_c \to \mathbb{C}$ is a locally trivial stratified fibration in the neighbourhood of $\bar{X}_c \cap H^\infty$.

Proof The first claim follows from the fact that the property of being reduced is an open property. For the second claim, we just take $Y = X_c$ in Proposition 6.1.1 □

Polar curves and topology of families of affine hypersurfaces. We stick from now on to the complex setting. Let $\{X_s\}_{s \in \delta}$ be a family of affine hypersurfaces $X_s \subset \mathbb{C}^{n+1}$, where δ is a small disk at the origin of $\mathbb{C}$. We assume (compare to §5.1) that the family $\{X_s\}_{s \in \delta}$ is *polynomial*, i.e. there is a polynomial $F : \mathbb{C}^{n+1} \times \mathbb{C} \to \mathbb{C}$ such that $X_s = \{x \in \mathbb{C}^{n+1} \mid F_s(x) = F(x, s) = 0\}$. Let $X = \cup_{s \in \delta} X_s$ be the total space of the family, which is itself a hypersurface in $\mathbb{C}^{n+1} \times \delta$. Let $\sigma : X \to \delta \subset \mathbb{C}$ denote the projection of X to the second factor of $\mathbb{C}^{n+1} \times \mathbb{C}$.

We endow the affine hypersurface X with the canonical (minimal) Whitney stratification S, cf. [Te3]. We recall that there exist finite Whitney stratifications of any algebraic set such that the closure of each stratum is algebraic, see [GWPL], and such that $X \setminus \mathrm{Sing}X$ is a stratum. The roughest, in the sense of Teissier, is the 'canonical' one. For instance, if X_s has no singularities and X_0 has at most isolated ones, then S consists of $X \setminus \mathrm{Sing}X_0$ and the point-strata $\mathrm{Sing}X_0$. For some linear form $l_H : \mathbb{C}^{n+1} \to \mathbb{C}$ corresponding to $H \in \check{\mathbb{P}}^n$, we shall use the same notation l_H for the application $\mathbb{C}^{n+1} \times \mathbb{C} \to \mathbb{C}, (x, s) \mapsto l(x)$, as well as for its restriction to X. Let $\Gamma_S(l_H, \sigma)$ denote the polar locus of the map $(l_H, \sigma) : X \to \mathbb{C}^2$ with respect to S.

We have the following version of the polar curve theorem for a family of hypersurfaces. Let us point out that Lemma 2.1.4 becomes a particular case of it, when the family X is given by the fibres $X_s := f^{-1}(s)$ of a polynomial $f : \mathbb{C}^{n+1} \to \mathbb{C}$.

Corollary 7.1.4 Let $X \subset \mathbb{C}^{n+1} \times \mathbb{C}$ be the total space of a polynomial family of affine hypersurfaces as above. There is a Zariski-open set $\Omega_\sigma \subset \check{\mathbb{P}}^n$ such that, for any $H \in \Omega_\sigma$, the polar locus $\Gamma_S(l_H, \sigma)$ is a curve or it is empty.

Notation 7.1.5 Similarly to the notation in Corollary 7.1.3(b), let $\Omega_{\sigma,0} \subset \Omega_\sigma$ denote the Zariski-open subset of hyperplanes which are also transversal to the canonical Whitney stratification of the projective hypersurface $\overline{X_0} \subset \mathbb{P}^{n+1}$. This supplementary condition insures that $\dim(\Gamma_S(l_H, \sigma) \cap X_0) \le 0$, for any $H \in \Omega_{\sigma,0}$.

Global polar numbers. One may define *global generic polar intersection multiplicities* $\alpha_{X_s}^{(i)}$ for a family of hypersurfaces[2] as in the particular case (6.3). We consider the germ of the family X at $\mathbb{C}^{n+1} \times \{0\}$ and the Zariski-open subset $\Omega_{\sigma,0}$. We get that for all $H \in \Omega_{\sigma,0}$, $\dim(\Gamma_S(l_H, \sigma) \cap X_0) \le 0$ and that for any $s \in \delta$ and small enough disk δ, the intersection multiplicity:

$$\alpha_{X_s}^{(n)} := \operatorname{mult}(\Gamma_S(l_H, \sigma), (X_s)_{\mathrm{reg}}) \tag{7.1}$$

is well defined as the sum of local intersection multiplicities at the points of intersection of the polar curve with the regular part $(X_s)_{\mathrm{reg}}$ of the hypersurface X_s, and moreover does not depend on the choice of $H \in \Omega_{\sigma,0}$.

The Whitney stratification S on X induces a Whitney stratification S_s on X_s, for all s within a small enough pointed disk. The strata of S_s are precisely the intersections of X_s with the strata of S, since X_s is transversal to S for small enough $|s| \ne 0$. It then follows that, for $H \in \Omega_{\sigma,0}$, the affine pencil l_H is general, not only for X_0, but also for all hypersurfaces X_s with small enough $|s| \ne 0$.

As we have seen in §6.1, the *geometric interpretation* of $\alpha_{X_s}^{(n)}$ is the number of Morse points of a generic linear function on $(X_s)_{\mathrm{reg}}$. This depends of course on the embedding of X_s into $\mathbb{C}^{n+1}$. If X_s is a general hypersurface (cf. Definition 4.1.1) of degree d, then, by Bezout's theorem, $\alpha_{X_s}^{(n)} = d(d-1)^n$.

We define the global polar intersection multiplicities $\alpha_{X_s}^{(i)}$ of lower order, as in §6.1. The idea is to consider successively general hyperplane slices of our family and apply to them the definition (7.1). It is similar to Teissier's construction of local polar multiplicities [Te1, Te2].

Namely, we take a general hyperplane $\mathcal{H} \in \Omega_{\sigma,0}$ and denote by $\alpha_{X_s}^{(n-1)}$ the global generic polar intersection multiplicity at $s \in \delta$ of the family of affine hypersurfaces $X' = X \cap \mathcal{H}$. In the last step we get $\alpha_{X_s}^{(0)} = \deg X_s$.

By the standard connectivity argument, the polar intersection multiplicities $\alpha_{X_s}^{(i)}$ do not depend on the choices of generic hyperplanes. They are also invariant up to linear changes of coordinates but not invariant up to nonlinear changes of coordinates (e.g. $\deg X_s$ is not invariant). The numbers $\alpha_{X_s}^{(i)}$ are constant on $\delta \setminus \{0\}$, provided that δ is small enough. We shall use the shorter notation $\alpha_s^{(i)}$ whenewer the family is fixed.

We have seen that the global polar intersection multiplicities enter in the description of the CW-complex structure (in particular, in the Euler characteristic) and that of the global Euler obstruction and of the MacPherson cycle of an affine space Y. We will show next that the global polar intersection multiplicities are key ingredients in an *equisingularity* condition at infinity for families of affine hypersurfaces and in a *Plücker formula* for affine hypersurfaces. As a by-product, we show how to study the evolution of the *integral of curvature* in families of affine hypersurfaces. In all those problems, the cornerstone is the *defect of intersection multiplicity* of a germ at 0 of a family $\{X_s\}_{s \in \delta}$, in other words the *asymptotic loss* of intersection multiplicity towards the 'ends' of X_0.

7.2 Asymptotic equisingularity

We define an equisingularity notion for a family $\{X_s\}_{s \in \delta}$ of affine hypersurfaces $X_s \subset \mathbb{C}^{n+1}$ with at most isolated singularities, such that to be controlled by numerical invariants and to imply the C^∞-triviality of this family in the neighbourhood of infinity. It turns out that this asymptotical equisingularity is different from the Whitney equisingularity. The following presentation is based on [Ti6, Ti4].

Equisingularity and topological triviality. As in the preceding section §7.1, let $F : \mathbb{C}^{n+1} \times \mathbb{C} \to \mathbb{C}$ be the polynomial defining the family of hypersurfaces $X_s = \{x \in \mathbb{C}^{n+1} \mid F_s(x) = F(x, s) = 0\}$. Let $X = \cup_{s \in \delta} X_s$ be the total space of the family, which is itself a hypersurface in $\mathbb{C}^{n+1} \times \delta$, where δ is a small disk centred at $0 \in \mathbb{C}$. Let $\sigma : X \to \delta \subset \mathbb{C}$ denote the projection of X to the second factor. We consider the total space of the projective compactification of the hypersurfaces:

$$\mathbb{X} := \{\tilde{F}_s(x_0, x) = 0\} \subset \mathbb{P}^{n+1} \times \delta$$

and the restriction to $\mathbb{X}$ of the projection $\hat{\sigma} : \mathbb{P}^{n+1} \times \delta \to \delta$, which is the proper extension of σ. As usual, $\tilde{F}_s$ denotes the polynomial obtained by homogenizing F_s by the new variable x_0, and $\mathbb{X}^\infty$ denotes the divisor at infinity $\mathbb{X} \cap \{x_0 = 0\}$. We have introduced in Definition 1.2.9 the subspace of *characteristic covectors at infinity* C^∞, which is an analytic subspace of the restriction over $\mathbb{X}^\infty$ of the cotangent bundle $T^*(\mathbb{P}^{n+1} \times \delta)$. Since C^∞ is conical, we may consider its projectivization $\mathbb{P}C^\infty$, and also denote by $\bar{\pi}$ the projection $\mathbb{P}T^*(\mathbb{P}^{n+1} \times \delta) \to$ $\mathbb{P}^{n+1} \times \delta$. For some subset $S \subset \mathbb{X}^\infty$, we denote $\mathbb{P}C^\infty(S) := \mathbb{P}C^\infty \cap \bar{\pi}^{-1}(S)$.

Definition 7.2.1 (equisingularity at infinity)
We say that the germ at $0 \in \delta$ of the family of hypersurfaces $\{X_s\}_{s\in\delta}$ is *equisingular at infinity* if $(p, d\hat{\sigma}) \notin \mathbb{P}C^\infty$ for all $p \in \mathbb{X}_0^\infty$.

This is a slight generalization of the 't-regularity' introduced in §1.2. The condition $(p, d\hat{\sigma}) \notin \mathbb{P}C^\infty$ is equivalent to the fact that the limit of the tangent spaces $T_x X_{\sigma(x)}$ as $x \to p$, whenever it exists in the appropriate Grassmannian, is not in the dual $\check{C}^\infty(p)$ of $C^\infty(p)$.

Definition 7.2.2 We say that the family $\{X_s\}_{s\in\mathbb{C}}$ is C^∞-*trivial at infinity* at $s_0 \in \mathbb{C}$ if there is a (large) ball $B_0 \subset \mathbb{C}^{n+1}$ centered at 0 and a disk $\delta \subset \mathbb{C}$ centered at s_0 such that, for any ball $B \supset B_0$ centered at 0, the restriction $\sigma_| : (X \setminus (B \times \mathbb{C})) \cap \sigma^{-1}(\delta) \to \delta$ is a C^∞-trivial fibration.

We have seen in §1.2, where we have investigated the fibres of a polynomial, that a family of hypersurfaces may fail to be topologically trivial because of the behaviour at infinity. We show in this more general setting that the triviality at infinity is insured by the equisingularity.

Theorem 7.2.3 *If the germ at 0 of the family $\{X_s\}_{s\in\mathbb{C}}$ is equisingular at infinity, then it is C^∞-trivial at infinity at 0.*

Proof The proof can be extracted from §1.2 We give just a brief account. We construct a C^∞ real nonnegative function ϕ on a neighbourhood of $\mathbb{X}_0^\infty$ by patching together the local equations of $\mathbb{X}^\infty$, such that the map σ is transversal to the positive levels of ϕ within some open neighbourhood V of the compact set $\mathbb{X}_0^\infty$. Consequently, there exists a bounded C^∞ vector field $\mathbf{v}$ which lifts the vector field $\frac{\partial}{\partial\sigma}$ over a small enough disk δ, to the open set $V \cap X \cap \sigma^{-1}(\delta)$ and is tangent to the positive levels of ϕ.

Since the restriction $\sigma_| : (\partial\bar{B} \times \delta) \cap X \to \mathbb{C}$ is a proper submersion, for big balls B and small enough disk δ, we may take a C^∞ vector field $\mathbf{w}$ on $((B' \setminus B) \times \delta) \cap X$, which lifts $\frac{\partial}{\partial\tau}$ and is tangent to $(\partial\bar{B} \times \delta) \cap X$. Then we glue it by a C^∞ partition of unity to the above defined vector field $\mathbf{v}$ and by integrating this new vector field, we get the desired C^∞-trivialization. $\qquad\square$

Corollary 7.2.4 If a family $\{X_s\}_{s\in\mathbb{C}}$ of nonsingular affine hypersurfaces is equisingular at infinity at $0 \in \mathbb{C}$, then $\sigma : X \to \mathbb{C}$ is a C^∞ locally trivial fibration at 0.

Proof Using the notations of the above proof and the same argument as for constructing the vector field $\mathbf{w}$ above, we construct this time a C^∞ vector field

$\mathbf{w}'$ on $(B \times \delta) \cap X$, which lifts $\frac{\partial}{\partial \sigma}$ and is tangent to $(\partial \bar{B} \times \delta) \cap X$, then glue it by a C^∞ partition of unity to the previously defined vector field $\mathbf{v}$. $\square$

Equisingularity and global polar invariants. We have defined in §7.1 the global generic polar intersection multiplicities $\alpha_s^{(i)} := \alpha_{X_s}^{(i)}$ for a family of hypersurfaces of dimension n, where the highest order one is:

$$\alpha_s^{(n)} := \mathrm{mult}(\Gamma_S(l_H, \sigma), (X_s)_{\mathrm{reg}}), \tag{7.2}$$

and $\Gamma_S(l_H, \sigma)$ denotes the polar curve relative to $H \in \Omega_{\sigma,0}$, with respect to the stratification S of X.

Definition 7.2.5 For $i \in \{0, \ldots, n\}$, we call *i-defect at infinity, at* $0 \in \mathbb{C}$ the following number:

$$\lambda_0^{(i)} := \alpha_s^{(i)} - \alpha_0^{(i)},$$

where $s \neq 0$ is close enough to 0.

From the CW-complex model explained in §6.1, p. 89, we get precise information on the evolution of the Euler characteristic of the hypersurfaces in the family, see also Exercise 7.2. But the central result that we want to prove here is that equisingularity at infinity is controlled by the constancy of the numbers $\alpha^{(i)}$.

Theorem 7.2.6 *Let* $F : \mathbb{C}^{n+1} \times \mathbb{C} \to \mathbb{C}$ *be a polynomial function defining a family of hypersurfaces* $X_s \subset \mathbb{C}^{n+1}$. *Assume that there is a compact set* $K \subset \mathbb{C}^{n+1}$ *such that* $\mathrm{Sing} X_s \subset K$ *for all* $s \in \delta$, *where* $\delta \subset \mathbb{C}$ *is some disk centered at 0.*

Then the family $\{X_s\}_{s \in \mathbb{C}}$ *is equisingular at infinity at 0 if and only if the generic polar intersection multiplicities* $\alpha_s^{(i)}$, $i = \overline{0, n}$, *are independent of s in some small disk* $\delta' \subset \delta$ *centred at 0.*

This reminds us of the well-known local equisingularity result due to Teissier, which characterizes the Whitney equisingularity by the constancy of the sectional Milnor–Teissier numbers μ^*. We refer to Teissier [Te1], [Te3], and Briançon–Speder [BrSp2], see also (1.14) and (1.15) in §1.2. The difference between the local setting and our affine global setting appears again to be essentially due to the genericity failure: general slices in the affine space are not general any more in the neighbourhood of infinity. If we try to use Whitney equisingularity along some stratum on the part at infinity X^∞, then the generic local polar invariants might not be global invariants of our family of affine hypersurfaces. We send the reader to §1.2 for the comparison to the

Whitney equisingularity in terms of integral closures, which holds in the current setting too.

Remark 7.2.7 The constancy of $\alpha_s^{(n)}$ does not imply that $\alpha_s^{(i)}$ is constant for $i < n$. This can be compared with the similar assertion in the local case, which has been proved by Briançon and Speder [BrSp1]: a μ-constant family of isolated hypersurface germs is not necessarily μ^*-constant. A simple example which we may use in the global affine setting is the following: $\{X_s\}_{s \in \mathbb{C}}$ is the family of fibres of the polynomial in three variables $f(x, y, z) = x + x^2 y$. We can easily see that $\alpha_s^{(2)}$ is constant, whereas $\alpha_s^{(1)}$ is not, since $\lambda_0^{(1)} = 1$.

Remark 7.2.8 Let $\{X_s\}_{s \in \delta}$ be a family of *smooth* hypersurfaces, where $\delta \subset \mathbb{C}$ is some disk. By using Theorem 7.2.6, the α^*-constancy implies that this family is C^∞ trivial over δ, provided this disk is small enough (see also Exercise 7.5). In particular, all the hypersurfaces have the same Euler characteristic.

It is, however, not true that the invariance of the Euler characteristic implies the invariance of α^*. We can show this fact by the example $f(x, y, z) = x + x^2 y z$. Up to homotopy type, the fibre $X_0 = f^{-1}(0)$ is the disjoint union of $\mathbb{C}^2$ and a torus $\mathbb{C}^* \times \mathbb{C}^*$, whereas the fibre $X_s = f^{-1}(s)$, for $s \neq 0$, is the disjoint union of $(\mathbb{C}^* \setminus \{s\}) \times \mathbb{C}^*$ and of the two coordinate axes in the plane $\{x = s\}$. The Euler characteristic of all fibres is therefore equal to one. On the other hand, by an easy computation, the defects at infinity $\lambda_{(0)}^2$ and $\lambda_{(0)}^1$ are both positive (see Example 7.2.10).

Proof of Theorem 7.2.6. '$\Leftarrow$'. We assume that the generic polar intersection multiplicities $\alpha_s^{(i)}$ are constant, for all s in some small disk centred at 0. This means in particular that $\deg F_s := \alpha_s^{(0)}$ is independent of s.

Extending Definition 1.2.10, we call the *singular locus at infinity* of σ the set $\mathrm{Sing}^\infty \sigma := \{p \in \mathbb{X}^\infty \mid (p, d\hat{\sigma}) \in \mathbb{P}C^\infty(p)\}$. Then the family $\{X_s\}_s$ is equisingular at infinity at 0 if and only if the compact analytic set $\mathrm{Sing}_0^\infty \sigma := \mathrm{Sing}^\infty \sigma \cap \hat{\sigma}^{-1}(0)$ is empty.

Suppose that the family $\{X_s\}_s$ is not equisingular at infinity. We will prove that the constancy of the numbers α_s^* implies $\mathrm{Sing}_0^\infty \sigma = \emptyset$, which gives a contradiction. The proof is in two steps: in the first one we show that, after a number $k \le n - 1$ of times of slicing the family $\{X_s\}_s$ by general hyperplanes, the singular locus at infinity at 0 reduces to dimension zero. The second step will then show that this contradicts our hypothesis $\lambda_0^{(n-k)} = 0$.

Step 1: reduction to $\dim \mathrm{Sing}^\infty \sigma = 0$.
Let us assume that $\dim \mathrm{Sing}^\infty \sigma > 0$. For any hyperplane $H \subset \mathbb{C}^{n+1}$, we have $\dim \mathrm{Sing}_0^\infty \sigma \ge \dim(\mathrm{Sing}_0^\infty \sigma \cap \overline{H}) \ge \dim \mathrm{Sing}_0^\infty \sigma - 1$, where $\overline{H}$ is

the closure of H in $\mathbb{P}^{n+1}$. The hyperplane slice $\mathrm{Sing}_0^\infty \sigma \cap \overline{H}$ is contained in the singular locus at infinity $\mathrm{Sing}_0^\infty \sigma'$ of the restriction $\sigma' := \sigma_{|X^{(1)}}$, where $X^{(1)} := X \cap (H \times \mathbb{C})$. It follows that, by hyperplane slicing, the dimension of the singular locus at infinity $\mathrm{Sing}_0^\infty \sigma$ cannot diminish by more than one. We may identify $\overline{H}$ with $\mathbb{P}^n$ and continue the slicing process. Let us show that, after slicing, the family of hypersurfaces $X_s^{(1)} := X_s \cap H \subset \mathbb{C}^n$ satisfies the hypothesis of our theorem, i.e. that $\mathrm{Sing} X_s \cap H$ consists of isolated points which do not tend to infinity as $s \to 0$. Let $\mathbb{X}$ be endowed with an algebraic Whitney stratification, which is moreover a Thom $(a_{\hat\sigma})$ stratification such that both $\mathbb{X}_0$ and $\mathbb{X}_0 \cap H^\infty$ are unions of strata. Then we may choose a hyperplane $H \in \check{\mathbb{P}}^n$ such that it is transversal within $H^\infty = \mathbb{P}^n$ to all the strata in $\mathbb{X}_0 \cap H^\infty$. The strata of $\mathbb{X}_0 \cap H^\infty$ are finitely many since this space is compact and the stratification is algebraic. By a Bertini type argument, the hyperplanes having this property are generic, i.e. varying in some Zariski-open subset Ω' of $\check{\mathbb{P}}^n$. Then such a generic hyperplane will cut the spaces X_s transversely in some fixed neighbourhood of infinity, for all s in some small enough disk. This proves our claim.

Let us now observe that, after repeating a finite number of times the slicing with generic hyperplanes, the singular locus at infinity has to become zero dimensional. Indeed, after slicing $n-1$ times, we obtain a family of plane curves $\{X_s^{(n-1)}\}_{s \in \mathbb{C}}$. If we consider generic slicing, then these curves are reduced. Moreover, they are of degree $\alpha_s^{(0)}$, which is constant, as we have seen above. We then show that we have $\dim \mathrm{Sing}^\infty \sigma^{(n-1)} = 0$. Indeed, there exists a Whitney stratification $\mathcal{W}$ of $\mathbb{X}^{(n-1)}$, which has $X^{(n-1)} \setminus \mathrm{Sing} X^{(n-1)}$ as a stratum and has a finite number of point-strata on $(\mathbb{X}^{n-1})^\infty$. Since $\deg F_s$ is constant and since the divisor at infinity $(\mathbb{X}^{(n-1)})^\infty$ is of dimension one, there is a finite number of points where $\hat\sigma^{(n-1)} : (\mathbb{X}^{(n-1)})^\infty \to \mathbb{C}$ is not a stratified submersion. The singular locus at infinity $\mathrm{Sing}^\infty \sigma^{(n-1)}$ is included in this finite set. Indeed, at the other points at infinity $p \in (\mathbb{X}^{(n-1)})^\infty$ we have $(p, d\hat\sigma) \notin \mathbb{P}\mathcal{C}^\infty(p)$, since our Whitney stratification $\mathcal{W}$ is Thom (a_{x_0})-regular, where $x_0 = 0$ is an equation of $(\mathbb{X}^{n-1})^\infty$ at p, by Theorem A1.1.7, cf. [BMM, Théorème 4.2.1] or [Ti4, Theorem 2.9]. Our argument is now complete.

Step 2: the case $\dim \mathrm{Sing}_0^\infty \sigma = 0$.
Let us fix *generic coordinates* x_i on $\mathbb{C}^{n+1}$, i.e. such that the hyperplanes defined by x_i belong to the Zariski-open subset $\Omega_{\sigma,0}$. We have the following result:

Lemma 7.2.9 Let $p \in \mathbb{X}^\infty$ be an isolated point of the singular locus at infinity $\mathrm{Sing}_0^\infty \sigma$. Then the family $\{X_s\}_s$ is equisingular in the neighbourhood p if and only if the polar curve $\Gamma(x_0, \sigma)_p^{(i)} = \emptyset$ for all i.

This is Theorem 2.1.7 stated in our slightly more general setting and the same proof applies, by just adapting the notations.

In our situation, since $\dim \mathrm{Sing}_0^\infty \sigma = 0$, any point $p \in \mathrm{Sing}_0^\infty \sigma$ is an isolated singularity, and by Lemma 7.2.9 we conclude that the polar curve $\Gamma(x_0, \sigma)_p^{(i)}$ at p is not empty. Now, we may apply Proposition 2.1.3 to our setting and deduce the following equality of germs of polar curves:

$$\Gamma(x_0, \sigma)_p^{(i)} = \bar{\Gamma}(x_i, \sigma)_p. \tag{7.3}$$

The later is used in the definition (7.2) of the intersection multiplicity $\alpha_s^{(n)}$ and since it is not empty, this implies that there is loss of intersection multiplicity $\alpha_s^{(n)}$ as $s \to 0$. Therefore we get that $\alpha_s^{(n)}$ is not constant, in other words $\lambda_0^{(n)} \neq 0$, which gives the desired contradiction and ends our proof.

Proof of '$\Rightarrow$'. We prove that $\lambda_0^{(n-j)} = 0, \forall j \in \{0, \dots, n\}$. Let us then start with $j = 0$, and suppose that $\alpha_s^{(n)}$ is not constant. This implies that for any $l \in \Omega_{\sigma,0}$, the closure of $\Gamma(l, \sigma)$ in $\mathbb{P}^{n+1} \times \mathbb{C}$ contains some point $p \in \mathbb{X}_0^\infty$. Then, by the above equality of germs of polar curves (7.3), the local polar locus $\Gamma(x_0, \sigma)_p^{(i)}$ at p is not empty. This contradicts the assumption $(p, d\hat{\sigma}) \notin \mathbb{P}C^\infty(p)$.

To continue with the case $j = 1$, we shall slice by a generic hyperplane H such that to preserve the equisingularity hypothesis. We endow $\mathbb{X}$ with a finite complex Whitney stratification such that $\mathbb{X}^\infty$ and $\mathbb{X}_0^\infty$ are unions of strata. There exists a Zariski-open set $\Omega' \subset \check{\mathbb{P}}^n$ such that, if $H \in \Omega'$, then H is transversal to all strata of $\mathbb{X}_0^\infty$. Slicing by H will preserve the hypothesis $d\hat{\sigma}' \notin \mathbb{P}C^\infty(p), \forall p \in \mathbb{X}_0^\infty$, since our Whitney stratification at infinity is Thom (a_{x_0})-regular, by the results mentioned in Step 1. We have also shown in Step 1 that the singularities of the family $X_s \cap H$ do not intersect some neighbourhood of infinity. $\qquad \square$

In this way we prove step-by-step descent that $\lambda_0^{(n-j)} = 0$ for $\forall j \in \{0, \dots, n-1\}$. It remains to show that $\lambda_0^{(0)} = 0$. Suppose that $\deg F_s$ is not constant at $s = 0$. Then $\mathbb{X}_0^\infty = H^\infty = \mathbb{P}^n$, since the homogenized $\tilde{F}_s$ of the polynomial F_s defining X_s is of the form $s\tilde{h}_s + x_0\tilde{g}_s$ for some polynomial functions $h_s, g_s : \mathbb{C}^{n+1} \to \mathbb{C}$ such that $\deg h_0 > \deg g_0$. Let $\hat{F} := \tilde{F}_s(x_0; x_1; \dots; x_n; 1) = s\hat{h}_s + x_0\hat{g}_s$. Without loss of generality, we assume that our point $p \in \mathbb{X}_0^\infty$ is such that $p \notin \mathbb{X}_0^\infty \cap \{\hat{h}_s = 0\}$ and its coordinate x_{n+1} is not zero. These conditions are satisfied by a Zariski-open subset $\mathcal{Z}$ of H^∞.

The equisingularity at infinity at p is characterized by the criterion (1.9) of the proof of Proposition 1.3.2, replacing the parameter t by s, namely:

$$\delta \left| \frac{\partial \hat{F}}{\partial s} \right| < \left\| \frac{\partial \hat{F}}{\partial x_1}, \dots, \frac{\partial \hat{F}}{\partial x_n} \right\|, \tag{7.4}$$

for some $\delta > 0$, in some neighbourhood $\mathcal{N}_p \subset \mathbb{X}$ of p. In our notations, this inequality amounts to:

$$\delta \left| \hat{h}_s + s\frac{\partial \hat{h}_s}{\partial s} + x_0\frac{\partial \hat{g}_s}{\partial s} \right| < \left\| s\frac{\partial \hat{h}_s}{\partial x} + x_0\frac{\partial \hat{g}_s}{\partial x} \right\|.$$

But this inequality is false for sequences of points in $\mathcal{N}_p$ such that $s \to 0$ and $x_0 \to 0$, since the derivatives of $\hat{h}_s$ and $\hat{g}_s$ are bounded near p. It follows that the singular locus at infinity $\mathrm{Sing}^\infty \sigma$ contains the Zariski-open set $\mathcal{Z} \subset \mathbb{X}_0^\infty$, which is a contradiction to the assumed equisingularity at infinity. $\square$

Example 7.2.10 $f : \mathbb{C}^3 \to \mathbb{C}, f(x,y,z) = x + x^2yz$.
We shall compute the generic polar intersection multiplicities and the defects at infinity in the neighbourhood of the value 0. As general linear form we may take $l = x + y + z$. Then $\Gamma(l,f) = \{x^2y - 2xy^2 - 1 = 0, y = z\}$ and this polar curve intersects transversely the fibre $X_0 = f^{-1}(0)$ in three points, hence $\alpha_0^{(2)} = 3$. The Euler characteristic of any fibre of f is 1 (see Remark 7.2.8). Since we have $\alpha_0^{(0)} = \deg f = 4$ and since $\alpha_0^{(0)} - \alpha_0^{(1)} + \alpha_0^{(2)} = \chi(f^{-1}(0)) = 1$, by Exercise 7.2, it follows that $\alpha_0^{(1)} = 6$.

Now let $t \neq 0$. We have $\alpha_t^0 = \deg f = 4$ and we want to find α_t^1. The function f restricted to the hyperplane $H = \{x + y + z = 0\}$ becomes $f_{|H} = x - x^3y - x^2y^2$. This is a polynomial in two variables of degree 4 and of degree 2 in y. One can easily compute that the homotopy type of a general fibre $f_{|H}^{-1}(s)$ is a bouquet of five circles, therefore $\chi(f_{|H}^{-1}(s)) = -4$. By using Exercise 7.2, we get $\alpha_s^{(1)} = 4 + 4 = 8$, for general s and $\alpha_s^{(2)} = 5$.

The defects at infinity are therefore $\lambda_0^{(1)} = \lambda_0^{(2)} = 2$. Referring to Exercise 7.3, we may also deduce that the polynomial f has nonisolated $\mathcal{W}$-singularities at infinity.

7.3 Plücker formula for affine hypersurfaces

The classical projective setting. In case of a projective hypersurface $V \subset \mathbb{P}^{n+1}$ of degree d, *Plücker's class formula* expresses the degree (or the 'class') $d^*(V) := \deg(\check{V})$ of the dual $\check{V}$ in terms of d and of certain invariants of the singularities of V. The one proved by Plücker himself in 1834 considers plane curves S with nodes and cusps:

$$d^*(S) = d(d-1) - 2(\#\text{nodes}) - 3(\#\text{cusps}).$$

Teissier generalized it in 1975 to the case of projective hypersurfaces with *isolated singularities* (see e.g. [Te4]). He found the following formula in terms of the sectional *Milnor–Teissier numbers* of isolated singularities (i.e. Milnor numbers $\mu_q^{\langle n \rangle}$ of the singular points q of V and Milnor numbers $\mu_q^{\langle n-1 \rangle}$ of generic hyperplane sections at $q \in \operatorname{Sing} V$):

$$d^*(V) = d(d-1)^n - \sum [\mu_q^{\langle n \rangle} + \mu_q^{\langle n-1 \rangle}] \tag{7.5}$$

and Laumon [Lau] proved it by a different method. Later Langevin [Lan2] showed the connection with the complex Gauss map (see §7.4) and provided the integral-geometric interpretation of it. Further generalizations, for *projective varieties* with isolated singularities, and then without conditions on singularities, were found notably by Kleiman, Pohl and Thorup. The reader may consult e.g. [Tho] for the references.

The affine setting. Let $Y \subset \mathbb{C}^{n+1}$ be an affine hypersurface of degree d. The *degree* of Y is by definition the intersection multiplicity $\operatorname{mult}(Y, L)$ of Y with a general affine line L. The *affine dual* of Y is by definition:

$$\check{Y} = \operatorname{closure}\{H \in \check{\mathbb{P}}^n \mid H = T_x Y_{\operatorname{reg}} \text{ for some } x \in Y_{\operatorname{reg}}\}$$

The degree $\deg(\check{Y})$ of the affine dual $\check{Y}$ is then equal to the number of tangent hyperplanes to Y_{reg} in a generic affine pencil of hyperplanes in $\mathbb{C}^{n+1}$. We call it the *affine class* of Y in analogy to the projective case and we denote it by $d^{\mathit{aff}}(Y)$. The affine pencils of hyperplanes (see §6) differ from the projective pencils especially in a neighbourhood of infinity, since, after projectivizing, the hyperplane at infinity H^∞ becomes a member of the pencil. Our hypersurface Y may be asymptotically tangent to H^∞ and so the pencil might be asymptotically nongeneric.

We prove a Plücker type formula for the affine class of Y, following [ST8]. The result expresses the asymptotic loss of intersection multiplicity in a family of constant degree *general hypersurfaces* tending to Y, cf. Definition 4.1.1. It was shown in §4.1 that any affine hypersurface Y may be deformed in such a family. As usual, X denotes the total space of the family.

Theorem 7.3.1 *Let $Y \subset \mathbb{C}^{n+1}$ be a hypersurface of degree d. Let $\{X_s\}_{s \in \delta}$ be a family of affine hypersurfaces of constant degree such that $Y = X_0$ and that X_s is general for all $s \neq 0$. Then the affine class of Y satisfies the following equality:*

$$d^{\mathit{aff}}(Y) = d(d-1)^n - \gamma_0(X) + \gamma_0^\infty(X), \tag{7.6}$$

where:

$$\gamma_0(X) := \operatorname{mult}_{\operatorname{Sing} X_0}(\Gamma_{\mathcal{S}}(l_H, \sigma), X_0) \tag{7.7}$$

is the polar intersection number at the points of $\text{Sing} X_0$ *through which the polar curve* $\Gamma_S(l_H, \sigma)$ *passes, and where:*

$$\gamma_0^\infty(X) = \lim_{R \to \infty} \lim_{s \to 0} \text{mult}(\Gamma_S(l_H, \sigma), X_s \cap \complement B_R) \tag{7.8}$$

is the loss of intersection multiplicity towards infinity.

Moreover, $\gamma_0(X)$ *and* $\gamma_0^\infty(X)$ *do not depend on* H *varying in some Zariski-open subset of* $\Omega_{\sigma,0}$.

Proof Let us first observe that the singular locus of the total space X is included into $\text{Sing} X_0$ and therefore the lower-dimensional strata of the Whitney stratification of X are included in $\text{Sing} X_0$.

Let us take $H \in \Omega_{\sigma,0}$ as in Notation 7.1.5. We have the following decomposition of the global intersection multiplicity $\alpha_s^{(n)}$ (cf. (7.1)):

$$\alpha_s^{(n)} = \alpha_0^{(n)} + \lambda_1 + \lambda_2. \tag{7.9}$$

The term $\alpha_0^{(n)}$ is by definition $\text{mult}(\Gamma_S(l_H, \sigma), (X_0)_{\text{reg}})$ and we have seen before that this does not depend on the choice of $H \in \Omega_{\sigma,0}$. According to the geometric interpretation of $\alpha_0^{(n)}$ (see the remark after (7.1)), this is precisely the affine class $d^{aff}(Y)$.

We define the term λ_1 of the sum (7.9) as the number of those intersection points of $\Gamma_S(l_H, \sigma)$ with X_s that tend to singular points of X_0. This is precisely the multiplicity $\text{mult}_{\text{Sing} X_0}(\Gamma_S(l_H, \sigma), X_0)$ that we have denoted by $\gamma_0(X)$ in the statement of our theorem. This multiplicity does not depend on the choice of H within some Zariski-open subset of $\Omega_{\sigma,0}$, by the standard arguments (local constancy and connectivity).

The term λ_2 from (7.9) counts the rest of the intersection points (with multiplicities), i.e. those that tend to infinity as $s \to 0$. This is the asymptotic loss of intersection points of the polar curve $\Gamma_S(l_H, \sigma)$ with X_s. In other words, we have $\lambda_2 = \gamma_0^\infty(X)$.

For the general hypersurfaces X_s we have $\alpha_s^{(n)} = d(d-1)^n$, as remarked after (7.1). It then follows from (7.9) and from the preceding discussion that $\gamma_0^\infty(X)$ is independent of $H \in \Omega_{\sigma,0}$. This ends our proof.   □

Case of isolated singularities. If $X_0 = Y \subset \mathbb{C}^{n+1}$ is a hypersurface with isolated singularities, then we may express the intersection multiplicity $\gamma_0(X)$ in terms of Milnor numbers as follows:

$$\text{mult}_{\text{Sing} X_0}(\Gamma_S(l_H, \sigma), X_0) = \sum_{q \in \text{Sing} X_0} [\mu_q^{\langle n-1 \rangle}(X_0) + \mu_q^{\langle n \rangle}(X_0)]. \tag{7.10}$$

This is a consequence of the following equality for the generic local polar multiplicity:

$$\text{mult}_q(\Gamma_{\mathcal{S}}(l_H, \sigma), X_0) = \mu_q^{\langle n \rangle}(X_0) + \mu_q^{\langle n-1 \rangle}(X_0) \quad (7.11)$$

proved by Teissier [Te2, Te3] for a germ (X_0, q) of a complex analytic hypersurface. It is actually well known that the local equality (7.11) is valid for any *smoothing* of X_0, see Exercise 7.6. In our case the local smoothing is embedded in the global smoothing $\sigma : X \to \mathbb{C}$ of Theorem 7.3.1.

Let us show how the affine class $d^{aff}(Y)$ can be expressed in terms of singularities at infinity of the projective closure $\bar{Y} \subset \mathbb{P}^{n+1}$. We assume that $\bar{Y}$ has isolated singularities only; nevertheless the part at infinity $\bar{Y} \cap H^\infty$ may have nonisolated singularities, of dimension not higher than one.

Proposition 7.3.2 Let Y be a hypersurface such that $\bar{Y}$ has at most isolated singularities. Then:

$$d^{aff}(Y) = d(d-1)^n - \sum_{q \in \text{Sing } Y} [\mu_q^{\langle n \rangle}(Y) + \mu_q^{\langle n-1 \rangle}(Y)] - \gamma^\infty(Y), \quad (7.12)$$

where:

$$\gamma^\infty(Y) = \sum_{p \in (\text{Sing } \bar{Y}) \cap H^\infty} \mu_p(\bar{Y}) + \\ \sum_{p \in \text{Sing } (\bar{Y} \cap \bar{\mathcal{H}} \cap H^\infty)} \mu_p(\bar{Y} \cap \bar{\mathcal{H}} \cap H^\infty) + (-1)^{n+1}[\chi^{n,d} - \chi(\bar{Y} \cap H^\infty)], \quad (7.13)$$

and where $\mathcal{H}$ is a general affine hyperplane and $\chi^{n,d}$ denotes the Euler characteristic of the general projective hypersurface of degree d in $\mathbb{P}^n$.

Proof We have by (6.23):

$$d^{aff}(Y) = (-1)^n \chi(Y) - (-1)^n \chi(Y \cap \mathcal{H}) - \sum_{q \in \text{Sing } Y} \mu_q^{\langle n-1 \rangle}(Y).$$

We consider a constant degree family of hypersurfaces such that X_s is general for $s \neq 0$ and such that $Y = X_0$. We observe that both $\bar{X}_s \cap \bar{\mathcal{H}}$ and $\bar{X}_s \cap \bar{\mathcal{H}} \cap H^\infty$ have at most isolated singularities for any s (see §2.2). We compute the difference of Euler characteristics:

$$\chi(X_0) - \chi(X_s) = \chi(\bar{X}_0) - \chi(\bar{X}_s) - \chi(\bar{X}_0 \cap H^\infty) + \chi(\bar{X}_s \cap H^\infty) =$$

$$(-1)^{n+1} \sum_{q \in \text{Sing } X_0} \mu_q^{\langle n \rangle}(X_0) + (-1)^{n+1} \sum_{p \in (\text{Sing } \bar{X}_0) \cap H^\infty} \mu_p(\bar{X}_0) + \chi^{n,d} - \chi(\bar{X}_0 \cap H^\infty)$$

and
$$\chi(X_s \cap \mathcal{H}) - \chi(X_0 \cap \mathcal{H}) =$$

$$[\chi(\bar{X}_s \cap \bar{\mathcal{H}}) - \chi(\bar{X}_0 \cap \bar{\mathcal{H}})] + [-\chi(\bar{X}_s \cap \bar{\mathcal{H}} \cap H^\infty) + \chi(\bar{X}_0 \cap \bar{\mathcal{H}} \cap H^\infty)] =$$

$$(-1)^{n-1} \sum_{q \in \mathrm{Sing}\,(\bar{X}_0 \cap \bar{\mathcal{H}})} \mu_q(\bar{X}_0 \cap \bar{\mathcal{H}}) + (-1)^{n-1} \sum_{p \in \mathrm{Sing}\,(\bar{X}_0 \cap \bar{\mathcal{H}} \cap H^\infty)} \mu_p(\bar{X}_0 \cap \bar{\mathcal{H}} \cap H^\infty).$$

We take the sum of these two last formulas. To finish our proof, we also use the genericity of the affine hypersurfaces X_s and $X_s \cap \bar{\mathcal{H}}$, which implies:

$$\chi(X_s) - \chi(X_s \cap \mathcal{H}) = (-1)^n(d-1)^{n+1} - (-1)^{n-1}(d-1)^n = (-1)^n d(d-1)^n.$$

$\square$

As a particular case of (7.12), the following formula holds for a hypersurface Y such that both $\bar{Y}$ and $\bar{Y} \cap H^\infty$ have isolated singularities:

$$d^{aff}(Y) = d(d-1)^n - \sum_{q \in \mathrm{Sing}\,Y}[\mu_q^{\langle n \rangle}(Y) + \mu_q^{\langle n-1 \rangle}(Y)] \atop - \sum_{p \in \mathrm{Sing}\,(\bar{Y} \cap H^\infty)}[\mu_p(\bar{Y}) + \mu_p(\bar{Y} \cap H^\infty)]. \tag{7.14}$$

The contribution from the affine singularities is contained in the first of the two sums and one recognizes the sectional Milnor–Teissier numbers from (7.5); see also Langevin's formula [Lan1] cited at (7.15).

The second sum is due to the 'singularities at infinity': the number $\mu_p(\bar{Y}) + \mu_p(\bar{Y} \cap H^\infty)$ is precisely the local polar number $\lambda_p = \mathrm{mult}_p(\Gamma(x_0, \sigma), \bar{X}_0)$ of the polar curve of the family $\{X_s\}_{s \in \delta}$ with respect to the local coordinate at infinity x_0.

The difference among $\mu_p(\bar{Y} \cap H^\infty)$ and the sectional Milnor–Teissier number $\mu_p^{\langle n-1 \rangle}(\bar{Y})$, for some $p \in \mathrm{Sing}\,(\bar{Y} \cap H^\infty)$, is the following: the later is the Milnor number of the *generic hyperplane section* $\bar{Y} \cap \mathcal{H}$, whereas in the former the hyperplane H^∞ is *not generic* at p relative to $\bar{Y}$. The reader should then compare with (7.11).[3]

Affine plane curves and the correction term at infinity. We get the following formula in the case of curves:

Corollary 7.3.3 Let $C \subset \mathbb{C}^2$ be a nonsingular complex affine curve of degree d. Then:

$$d^{aff}(Y) = d^2 - 2d + r - \sum_{p \in \mathrm{Sing}\,(\bar{C} \cap H^\infty)} \mu_p(\bar{C}),$$

where r is the number of points in the set $\bar{C} \cap H^\infty$

Proof We may apply formula (7.14) since $\bar{C} \cap H^\infty$ consists of finitely many points, also called 'asymptotic directions of C'. Then the sum of Milnor numbers $\sum_{p \in \mathrm{Sing}(\bar{C} \cap H^\infty)} \mu_p(\bar{C} \cap H^\infty)$ is precisely $d - r$. $\qquad\square$

In other words, the above formula tells that the affine class of a smooth plane curve C is equal to that of a general curve (which is $d(d - 1)$) diminished by the number of tangencies (with multiplicities) between $\bar{C}$ and the line at infinity (which is equal to $d - r$), and by the sum of the Milnor numbers of $\bar{C}$.

Example 7.3.4 Let $f : \mathbb{C}^3 \to \mathbb{C}, f(x, y, z) = x + x^2yz$, and consider the family $X_s = \{f = s\}$. The generic polar intersection multiplicities and the defects at infinity in the neighbourhood of the value 0 are given in Example 7.2.10; from those results we may extract the folowing data: $\alpha_s^{(2)} = 5, \alpha_s^{(1)} = 8, \alpha_s^{(0)} = 4$ for $s \neq 0$, and $\alpha_0^{(2)} = 3, \alpha_0^{(1)} = 6, \alpha_0^{(0)} = 4$. We get: $d^{\mathit{aff}}(X_s) = 5$ if $s \neq 0$ and $d^{\mathit{aff}}(X_0) = 3$.

Therefore the affine class of X_s is not constant in the family, even if X_s is nonsingular and $\chi(X_s) = 1$ for all $s \in \mathbb{C}$ (see [Ti6]). Note that in our case the variation of the affine class is equal to the vanishing polar multiplicity at infinity $\gamma_0^\infty(X)$, as interpreted in (7.13). The family is clearly not topologically trivial since the number of connected components of X_s change at $s = 0$.

Example 7.3.5 [ST8] Consider the double parameter family of surfaces in $\mathbb{C}^3$, $X_{s,t} = \{f_s = x^4 + sz^4 + z^2y + z = t\}$. Then, for all s, f_s has a generic fibre that is homotopy equivalent to a bouquet of three 2-spheres. There are no affine critical points and $t = 0$ is the only atypical value of f_s.

In order to compute the affine class of $X_{s,t}$ we use the formulas (7.12) for $s = 0$ and (7.14) for $s \neq 0$. We have the following information and notations, also collected in Tables 7.1 and 7.2 as input data for the formulas:
(a). $\bar{X}_{s,t}$ has isolated singularities at infinity in $p := ([0 : 1 : 0], 0)$ for all s and in $q := ([1 : 0 : 0], 0)$ for $s = 0$. The μs are listed in Table 7.1.
(b). The singularities of $\bar{X}_{s,t} \cap H^\infty \subset \mathbb{P}^2$ are a single smooth line $\{x^4 = 0\}$ for $s = 0$ and the isolated point p with $\tilde{E}_7$ singularity for $s \neq 0$.

Table 7.1.

(s, t)	$\mu_p(\bar{X}_{s,t}) + \mu_q(\bar{X}_{s,t})$	$\mu(\bar{X}_{s,t} \cap \bar{\mathcal{H}} \cap H^\infty)$
$(0, 0)$	$18 + 3$	3
$(0, t)$	$15 + 3$	3
$(s, 0)$	$18 + 0$	$-$
(s, t)	$15 + 0$	$-$

Table 7.2.

(s,t)	$(-1)^{n+1}\Delta\chi$	$\alpha_{s,t}^{(2)}$	$\chi(X_{s,t})$
$(0,0)$	$4+2$	$36-30=6$	$6-6=0$
$(0,t)$	$4+2$	$36-27=9$	$9-6=3$
$(s,0)$	9	$36-27=9$	$9-9=0$
(s,t)	9	$36-24=12$	$12-9=3$

(c). The space $\bar{X}_{s,t} \cap \bar{\mathcal{H}} \cap H^\infty$ has a single singularity of type A_3 for $s=0$ and is smooth if $s \neq 0$.

We use the notation $\Delta\chi = \chi^{2,4} - \chi(\bar{X}_{s,t} \cap H^\infty)$, where $\chi^{2,4} = -4$ by the definition of $\chi^{n,d}$ given in Proposition 7.3.2. By the notations $(0,t)$, $(s,0)$ and (s,t) we mean $s,t \neq 0$.

We observe that, for each fixed t, the family $X_{s,t}$ has constant degree and Euler characteristic, but nonconstant affine class. Actually, by using a coordinate change in the variable y, it turns out that this family is topologically trivial.

7.4 Curvature loss at infinity and the Gauss–Bonnet defect

We now study the influence of the position of Y at infinity upon the *total curvature* of the affine hypersurface $Y \subset \mathbb{C}^{n+1}$ following [ST8].

The curvature denoted by K means the *Lipschitz–Killing curvature* of the real analytic 2-codimensional space Y, with respect to the metric induced by the flat Euclidean metric of $\mathbb{C}^{n+1}$. Let dv denote the associated volume form. The integral of the curvature $\int_Y K dv$ will be called 'total curvature' of Y.

We shall take two standpoints for computing the total curvature of Y: comparing it with the Euler characteristic $\chi(Y)$, and comparing it to the total curvature of a general hypersurface, after embedding Y into a family. Let us observe that computing the total curvature of the projective closure $\bar{Y}$ would not help, since the metrics on $\mathbb{P}^{n+1}$ and $\mathbb{C}^{n+1}$ are different.

The first approach goes back to extrinsic proofs of the Gauss–Bonnet theorem. The failure of this celebrated theorem in case of *open surfaces* is a theme which has been under constant attention ever since Cohn–Vossen's pioneering work [Co] in 1935.

The second approach recalls the work by Langevin [Lan1, Lan2] and Griffiths [Gri] in the late 1970s on the influence of an isolated singularity upon the total curvature of the local Milnor fibre in case of analytic hypersurface germs.

Langevin studied the total curvature of the local Milnor fibre at an isolated hypersurface singularity defined by the germ of a holomorphic function g : $(\mathbb{C}^{n+1}, 0) \to (\mathbb{C}, 0)$.[4] The next local formula due to Langevin [Lan1, Théorème 1] shows that the 'loss of total curvature' at an isolated singularity is measured by the sum of the first two sectional Milnor–Teissier numbers (compare with the results and comments in §7.3)

$$\lim_{\varepsilon \to 0} \lim_{t \to 0} \int_{g^{-1}(t) \cap B_\varepsilon} |K| \mathrm{d}v = \omega_n (\mu_0^{\langle n \rangle} + \mu_0^{\langle n-1 \rangle}). \tag{7.15}$$

In our global affine setting, it turns out that there is an identification, up to a constant, of the total curvature to the affine class of Y (defined before in §7.3). Let us first give some background on the total curvature.

Real submanifolds. For a real oriented hypersurface of $\mathbb{R}^N$ we have a well-defined *Gauss map*. One defines the *Gauss–Kronecker curvature* $K(x)$ as the Jacobian of the Gauss map at $x \in \mathbb{R}^N$. For a submanifold V in $\mathbb{R}^N$, Fenchel [Fen] computes the curvature as follows. For a given point x on V we consider a unit normal vector $\mathbf{n}$, projects V orthogonally to the affine subspace W generated by the affine tangent space to V and this normal vector. The projection of V to W is a hypersurface, which has a well-defined Gauss–Kronecker curvature $K(x, \mathbf{n})$. The Lipschitz–Killing curvature $K(x)$ of V in x is defined (see e.g.[ChL, pp. 246–247]) as the integral of these curvatures over all normal directions, up to a universal constant u, namely: $K(x) = u \int_{N_x V} K(x, \mathbf{n}) \mathrm{d}\mathbf{n}$.

The classical *Gauss–Bonnet theorem* says that, if V is compact and of even dimension $2n$, then the total curvature is equal, modulo an universal constant, to the Euler characteristic

$$\omega_n^{-1} \int_V K \mathrm{d}v = \chi(V),$$

where $\mathrm{d}v$ denotes the restriction of the canonical volume form and where $\omega_n = \frac{(2\pi)^n}{1 \cdot 3 \cdots (2n-1)}$ is half the volume of the sphere S^{2n}.

Complex hypersurfaces. Langevin [Lan1, Lan2] studied the integral of curvature of the Milnor fibre of a complex hypersurface germ with isolated singularity $X \subset \mathbb{C}^{n+1}$. He uses Milnor's approach [Mi1] as follows: *compute the total curvature from the number of critical points of an orthogonal projection on a generic line.* We recall below some results and fix the notations.

The curvature $K(x)$ of a smooth complex hypersurface is the Lipschitz–Killing curvature of Y_{reg} as a 2-codimensional submanifold of $\mathbb{R}^{2n+2}$, where Y_{reg} denotes the regular part of Y. A computation due to Milnor allows us to express the Lipschitz–Killing curvature of Y in terms of the complex Gauss map $\nu_{\mathbb{C}} : Y_{\mathrm{reg}} \to \mathbb{P}_{\mathbb{C}}^n$ which sends a point $x \in Y_{\mathrm{reg}}$ to the complex tangent space of Y_{reg} at x, cf. [Lan1, p. 11]:

$$(-1)^n K(x) = |K(x)| = \frac{2 \cdot 4 \cdots 2n}{1 \cdot 3 \cdots (2n-1)} |\mathrm{Jac}\, \nu_{\mathbb{C}}|^2. \qquad (7.16)$$

In the complex case the curvature K is well known to have the constant sign $(-1)^n$. Using (7.16), we can prove an *exchange formula*, as follows.[5] Let H be a hyperplane in $\mathbb{P}^n$, defined by a linear form $l_H : \mathbb{C}^{n+1} \to \mathbb{C}$. For almost all $H \in \check{\mathbb{P}}^n$ the restriction of l_H to Y_{reg} has only complex Morse critical points.

Let $\alpha_Y(l_H)$ be the number of those critical points (which is finite, since Y is algebraic). On the complement of the zero set of its Jacobian, the complex Gauss map is a local diffeomorphism with locally constant degree $\alpha_Y(l_H)$. This is precisely the number $\alpha_Y^{(n)}$ defined in §6.1, (6.3), and does not depend on l_H running in some Zariski-open set of the dual projective space $\check{\mathbb{P}}^n$. From the above discussion and from Langevin's [Lan2, Theorem A.III.3], we may draw the following result:

Lemma 7.4.1 (Théorème d'échange) *Let* $Y \subset \mathbb{C}^{n+1}$ *be any affine hypersurface. Then*

$$\int_Y |K| \mathrm{d}v = \frac{2 \cdot 4 \cdots 2n}{1 \cdot 3 \cdots (2n-1)} \int_{\check{\mathbb{P}}^n} \alpha_Y(l_H)\, \mathrm{d}H = \omega_n \alpha_Y^{(n)}.$$

By definition, the integral $\int_Y |K| \mathrm{d}v$ is taken over Y_{reg}; the notation makes sense since Y differs from Y_{reg} by a set of measure zero.

The vanishing curvature. The above lemma establishes an equality, up to the constant ω_n, among the total curvature and the affine class of Y. We therefore get the following interpretation of Theorem 7.3.1:

Corollary 7.4.2 [ST8] *Let* $Y \subset \mathbb{C}^{n+1}$ *be any hypersurface. Let* $\{X_s\}_{s \in \delta}$ *be a one-parameter deformation of* $X_0 := Y$ *such that* X_s *is nonsingular for all* $s \neq 0$. *Let* X *denote the total space of this family. Then the following limit exists*

$$\lim_{s \to 0} \omega_n^{-1} \int_{X_s} |K| \mathrm{d}v = \omega_n^{-1} \int_{X_0} |K| \mathrm{d}v + \gamma_0(X) + \gamma_0^\infty(X). \qquad (7.17)$$

We may identify $\gamma_0^\infty(X)$ as the 'loss of total curvature' at infinity, or the *vanishing curvature* at infinity:

$$\gamma_0^\infty(X) := \omega_n^{-1} \lim_{R \to \infty} \lim_{s \to 0} \int_{X_s \cap \mathbb{C}B_R} |K| \mathrm{d}v. \qquad (7.18)$$

We may therefore translate into the curvature language all the results in §7.3, replacing the affine class $d^{aff}(Y)$ by $(-1)^n \omega_n^{-1} \int_Y K dv$.

Gauss–Bonnet defect. In order to measure the failure of the Gauss–Bonnet theorem in case of arbitrary singular affine hypersurfaces, we compute the *Gauss–Bonnet defect* of Y:

$$GB(Y) := \omega_n^{-1} \int_Y K dv - \chi(Y).$$

This integer may be interpreted as the correction term due to the 'boundary at infinity' of Y. Indeed, let us assume that Y has isolated singularities. Let $B_R \subset \mathbb{C}^{n+1}$ be a ball centred at the origin and denote $Y_R := Y \cap B_R$ and $\partial Y_R := Y \cap \partial \bar{B}_R$. Since Y has isolated singularities and is affine, the intersection $Y \cap \partial \bar{B}_R$ is transversal and Y_R is diffeomorphic to Y, for large enough radius R. By applying Griffith's Gauss–Bonnet formula for the manifold with boundary $\bar{Y}_R$, see [Gri, p. 479], we get: $\omega_n^{-1} \int_Y K dv - c \int_{\partial Y_R} k ds = \chi(Y_R)$, where k is the generalized *geodesic curvature* of ∂Y_R and c is a universal constant which we do not specify here. It then follows:

$$GB(Y) = \lim_{R \to \infty} c \int_{\partial Y_R} k ds.$$

Nevertheless, it is the interpretation of the total curvature in terms of the affine class which allows us to compute the Gauss–Bonnet defect too, at least in certain cases, as follows:

Proposition 7.4.3 Let Y be a hypersurface of degree d with isolated singularities such that $\bar{Y} \cap H^\infty$ has at most one-dimensional singularities. Then:

$$GB(Y) = (-1)^n (d-1)^n - 1 + (-1)^{n+1} \sum_{q \in \text{Sing } Y} \mu_q^{\langle n-1 \rangle}(Y) + \tag{7.19}$$

$$(-1)^{n+1} \left[\sum_{q \in \text{Sing} (\bar{Y} \cap \bar{\mathcal{H}})} \mu_q(\bar{Y} \cap \bar{\mathcal{H}}) + \sum_{p \in \text{Sing} (\bar{Y} \cap \bar{\mathcal{H}} \cap H^\infty)} \mu_p(\bar{Y} \cap \bar{\mathcal{H}} \cap H^\infty) \right].$$

Proof We have by Lemma 7.4.1 and by (6.23):

$$GB(Y) = (-1)^{n-1} \sum_{q \in \text{Sing } Y} \mu_q^{\langle n-1 \rangle}(Y) - \chi(Y \cap \mathcal{H}).$$

Consider a constant degree family of hypersurfaces such that X_s is general for $s \neq 0$ and that $Y = X_0$. We observe that both $\bar{X}_s \cap \bar{\mathcal{H}}$ and $\bar{X}_s \cap \bar{\mathcal{H}} \cap H^\infty$ have at most isolated singularities, for any s. Then we may compute the difference $\chi(X_s \cap \mathcal{H}) - \chi(X_0 \cap \mathcal{H})$ as in the proof of Proposition 7.3.2. Combining this with the equality $\chi(X_s \cap \mathcal{H}) = 1 + (-1)^{n-1}(d-1)^n$ for general hypersurfaces, we get our formula.   □

Plane affine curves. In case of a nonsingular complex affine plane curve C of degree d, we get the following Gauss–Bonnet defect:

$$GB(C) = -d.$$

We therefore have a precise evaluation in this particular case of a well-known more general result due to Cohn–Vossen [Co], which tells that $GB(M) \leq 0$ if M is a complete, finitely connected Riemann surface, having absolutely integrable Gauss curvature.

An expression of the total curvature of such a curve C can also be extracted from Corollary 7.3.3.

Exercises

7.1 Let $X = \mathbb{C}^{n+1}$ and let $f : \mathbb{C}^{n+1} \to \mathbb{C}$ be a polynomial function. Let X_c have an isolated singularity at $p \in X_c$ and let $H \in \Omega'_{f,c}$. Show that $p \in \Gamma(l_H, f)$. In case X_c is singular and p is an isolated stratified singularity, do we have $p \in \Gamma(l_H, f)$ for generic H?

7.2 Let $\{X_s\}_{s \in \mathbb{C}}$ be the germ at $0 \in \mathbb{C}$ of a polynomial family of hypersurfaces in $\mathbb{C}^{n+1}$. Suppose that, for all s in some disk $\delta \subset \mathbb{C}$ centered at 0, the hypersurfaces $X_s := \{F_s(x) = 0\}$ have at most isolated singularities, and they do not tend to infinity as s tends to zero. Let $\mu(X_0)$ denote the sum of the Milnor numbers of the isolated singularities of X_0.

Show that X_0 is homotopy equivalent to a generic hyperplane section $X_0 \cap \mathcal{H}$ to which one attaches $\alpha_0^{(n)} - \mu(X_0)$ cells of dimension n. Next prove that X_0 is homotopy equivalent to the CW-complex obtained by successively attaching to $\deg X_0$ points a number of $\alpha_0^{(1)}$ cells of dimension 1, then $\alpha_0^{(2)}$ cells of dimension 2, ..., $\alpha_c^{(n-1)}$ cells of dimension $n-1$ and finally $\alpha_c^{(n)} - \mu(X_0)$ cells of dimension n. In particular, $\chi(X_0) = (-1)^{n+1}\mu(X_0) + \sum_{i=0}^{n}(-1)^i \alpha_0^{(i)}$.

Supposing in addition that X_s is nonsingular, $\forall s \in \delta \setminus \{0\}$, show the equality:

$$\chi(X_s) - \chi(X_0) = (-1)^n \mu(X_0) + \sum_{i=0}^{n} (-1)^i \lambda_0^{(i)}.$$

7.3 Let $\{X_s\}_{s \in \mathbb{C}}$ be the family of fibres of a polynomial $f : \mathbb{C}^{n+1} \to \mathbb{C}$. Assume that f is a W-type polynomial, according to Definition 2.2.2. We have defined in §3.3, (3.23) the number $\lambda_p \geq 0$, which measures a certain local defect at infinity, at some point $p \in \mathbb{X}^\infty$. (We have actually given several interpretations of λ_p in §3.3.) Then show that $\lambda_0^{(n)} = \sum_{p \in \overline{X_0} \cap \mathbb{X}^\infty} \lambda_p$ and $\lambda_0^{(i)} = 0$ for $i \leq n-1$.

7.4 In Theorem 7.2.6, assume that $n = 1$ and that $\deg X_s = \text{constant}$, for any $s \in \delta$. Then show that the constancy of $\alpha_s^{(1)}$ at $0 \in \mathbb{C}$ is equivalent to the topological triviality at infinity of the family $\{X_s\}_{s \in \mathbb{C}}$ in the neighbourhood of 0.

Show that this also holds in higher dimensions for the following more general situation: $\hat{\sigma}$ has isolated stratified singularities at infinity with respect to some stratification of $\mathbb{X}$, which is a *partial Thom stratification at infinity* (cf. Appendix A1.1).

7.5 In Theorem 7.2.6, assume that the family $\{X_s\}_{s \in \delta}$ is nonsingular. Show that the α^*-constancy at $s = 0$ implies that this family is C^∞ trivial over some small enough disk δ.

7.6 Let X_s be a smoothing of a germ of a hypersurface $X_0 \subset \mathbb{C}^{n+1}$ with isolated singularity at q, i.e. $X_s := \{F(x, s) = 0\}$ for all s in some small disk δ centred at 0 and $X_s \cap B\varepsilon$ is nonsingular for all $s \neq 0$ and for a small enough ball $B\varepsilon$ centred at q.

7.7 Show that, if $Y \subset \mathbb{C}^{n+1}$ is a nonsingular affine hypersurface of degree d such that $\overline{Y} \cap H^\infty$ has at most isolated singularities, then the Gauss–Bonnet defect depends only on n and d. More precisely $GB(Y) = (-1)^n (d-1)^n - 1$.

8

Monodromy of polynomials

A *global geometric monodromy* is a representation

$$\rho : \pi_1(\mathbb{C} \setminus \mathrm{Atyp} f) \to \mathrm{Diff}\,(G),$$

where Diff (G) is the group of C^∞-diffeomorphisms of the typical fibre G of the polynomial function $f : \mathbb{C}^n \to \mathbb{C}$.

Some of the methods employed in the study of the monodromy, usually under certain restrictions on f, are: Newton polyhedra [LS], Hodge theory of families of algebraic hypersurfaces [GN, Di2], Fourier transform of $\mathcal{D}$-modules [Sab], resolution of singularities [MW], [ACD], [GLM3], relative monodromy in case of two variables [Hà].

We discuss here our viewpoint of studying the monodromy of a polynomial function f via the *relative monodromy*. Knowing the relative monodromy reduces the problem to less variables. We shall first construct a *model* of the typical fibre, by using the tomographic method (see §6.1), and next *geometric monodromies* along loops in $\mathbb{C} \setminus \mathrm{Atyp} f$ acting on this model.[1] The presentation is based on [Ti1, ST3, ST4, ST5].

8.1 Models of fibres and a global geometric monodromy

Let $f : \mathbb{C}^n \to \mathbb{C}$ be a polynomial function of degree d. We recall some notations used before, e.g. in §1.1, and set some more. We consider the closure of the graph of f, this time in $\mathbb{P}^n \times \mathbb{P}^1$, namely:

$$\mathbb{X} = \{[x_0 : x_1 : \cdots x_n] \in \mathbb{P}^n, [s : t] \in \mathbb{P}^1 \mid s\tilde{f} - tx_0^d = 0\} \subset \mathbb{P}^n \times \mathbb{P}^1,$$

where $\tilde{f}$ is the homogenized of the polynomial f, x_0 denotes the variable at infinity and $\mathbb{X}^\infty := \mathbb{X} \cap \{x_0 = 0\}$ is the hyperplane at infinity of $\mathbb{X}$. Denote by $\tau : X \to \mathbb{P}^1$ the second projection. The map f is then the composition $\mathbb{C}^n \overset{i}{\hookrightarrow} X \overset{\tau}{\to} \mathbb{C}$, where i is the natural embedding $i : \mathbb{C}^n \simeq \text{Graph}(f) \hookrightarrow X$. The study of f is therefore equivalent to the study of the restriction of τ to the graph of f.

There is a finite set $\Lambda \subset \mathbb{P}^1$, $\Lambda \supset \text{Atyp}f$, such that the restrictions $\tau_| :$ $X \setminus \tau^{-1}(\Lambda) \to \mathbb{P}^1 \setminus \Lambda$, resp. $\tau_| : X \setminus (\tau^{-1}(\Lambda) \cup X^\infty) \to \mathbb{P}^1 \setminus \Lambda$, are C^0 and C^∞ locally trivial fibrations, respectively. The point at infinity $[0 : 1]$ of the embedding $\mathbb{C} \subset \mathbb{P}^1$ belongs to Λ by definition.

We consider, as in § 2.1, the polar locus $\Gamma(l_H, f) := \text{closure}[\text{Sing}(l_H, f) \setminus \text{Sing}f] \subset \mathbb{C}^n$ of f with respect to a linear form $l_H : \mathbb{C}^n \to \mathbb{C}$ associated to a projective hyperplane $H \in \check{\mathbb{P}}^{n-1}$. Let us also denote here by $\Delta := (l_H, f)(\Gamma(l_H, f))$ the polar image in $\mathbb{C}^2$. We prove the following key lemma, which is another supplement to Corollary 7.1.3 of the polar curve theorem 7.1.2:

Lemma 8.1.1 There is a Zariski-open set $\Omega''_f \subset \check{\mathbb{P}}^{n-1}$ such that, for any $H \in \Omega''_f$, the global polar locus $\Gamma(l_H, f)$ is a reduced curve or it is empty and the map $(l_H, f) : \mathbb{C}^n \to \mathbb{C} \times \mathbb{C}$ is a C^∞-trivial fibration above $(\mathbb{C} \times (\mathbb{C} \setminus \Lambda_H)) \setminus \Delta$, for some finite set $\Lambda_H \subset \mathbb{C}$.

Proof The first assertion is Corollary 7.1.3(a), so we only prove the claim about the map (l_H, f). We fix a complex algebraic Whitney stratification $\mathcal{W} := \{W_i\}_{i \in I}$ of $X' := X \cap (\mathbb{P}^n \times \mathbb{C})$ with finitely many strata such that $\mathbb{C}^n \subset X'$ is a stratum. The embedding of $\mathbb{C}$ into $\mathbb{P}^1$ is via the chart $s \neq 0$ and the embedding of $\mathbb{C}^n$ into X' is the graph embedding i.

We take two copies of $\mathbb{C}$, denoted $\mathbb{C}_r$ and $\mathbb{C}_t$, of variables r, respectively t. Let $\mathbb{H} := \{(r, [x_0 : x], t) \in \mathbb{C}_r \times \mathbb{P}^n \times \mathbb{C}_t \mid l_H(x) - rx_0 = 0\}$ (notice that t is free). Consider X' as subset of $\mathbb{P}^n \times \mathbb{C}_t$. Note that $\mathbb{H}$ is nonsingular, whereas X' can have singularities on X^∞. Define the space:

$$\mathbb{Y} := (\mathbb{C}_r \times X') \cap \mathbb{H} \subset \mathbb{C}_r \times \mathbb{P}^n \times \mathbb{C}_t \qquad (8.1)$$

and consider the projection $(r, t) : \mathbb{Y} \to \mathbb{C}_r \times \mathbb{C}_t$. This is a proper extension of the map (l_H, f). It follows that $\mathbb{Y} \setminus \mathbb{Y}^\infty$ is nonsingular (since it can be identified with $\mathbb{C}^n$), where $\mathbb{Y}^\infty := \mathbb{Y} \cap (\mathbb{C}_r \times \{x_0 = 0\} \times \mathbb{C}_t)$, and that the critical locus of the restriction $(r, t)_{|\mathbb{Y} \setminus \mathbb{Y}^\infty}$ is $\text{Sing}(l_H, f)$. It remains to understand the situation at infinity.

We first return to the space X' and the map $\tau : X' \to \mathbb{C}$. Denote by $\text{Sing}_{\mathcal{W}}\tau := \cup_{W_i \in \mathcal{W}}\text{Sing}\,\tau_{|W_i}$ the singular locus of τ with respect to the stratification $\mathcal{W}$. It is a closed analytic subset of X'.

For a fixed stratum $W_i \subset X' \cap X^\infty \subset \mathbb{P}^{n-1} \times \mathbb{C}_t$, we consider the projectivized relative conormal $\mathbb{P}T^*_{\tau|W_i} \subset \mathbb{P}^{n-1} \times \mathbb{C}_t \times \check{\mathbb{P}}^{n-1}$.

We remark that, if τ is not constant on $\mathcal{W}_i$, then $\dim \mathbb{P}T^*_{\tau|\mathcal{W}_i} = n - 1$. By using a Bertini type argument for the projection $\mathbb{P}T^*_{\tau|\mathcal{W}_i} \to \check{\mathbb{P}}^{n-1}$, it follows that there exists a Zariski-open set $\Omega'' \subset \check{\mathbb{P}}^{n-1}$ such that, for any $H \in \Omega''$, the map $\tau_{|\mathcal{W}_i}$ restricted to $\mathcal{W}_i \cap (H \times \mathbb{C}_t)$ is a submersion at all points except of the singular locus $\mathrm{Sing}_{\mathcal{W}}\tau$ and except of at most a finite set A_H, for any $i \in I$. We may and shall assume, after eventually intersecting with some Zariski-open set, that the set Ω'' also verifies the first claim of our lemma.

Now let us construct a Whitney stratification of $\mathbb{Y}$. We take $\mathbb{Y} \setminus \mathbb{Y}^\infty \simeq \mathbb{C}^n$ as a stratum and start to stratify $\mathbb{Y}^\infty$. Consider the product stratification $\mathbb{C}_r \times \mathcal{W}'$ of $\mathbb{C}_r \times \mathbb{X}'$, where $\mathcal{W}' \subset \mathbb{X}' \cap \mathbb{X}^\infty$. This is a Whitney stratification too. The intersection of $\mathbb{C}_r \times \mathcal{W}_i$ with $\mathbb{H}$ is transverse within $\mathbb{C}_r \times \mathbb{P}^n \times \mathbb{C}_t$ if and only if the intersection $(\mathbb{C}_r \times \mathcal{W}_i) \cap (\mathbb{C}_r \times H \times \mathbb{C}_t)$ is transverse within $\mathbb{C}_r \times \mathbb{P}^{n-1} \times \mathbb{C}_t$. The latter is indeed transverse for any i, at all points except of the set $\mathbb{C}_r \times A_H$, for $H \in \Omega''$. Leaving aside the exceptional set $\mathbb{C}_r \times A_H$, the transverse slices become Whitney strata of $\mathbb{Y}$ at infinity, since Whitney property is preserved by slicing transversely.

Next we have to see whether the fibres of the map (r, t) are transverse to the above defined Whitney strata of $\mathbb{Y}$. Namely, the map $(\mathbb{C}_r \times \mathcal{W}_i) \cap (\mathbb{C}_r \times H \times \mathbb{C}_t) \overset{(r,t)}{\to} \mathbb{C}_r \times \mathbb{C}_t$ is a submersion whenever the second projection $\mathcal{W}_i \cap (H \times \mathbb{C}_t) \overset{t}{\to} \mathbb{C}_t$ is a submersion. This situation was treated before, since the map t coincides with the map τ. The result is that, if $H \in \Omega''$, then the fibres of (r, t) are transverse to the Whitney stratification of $\mathbb{Y}$ except at the points of the following set:

$$\Sigma_H := \Gamma(l_H, f) \cup (\mathbb{C}_r \times (\mathrm{Sing}_{\mathcal{W}}\tau \cup A_H)).$$

Note that $t(\mathbb{C}_r \times \mathrm{Sing}_{\mathcal{W}}\tau)$ is a finite set. Moreover, for $H \in \Omega''$, $\Gamma(l_H, f) \cup (\mathbb{C}_r \times A_H)$ is a curve (or empty). Denote by $\Delta_H(r, t) \subset \mathbb{C}_r \times \mathbb{C}_t$ the image of Σ_H by (r, t). We have thus proved that the map:

$$(r, t)_| : \mathbb{Y} \setminus (r, t)^{-1}(\Delta_H(r, t)) \to (\mathbb{C}_r \times \mathbb{C}_t) \setminus \Delta_H(r, t) \qquad (8.2)$$

is a locally trivial, stratified fibration. Its restriction to the stratum $\mathbb{C}^n$, is what we were looking for. We may therefore set $\Omega''_f := \Omega''$ and define Λ_H as $t(\mathbb{C}_r \times (\mathrm{Sing}_{\mathcal{W}}\tau \cup A_H))$, noting that this contains the set of atypical values $\mathrm{Atyp} f$. $\qquad \square$

Remark 8.1.2 Corollary 7.1.3(b) and Lemma 8.1.1 show that, for any $c \in \mathbb{C}$, there exists a Zariski-open subset $\Omega''_{f,c} \subset \Omega''_f$ such that, if $H \in \Omega''_{f,c}$, then the restriction $l_{H|f^{-1}(t)} : f^{-1}(t) \to \mathbb{C}$ is a locally trivial stratified fibration in the neighbourhood of $\overline{f^{-1}(t)} \cap \mathbb{X}^\infty$ and has only stratified Morse critical points outside the point-strata of the stratification $\mathcal{W}$, and that these properties hold for any t within a small enough neighbourhood of c in $\mathbb{C}$. For the special infinity point $c = [0 : 1] \in \mathbb{P}^1$, this is not well defined, but we may set by definition $\Omega''_\infty := \Omega''_f$.

Tomographic model of the fibre. We recall from § 6.1 and § 7.1 the description of the 'tomographic' model of a fibre $F_c := f^{-1}(c)$. For $H \in \Omega''_{f,c}$ the singularities of the linear function $l_H : F_c \to \mathbb{C}$ are the intersection points $\Gamma(l_H, f) \cap F_c = \{c_1, \ldots, c_k\}$. As we have already seen in § 1, by the Lefschetz–Morse–Smale theory and its generalization by Milnor to holomorphic functions with isolated singularities, the nonsingular hypersurface F_c is built from a general slice $F_c \cap l_H^{-1}(s)$ by attaching a number of cells of dimension $n - 1$. In case F_c is singular, this result is still true by Lê's theorem [Lê5] for holomorphic functions with isolated singularities on singular hypersurfaces.[*]

To each singular point c_i, there corresponds an attaching of a number of $(n-1)$-cells equal to the $(n-2)$th Betti number of the local Milnor fibre M_{H,c_i} of the germ $l_H : (F_c, c_i) \to \mathbb{C}$.

Let us be more explicit in case when F_c is nonsingular or it has only isolated singularities. By Milnor's theorem, the Milnor fibre M_{H,c_i} is homotopically a bouquet of spheres and by Lê's theorem [Lê5], their number is equal to the intersection multiplicity $\mathrm{mult}_{c_i}(\Gamma(l_H, f), \{l_H = l_H(c_i)\})$. If the hyperplane H is generic (i.e. not in the conormal of F_c at c_i), then this is in turn equal to the multiplicity $\mathrm{mult}_{c_i} \Gamma(l_H, f)$ of the polar curve at c_i. We deduce, see also Exercise 7.2, that if F_c has at most an isolated singularity at c_i, then:

$$\mathrm{rank}\, H_{n-2}(M_{H,c_i}, \mathbb{Z}) = \mathrm{mult}_{c_i}(\Gamma(l_H, f), F_c) - \mu_{c_i}(F_c), \qquad (8.3)$$

and the following model, up to homotopy type: F_c is built from some generic hyperplane section $F_c \cap l_H^{-1}(s)$, $s \notin \Lambda_H$, to which one attaches $\alpha_c^{(n-1)} - \mu_{F_c}$ cells of dimension $n - 1$, where μ_{F_c} is the sum of the Milnor numbers of the singularities of the hypersurface F_c and $\alpha_c^{(n-1)}$ is the generic global polar intersection multiplicity $\mathrm{mult}(\Gamma(l_H, f), F_c)$. We have defined for example in §7.2 the set of generic global polar intersection multiplicities:

$$\alpha_c^{(n-1)}, \alpha_c^{(n-2)}, \ldots, \alpha_c^{(1)}, \alpha_c^{(0)}.$$

The *generic skeleton* $\mathrm{Sk}_c(f)$ of F_c is the CW-complex obtained in the above process: starting with $\alpha_c^{(0)}$ points, one successively attaches $\alpha_c^{(1)}$ cells of dimension

---

[*] For more general spaces, see the proof of Theorem 6.1.6.

1, then $\alpha_c^{(2)}$ cells of dimension 2, ..., $\alpha_c^{(n-2)}$ cells of dimension $n-2$ and finally $\alpha_c^{(n-1)} - \mu_{F_c}$ cells of dimension $n-1$. Up to homotopy type, the generic skeleton does not depend on the choices of generic hyperplanes.

A global geometric monodromy. We pursue and construct now a geometric monodromy of a polynomial function $f : \mathbb{C}^n \to \mathbb{C}$ by using the key Lemma 8.1.1 and the generic skeleton $\mathrm{Sk}_c(f)$ of a typical fibre of F_c.

Let $H \in \Omega_f''$ and $\Lambda_H \subset \mathbb{C}$, in the notations of Lemma 8.1.1. For any simple loop within $\mathbb{C} \setminus \Lambda_H$, we may define a 'rough' geometric monodromy as follows. By the proof of Lemma 8.1.1, the map $\tau : \mathbb{X} \setminus \tau^{-1}(\Lambda_H) \to \mathbb{C} \setminus \Lambda_H$ is a locally trivial stratified fibration. In particular, the restriction of τ to the open stratum $\mathbb{C}^n$ is a C^∞ locally trivial fibration. We may produce a trivializing vector field tangent to the strata at infinity, as in the proof of the Thom–Mather isotopy lemma (cf. e.g. [Ve]). We therefore get a geometric monodromy representation $\rho : \pi_1(\mathbb{P}^1 \setminus \Lambda_H \cup \{\infty\}, c) \to \mathrm{Diff}(G)$. This induces an algebraic monodromy representation $\rho_{\mathrm{alg}} : \pi_1(\mathbb{P}^1 \setminus \Lambda_H \cup \{\infty\}) \to H_*(G, \mathbb{Z})$.

We would like to refine the construction of the trivializing vector field, hence of the geometric monodromy, so as to get more information on the algebraic monodromy. We therefore focus on defining a geometric monodromy of a general fibre of f along a small circle included in $\mathbb{P}^1 \setminus \mathrm{Atyp}f \cup \{\infty\}$. We refer to Lemma 8.1.1 and to the notations in its proof.

Let us take a small closed disk D_a at $a \in \mathrm{Atyp}f \cup \{\infty\}$ such that $\Lambda_H \cap D_a = \{a\}$, for some $H \in \Omega_{f,a}''$. Then take a lift of the unitary vector field **u** on the circle ∂D_a to a vector field **w** in the tube $f^{-1}(\partial D_a)$ such that **w** is tangent to $t^{-1}(\partial D_a) \cap \mathbb{Y}^\infty$ in a stratified sense and tangent to $\Gamma(l_H, f) \cap f^{-1}(\partial D_a)$, for some general $H \in \Omega_{f,a}''$. Note that the set $\Gamma(l_H, f) \cap f^{-1}(\partial D_a)$ is a finite union of (knotted) circles.

We may produce a more refined construction of a vector field **w** by lifting **u** in two steps, as follows:

$$f^{-1}(\partial D_a) \xrightarrow{(l,f)} \mathbb{C} \times \partial D_a \xrightarrow{\mathrm{pr}_2} \partial D_a. \tag{8.4}$$

In the local setting, this leads to the *carrousel monodromy* introduced by Lê D.T. [Lê2, Lê4].[2] In the global setting, we may decompose the monodromy flow in regions where the local carrousel construction holds. There is a 'carrousel' associated to each point $q \in \Gamma(l_H, f) \cap \tau^{-1}(a)$.

Moreover, we may construct a similar 'carrousel' at some point at infinity $q \in \mathbb{X}_a^\infty \cap \overline{\Gamma}(l_H, f)$, where $\overline{\Gamma}(l_H, f)$ denotes the closure of $\Gamma(l_H, f)$ in $\mathbb{X}$, as follows. We have defined at (8.1) the space $\mathbb{Y} \subset \mathbb{C}_r \times \mathbb{P}^n \times \mathbb{C}_t$ with projection $(r, t) : \mathbb{Y} \to \mathbb{C}_r \times \mathbb{C}_t$. We now consider the closure $\hat{\mathbb{Y}}$ of $\mathbb{Y}$ in $\mathbb{P}_r^1 \times \mathbb{P}^n \times \mathbb{C}_t$

and denote by $\hat{r} : \hat{\mathbb{Y}} \to \mathbb{P}_r^1$ the projection. Let δ_i be a small open disk centered at d_i, where $\{d_1, \ldots, d_k\} \in \mathbb{P}_r^1$ is the image by $\hat{r}$ of the set $\overline{\Gamma}(l_H, f) \cap t^{-1}(a)$.

We first lift the vector field $\mathbf{u}$ by pr_2 to a vector field $\mathbf{v}$ on $\mathbb{C}_r \times \partial D_a$ such that, for any $i = \overline{1,k}$, $\mathbf{v}$ is the carrousel vector field on $\delta_i \times \partial D_a$, for small enough disks D_a and δ_i. This means that the lift $\mathbf{v}$ of $\mathbf{u}$ to $\mathbb{C}_r \times \partial D_a$ is tangent to the discriminant $\Delta(l_H, f)$ and that $\mathbf{v}$ is by definition the identical lift of $\mathbf{u}$ by the projection $\mathrm{pr}_2 : \{b\} \times \partial D_a \to \partial D_a$ for any point $b \in \mathbb{P}_r^1 \setminus \cup_{i=1}^k \delta_i$.

This carrousel vector field $\mathbf{v}$ is now lifted to $f^{-1}(\partial D_a)$ via the stratified fibration (8.2) and yields, by integration, a geometric monodromy, denoted by h_a. By its definition, for any point $b \in \mathbb{P}^1 \setminus \cup_{i=1}^k \delta_i$, h_a acts on the slice $l_H^{-1}(b)$ as the monodromy of the slice fibration:

$$f^{-1}(\partial D_a) \cap l_H^{-1}(b) \to \partial D_a.$$

For some fixed $c \in \partial D_a$, the action of the monodromy h_a on the pair $(F_c, F_c \cap l_H^{-1}(b))$, will be called *relative geometric monodromy* and will be denoted by h_a^{rel}.

Localization of the monodromy. After [Ti4, §4], we say that the *variation of topology* of the fibres of f at $a \in \mathbb{C}$ is *localizable* if there is a finite set $\{a_1, \ldots, a_k\} \in \mathbb{X}_a$, such that for any system of small enough balls $B_i \subset \mathbb{X}$ centered at a_i, there exist a disk D_a such that the restriction $f_| : (\mathbb{C}^n \setminus \cup_{i=1}^k B_i) \cap f^{-1}(D_a) \to D_a$ is a trivial fibration.

In this case, the geometric monodromy h_a is trivial on the complement $F_c \setminus \cup_{i=1}^k B_i$ of the balls, for some $c \in \partial D_a$. We also say that the monodromy h_a is localizable at the points a_i.

In principle, the change of topology from a typical fibre to an atypical one cannot be localized for any polynomial $f : \mathbb{C}^n \to \mathbb{C}$, since its singularities at infinity may be nonisolated in the meaning of Definition 2.2.2. Assuming that our polynomial has isolated t-singularities, we get the following direct consequence of the proof of the bouquet theorem, 3.2.1, and of the above definition. We use the notations in Definition 2.2.2.

Proposition 8.1.3 (localization of the monodromy) If f is of $\mathcal{T}$-type, or more generally, if f has t-singularities at the fibre F_a, then the variation of topology of the fibres of f at a is localizable at the t-singularities of the fibre F_a, i.e. at the finite set $F_a \cap (\mathrm{Sing} f \cup \mathrm{Sing}^\infty f)$. In particular, the monodromy h_a is localizable at the t-singularities.

We next show that certain localization results hold without any assumption on f in case of the relative monodromy h_a^{rel}. For more details on the case of isolated t-singularities, we refer to Proposition 8.2.3.

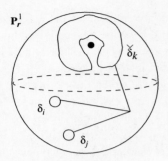

Figure 8.1. Disks and truncated disks in $\mathbb{P}_r^1$

First localization of the relative monodromy. As before, $d_1, d_2, \ldots, d_k$ are the points where the closure $\bar{\Delta}(l_H, f) \subset \mathbb{P}_r^1 \times \mathbb{C}_t$ of the discriminant $\Delta(l_H, f)$ cuts the projective line $\{t = a\} \subset \mathbb{P}_r^1 \times \mathbb{C}_t$. We fix points $s_i \in \partial \delta_i$, for $i \in \{1, \ldots k\}$ at the boundaries of the small disks δ_i. We denote $\check{\delta}_i := \delta_i$ if $d_i \in \mathbb{C}_r$ and $\check{\delta}_i := \delta_i \setminus \rho_i$ if $d_i = \infty$, where ρ_i is a radius segment from d_i to some point on $\partial \delta_i$ different from s_i (see Figure 8.1).

By construction, $\check{\delta}_i \times \{c\}$ contains all the points of $\bar{\Delta}_i(l_H, f) \cap \mathbb{P}_r^1 \times \{c\}$, where $c \in \partial D_a$ and $\bar{\Delta}_i(l_H, f)$ denotes the germ of $\bar{\Delta}(l_H, f)$ at $(d_i, a) \in \mathbb{P}_r^1 \times \mathbb{C}_t$. The structure of the generic skeleton $\mathrm{Sk}_c(f)$ implies that the relative homology $H_*(F_c, F_c \cap l_H^{-1}(b))$ is concentrated in dimension $n - 1$, where $b \in \mathbb{P}_r^1 \setminus \cup_{i=1}^k \delta_i$.

We get the following localization result:

Lemma 8.1.4 Let $f : \mathbb{C}^n \to \mathbb{C}$ be a polynomial function and $a \in \mathbb{P}_t^1$. The relative homology splits into a direct sum:

$$H_{n-1}(F_c, F_c \cap l_H^{-1}(b)) = \oplus_{i=1}^k H_{n-1}(F_c \cap l_H^{-1}(\check{\delta}_i), F_c \cap l_H^{-1}(s_i)).$$

Suppose in addition that $a \in \mathbb{C}_t$ and that $\bar{\Gamma}(l_H, f) \cap \mathbb{X}_a^\infty = \emptyset$. Then the relative monodromy splits accordingly:

$$(h_a^{\mathrm{rel}})_* = \oplus_{i=1}^k (h_a^{\mathrm{rel}})_{*,i},$$

where $(h_a^{\mathrm{rel}})_{*,i}$ denotes the monodromy acting on $H_{n-1}(F_c \cap l_H^{-1}(\delta_i), F_c \cap l_H^{-1}(s_i))$.

Proof We get the claimed homology splitting by applying an excision, using the fact that the projection $l_H : F_c \to \mathbb{C}$ is locally trivial over $\mathbb{P}_r^1 \setminus \cup_{i=1}^k \check{\delta}_i$.

Let us fix now $a \in \mathbb{C}_t$ and remark that in general the carrousel vector field does not preserve the truncated disk $\check{\delta}_i$ around infinity. Actually, by construction, the geometric monodromy h_a^{rel} acts on the pair of *spaces* $(F_c \cap l_H^{-1}(\check{\delta}_i), F_c \cap l_H^{-1}(s_i))$ if and only if $d_i \in \mathbb{C}$. The supplementary hypothesis implies that all the points d_i are different from ∞, see also Exercise 8.2, and so we get the splitting of our relative monodromy since in the direct sum splitting there is no disk δ_i around infinity. $\qquad\square$

Second localization of relative monodromy. We further localize the relative monodromy in the source space. Let $a \in \mathbb{C}$. By Remark 8.1.2, if $H \in \Omega''_{f,a}$, then there exists a small enough disk D_a such that, for any $t \in D_a$, the map $l_{H|} : F_t \to \mathbb{C}$ is a locally trivial stratified fibration in the neighbourhood of $\mathbb{X}_a^\infty$ and has only stratified Morse critical points at $\Gamma(l_H, f) \cap F_a$.

Let $c \in \partial D_a$ and let $\{p_{i,j}\}_{j \in R_i}$ denote the set of points $\Gamma(l_H, f) \cap F_c \cap l_H^{-1}(\check{\delta}_i)$. Take Milnor disks and balls for the germs $l_H : (F_c, p_{i,j}) \to \mathbb{C}$, which means small enough disks $D_{i,j} \subset \check{\delta}_i$ centered at $l_H(p_{i,j})$ and small balls $B_{i,j} \subset F_c \cap l_H^{-1}(\check{\delta}_i)$ centered at $p_{i,j}$. We assume that $D_{i,j_1} = D_{i,j_2}$ if $l_H(p_{i,j_1}) = l_H(p_{i,j_2})$. Note again that these are germs of complex Morse singularities. With these notations, we have the following result:

Proposition 8.1.5 Let $a \in \mathbb{P}^1$ and let i be fixed. The homology Milnor datum $H_{n-1}(F_c \cap l_H^{-1}(\check{\delta}_i), F_c \cap l_H^{-1}(s_i))$ splits into the following direct sum

$$\oplus_{j \in l_i} H_{n-1}(F_c \cap l_H^{-1}(D_{i,j}) \cap B_{i,j}, F_c \cap l_H^{-1}(s_{i,j}) \cap B_{i,j}),$$

where $s_{i,j} \in \partial D_{i,j}$ are some fixed points. The restriction of the relative monodromy h_a^{rel} acts on the set of all pairs $(F_c \cap l_H^{-1}(D_{i,j}) \cap B_{i,j}, F_c \cap l_H^{-1}(s_{i,j}) \cap B_{i,j})$ by permutations, more precisely h_a^{rel} is a diffeomorphism from one pair to another one.

Proof We get by excision the following splitting:

$$H_{n-1}(F_c \cap l_H^{-1}(\check{\delta}_i), F_c \cap l_H^{-1}(s_i)) \simeq \oplus H_{n-1}(F_c \cap l_H^{-1}(D_{i,j}), F_c \cap l_H^{-1}(s_{i,j})),$$

where the direct sum is taken over the *distinct* disks $D_{i,j}$, with fixed i. It also follows by the above arguments that, for any fixed $j_0 \in R_i$, the map:

$$l_H : F_c \cap l_H^{-1}(D_{i,j_0}) \setminus \cup B_{i,j} \to D_{i,j_0},$$

where the union is over all $j \in R_i$ such that $D_{i,j} = D_{i,j_0}$, is a trivial fibration. Indeed, this map has no singularities at infinity and its fibres are transverse to the spheres $\partial B_{i,j}$. The splitting of the homology follows. The second statement is a consequence of the construction of the carrousel monodromy, more specifically of the fact that the lift of the carrousel vector field is tangent to the polar curve.	□

The relative monodromy h_a^{rel} does not act in general on each term of the direct sum of Proposition 8.1.5. The interactions among the relative cycles are prescribed by the global geometric carrousel monodromy. It may happen, for instance, that one or more branches of the discriminant locus at d_i are multiple, being the images of more than one branch of the polar curve.

Remark 8.1.6 Let $H \in \Omega''_{f,a}$ as in Remark 8.1.2. Let $p \in F_a \cap \Gamma(l_H, f)$ such that $p \notin \text{Sing}\, F_a$. Since p is a Morse point of the restriction of l_H to F_a, the polar curve is nonsingular at p and it cuts the nearby fibre F_c at a single point $p_{i,1}$ within a small neighbourhood of p. The construction of the carrousel monodromy then tells that the relative monodromy h_a^{rel} acts on the pair $(F_c \cap l_H^{-1}(D_{i,1}) \cap B_{i,1}, F_c \cap l_H^{-1}(s_{i,1}) \cap B_{i,1})$ as the identity.

Proposition 8.1.7 Let $a \in \mathbb{C}$, $d_i \in \mathbb{C}_r$ for some fixed $i \in \{1, \ldots, k\}$ and $H \in \Omega''_{f,a}$. Then there is the following splitting:

$$(h_a^{\text{rel}})_{*,i} = \oplus_{j=1}^{g_i} (h_{b_j}^{\text{rel}})_*,$$

where $\{b_1, \ldots, b_{g_i}\} := \Gamma(l_H, f) \cap F_a \cap l_H^{-1}(\delta_i)$ and $(h_{b_j}^{\text{rel}})_*$ denotes the local relative monodromy of f with respect to l_H, at the space germ $(\mathbb{C}^n, b_j)$.

Proof By Remark 8.1.2, the map $l_H : F_c \cap l_H^{-1}(\delta_i) \to \delta_i$ is topologically trivial at infinity, for all c close enough to a. The carrousel monodromy shows that it is also trivial away from some small enough polydisks $P_j := \delta_i \times B_j$ centered at b_j, for $j \in \{1, \ldots, g_i\}$. We obtain the splitting:

$$H_{n-1}(F_c \cap l_H^{-1}(\delta_i), F_c \cap l_H^{-1}(s_i)) = \oplus_{j=1}^{g_i} H_{n-1}(P_j \cap F_c \cap l_H^{-1}(\delta_i), P_j \cap F_c \cap l_H^{-1}(s_i))$$

and the monodromy acts on each term of the direct sum. These are just local relative homology groups of f with respect to l_H, at the space germ $(\mathbb{C}^n, b_j)$.	□

Proposition 8.1.8 Let $H \in \Omega''_{f,a}$. If f has isolated t-singularities at a, then the carrousel monodromy can be defined such that its restriction $h_{a|H_t}$ to a general hyperplane $H_t := \{l_H = t\}$ is the identity.

Proof As in the proof of Theorem 7.2.6, the t-singularities at infinity of f will be denoted by $\text{Sing}^\infty f$. By the proof of Lemma 8.1.1 (see also [Ti4, §5.]), for $H \in \Omega''_{f,a}$ and for some general value of t we have:

$$\dim \text{Sing}^\infty f_{|H_t} \cap \mathbb{X}_a \leq \dim \text{Sing}^\infty f \cap \mathbb{X}_a - 1. \tag{8.5}$$

Since in our case f has isolated t-singularities, we get that the hyperplane $\bar{H}_t$ avoids the singular set $(\mathrm{Sing} f \cup \mathrm{Sing}^{\infty} f) \cap \mathbb{X}_a$ and is stratified transverse to $\mathbb{X}_a$. It follows that the restriction $f_{|H_t}$ has no t-singularities on the fibre $\mathbb{X}_a$, hence this fibre is typical for $f_{|H_r}$. Therefore the monodromy of $f_{|H_r}$ around a is isotopic to the identity. $\qquad\qquad\square$

8.2 Relative monodromy and zeta function

Zeta function formulae have been proved for special classes of polynomials: in terms of Newton polyhedra (under nondegeneracy conditions, cf. [LS]), in terms of the projective compactification of f (when this is nonsingular, cf. [GN]), or using resolution of singularities and the fact that the zeta function is a constructible function (see e.g. [GLM3]).

We show here how to deduce the zeta function of the monodromy of f around an atypical value $a \in \mathbb{P}^1$ from the zeta function of the relative monodromy.

Definition 8.2.1 Let h_* be the algebraic monodromy associated to a fibration $F \hookrightarrow E \to S^1$ with bounded and finitely dimensional homology of the fibre $H_*(F, \mathbb{C})$. Let H_j denote the jth homology group. The *zeta function* of h_* is the following rational function:

$$\zeta_{h_*}(t) = \prod_{j \geq 0} \det[(\mathrm{id} - t h_*) : H_j \to H_j]^{(-1)^{j+1}}.$$

With this sign convention, the degree of the zeta-function is equal to $-\chi(M)$.

The *Lefschetz number of the monodromy* h_* is the following integer:

$$L(h_*) = \sum_{j \geq 0} (-1)^j \mathrm{trace}[h_* : H_j \to H_j].$$

We refer to [Mi2, A'C2, Lê2, Ti1, ST1, GLM2] for some aspects of the local zeta function. We shall currently use the multiplicativity of the zeta function of the monodromy acting on exact sequences, and the fact that the derivative at $t = 0$ of the zeta function is equal to the Lefschetz number of the monodromy $L(h_*)$.

We use the same notations as in the previous section. Let F' be a general fibre of the map (l_H, f), see Lemma 8.1.1, and let $\zeta_i^{\mathrm{rel}}(t)$ denote the zeta-function of the relative monodromy $(h_a^{\mathrm{rel}})_{*,i}$, in the notations of Lemma 8.1.4. As a consequence of the localization of the relative monodromy, we have the following general zeta function formula for the monodromy $(h_a)_*$, with no restrictions on f:

Theorem 8.2.2 *Let* $f : \mathbb{C}^n \to \mathbb{C}$ *be any polynomial function,* $a \in \mathbb{C}$ *and* $H \subset \Omega''_{f,a}$. *Let* $\{d_1, d_2, \ldots, d_k\} := \bar{\Delta}(l_H, f) \cap \mathbb{P}^1_r \times \{t = a\}$ *and assume that* $d_k = \infty$. *Then the zeta function of the monodromy* h_a *decomposes into a product as follows:*

$$\zeta_{h_a}(t) = \zeta_{h_{a|F'}}(t) \cdot \zeta_{\delta^*_k}(t) \cdot \prod_{i=1}^{k-1} \zeta_i^{\mathrm{rel}}(t),$$

where δ^*_k *is the pointed disk and* $\zeta_{\delta^*_k}$ *is the zeta function of the monodromy[3] acting on the space* $F_c \cap l_H^{-1}(\delta^*_k)$.

Proof We decompose $\mathbb{C} = \mathbb{P}^1 \setminus \infty$ as the union $D \cup \delta^*_k$ of a big closed disk D and the pointed disk around infinity δ^*_k, their intersection $A := D \cap \delta^*_k$ being a thickened circle. The carrousel monodromy acts on these subspaces and the zeta function of the monodromy on $F_c \cap l_H^{-1}(A)$ is equal to 1 by general reasons (see e.g. [A'C2, Ti1]). A Mayer–Vietoris argument then yields the equality:

$$\zeta_{h_a}(t) = \zeta_D(t) \cdot \zeta_{\delta^*_k}(t),$$

where ζ_D is the zeta function of the monodromy acting on the space $F_c \cap l_H^{-1}(D)$.

Next, D retracts to the union $\cup_i (\delta_i \cup \gamma_i)$ of the disks $\delta_i \subset \mathbb{C}$ and a collection of simple nonintersecting paths γ_i connecting each δ_i to some exterior point $p \in \partial D$. Excising now $l_H^{-1}(\cup_i \gamma_i)$ in the relative homology $H_*(F_c \cap l_H^{-1}(D), F_c \cap l_H^{-1}(p))$, where $F_c \cap l_H^{-1}(p)$ is a general fibre F' of the map (l_H, f), and using the proof of the second part of Lemma 8.1.4, we get the formula:

$$\zeta_D(t) = \zeta_{h_{a|F'}}(t) \cdot \prod_{d_i \subset \mathbb{C}} \zeta_i^{\mathrm{rel}}(t).$$

We obtain the claimed relation by comparing this with the previous formula for ζ_{h_a}. $\qquad\qquad\qquad\qquad\qquad\qquad\qquad\qquad\qquad\qquad\qquad\qquad\qquad\square$

Isolated *t***-singularities.** Proposition 8.1.3 tells that, if f has isolated t-singularities at the fibre F_a, then the variation of topology is localizable. Let $\{a_s \mid s = 1, \ldots, v\} \subset \mathbb{X}_a$ be the t-singularities of f at the fibre F_a, where some of them may be at infinity. By definition, at each point a_s there is a complete system of ball neighbourhoods $B_{s,\varepsilon}$ within the space $\mathbb{X}$, and of disks D_δ centered at $a \in \mathbb{C}$, such that the map

$$f_| : (\mathbb{C}^n \setminus \cup_s \bar{B}_{s,\varepsilon}) \cap f^{-1}(D_\delta) \to D_\delta$$

is a trivial fibration, for small enough $\varepsilon > 0, 0 < \delta \ll \varepsilon$. Moreover, for any s, the restriction

$$f_| : \mathbb{C}^n \cap B_{s,\varepsilon} \cap f^{-1}(D_\delta \setminus \{a\}) \to D_\delta \setminus \{a\} \tag{8.6}$$

is locally trivial, by Milnor's fibration theorem 3.1.1, and Lê's fibration theorem A1.1.2, respectively. The carrousel monodromy acts on the fibres of these fibrations and on their boundaries. See also §3.2 for related facts. Let us denote by $\zeta_{h_a,s}(t)$ the zeta function of the monodromy induced by the local fibration (8.6) on the pair $(F_{D_\delta} \cap B_{s,\varepsilon}, F_c \cap B_{s,\varepsilon})$, where $c \in \partial D_\delta$. Under these notations we have:

Proposition 8.2.3 If f has isolated t-singularities at F_a, then:

$$\zeta_{h_a}(t) = (1 - t)^{-\chi(F_a)} \prod_{s=1}^{\nu} \zeta_{h_a,s}^{-1}(t). \tag{8.7}$$

Proof The carrousel monodromy acts on the homology exact sequence of the pair (F_{D_δ}, F_c); we get that $\zeta_{h_a}(t) = \zeta_{F_{D_\delta}}(t) \cdot \zeta_{(F_{D_\delta}, F_c)}^{-1}(t)$. Since by excision we have the direct sum decomposition:

$$H_\star(F_{D_\delta}, F_c) \simeq \oplus_s H_\star(F_{D_\delta} \cap B_{s,\varepsilon}, F_c \cap B_{s,\varepsilon}),$$

we get:

$$\zeta_{(F_{D_\delta}, F_c)}(t) = \prod_{s=1}^{\nu} \zeta_{h_a,s}(t). \tag{8.8}$$

Since the monodromy acts on F_{D_δ} as the identity and since $\chi(F_{D_\delta}) = \chi(F_a)$. (Note also that the relative homology $H_\star(F_{D_\delta}, F_c)$ is concentrated in dimension $n - 1$, by Theorem 3.2.1.)

Finally, for the affine isolated singularity $a_s \in F_a$, the space $F_{D_\delta} \cap B_{s,\varepsilon}$ is contractible and therefore in the formula (8.8) we may replace the factor $\zeta_{h_a,s}(t)$ by $(1 - t)^{-1} \zeta_{a_s}^{-1}(t)$. □

Lefschetz number of the monodromy. A'Campo's local result [A'C1] says that the Lefschetz number of the monodromy of the germ of a singular holomorphic function is zero.[4] In the global affine setting we have the following result:

Proposition 8.2.4 Let $f : \mathbb{C}^n \to \mathbb{C}$ be a polynomial with isolated t-singularities at the fibre F_a. Then the Lefschetz number of the monodromy $L(h_a)$ satisfies the relation:

$$L(h_a) = \chi(F_a \setminus \mathrm{Sing} f) + \sum_{a_s \in \mathbb{X}_a^\infty} L(h_{a|F_c \cap B_{s,\varepsilon}}),$$

where $h_{a|F_c \cap B_{s,\varepsilon}}$ is the monodromy acting on the local nearby fibre $F_c \cap B_{s,\varepsilon}$.

Proof We apply the Lefschetz number to the exact sequence of the pair $(F_c, F_c \setminus \cup_s B_{s,\varepsilon})$. By its definition the monodromy h_a acts as the identity on $F_c \setminus \cup_s B_{s,\varepsilon}$ and so the Lefschetz number on this subspace is equal to its Euler characteristic. We get:

$$L(h_a) = \chi(F_c \setminus \cup_s B_{s,\varepsilon}) + L(h_{a|(F_c, F_c \setminus \cup_s B_{s,\varepsilon})}).$$

We also have the diffeomorphism $F_c \setminus \cup_s B_{s,\varepsilon} \simeq F_a \setminus \cup_s B_{s,\varepsilon}$ and we deduce $\chi(F_c \setminus \cup_s B_{s,\varepsilon}) = \chi(F_a \setminus \text{Sing} f)$. By excision, we get the direct sum splitting: $H_*(F_c, F_c \setminus \cup_s B_{s,\varepsilon}) \simeq \oplus_s H_*(F_c \cap B_{s,\varepsilon}, F_c \cap \partial B_{s,\varepsilon})$ on which the monodromy acts. Moreover, the monodromy h_a is trivial on all links $F_c \cap \partial B_{s,\varepsilon}$ and these have zero Euler characteristics since they are odd real algebraic manifolds. It then follows that:

$$L(h_{a|(F_c \cap B_{s,\varepsilon}, F_c \cap \partial B_{s,\varepsilon})}) = L(h_{a|F_c \cap B_{s,\varepsilon}}).$$

Finally, A'Campo's theorem [A'C1] tells that the Lefschetz number of the monodromy at the affine points $a_s \in F_a$ is zero. Collecting all these formulas we get the claimed formula. □

Let us point out that unlike the local case where A'Campo's theorem holds, the Lefschetz number $L(h_{a|F_c \cap B_{s,\varepsilon}})$ may be nonzero at t-singularities at infinity (see also some examples below).

Computing the relative monodromy at infinity. We consider the case of an isolated singularity at infinity with $\lambda = 1$. Then the local relative monodromy action on the single relative cycle can be either $+\text{id}$ or $-\text{id}$. When the dimension is fixed, it appears that both cases are possible. This has to be contrasted to the affine local setting, where the monodromy of a $(n-1)$-cycle of a Morse singularity is equivalent to $(-1)^n \text{id}$, hence depends only on the dimension. We illustrate by the following examples what is the relative monodromy in the neighbourhood of infinity and present the two possible cases for the relative monodromy of a $\lambda = 1$ singularity.

Example 8.2.5 $f : \mathbb{C}^2 \to \mathbb{C}, f(x,y) = x + x^2 y$, as in Example 1.1.2, cf. [Br2], is the simplest polynomial with a noncritical atypical fibre, no critical fibres and a single $\mathcal{B}$-type singularity at infinity $q = ([0:1], 0) \in \mathbb{X}_0^\infty$, with $\lambda_q = 1$.

We may take $l = x + y$ as a general function. The polar curve $\Gamma(l,f)$ intersects transversely a general fibre F_t at 3 points and the atypical fibre F_0 at 2 points, a_1 and a_2. We get $\alpha_t^{(1)} = 3$, $\alpha_0^{(1)} = 2$ and $\alpha_t^{(0)} = \alpha_0^{(0)} = 3$. Therefore $\chi(F_c) = 0$, $\chi(F_0) = 1$. By Theorem 8.2.2, $\zeta_{h_0}(t) = \zeta_{F'}(t) \cdot \zeta_{\delta*}(t) \cdot \prod_{i=1}^2 \zeta_i^{\text{rel}}(t)$, where F' is 3 points. By Remark 8.1.6, $\zeta_1^{\text{rel}}(t) = \zeta_2^{\text{rel}}(t) = (1 - t)$.

We use the setting and notations of Proposition 8.1.5. The carrousel δ^* centered at $\infty \in \mathbb{P}^1$ contains a small disk D centered at the unique point

$p := l(\Gamma(l,f) \cap F_c \cap l^{-1}(\delta^*))$. Let B be a small enough Milnor ball centered at $\Gamma(l,f) \cap F_c \cap l^{-1}(\delta^*)$, and let $s \in \partial D$. Then $\zeta_{\delta^*}(t)$ is equal to $\zeta_{D,s}(t)$, the zeta function of the carrousel monodromy acting on $H_1(F_c \cap l^{-1}(D) \cap B, F_c \cap l^{-1}(s) \cap B) \simeq H_1(I, \partial I)$. The carrousel monodromy of the space $F_c \cap l^{-1}(s) \cap B$ is isotopic to the monodromy of this space along a loop around both the center of δ^* and the center of D. We decompose it into a simple loop around $\infty \in \delta$ followed by a small loop around p. Since p is an A_1-singularity of l restricted to F_c, the l-monodromy of $H_1(I, \partial I)$ around it is $-\mathrm{id}$. The monodromy around the point ∞ is also $-\mathrm{id}$, by the following reason: the compactified curve $\bar{F}_c$ has an A_2 singularity at $[0 : 1]$, if $c \neq 0$ (and an A_3 singularity if $c = 0$). The loop in F_c around this point is the complex link monodromy of the germ of $\bar{F}_c$ (i.e. the monodromy of a general linear function on $\bar{F}_c$) and this is $-\mathrm{id}$ by Exercise 8.3. It follows that $\zeta_{\delta^*}(t) = (1 - t)$. Since $\zeta_{h_{0|F'}}(t) = (1 - t)^{-3}$, we finally get $\zeta_{h_0}(t) = 1$. Actually the monodromy h_a is isotopic to the identity.

Example 8.2.6 [SiSm, ST5] $f : \mathbb{C}^2 \to \mathbb{C}, f(x,y) = x^2y^2 + xy + x$.
There is a Morse singularity at $(0, -1)$, on the fibre F_0 and a singularity at infinity at $[0 : 1] \in \mathbb{P}^1$ for the fibre $F_{-\frac{1}{4}}$. Hence $\mu = 1$ and $\lambda = 1$. The bouquet theorem tells that the general fibre is homotopy equivalent to $S^1 \vee S^1$. We may take $l = x + y$ as a general linear form. Then $\Gamma(l,f) = \{2xy^2 + y + 1 - 2x^2y - x = 0\}$ and its intersection with F_t is 4 points, if $t = 0$ or $t = -\frac{1}{4}$ and 5 points for the other values of t. The fibre $F' = F_c \cap \{x + y = s\}$ is 4 points, for generic s. We get $\alpha_c^{(1)} = 5$, $\alpha_0^{(1)} = 3$ and $\alpha_{-\frac{1}{4}}^{(1)} = 4$. We now compute the zeta-function of the monodromy $h_{-\frac{1}{4}}$. In the carrousel disk δ^* centered at $q_1 = \Gamma(l,f) \cap F_{-\frac{1}{4}}$, the situation is similar to that in the first example. The only difference is that the complex link monodromy in cause is this time the one of an A_3 curve singularity. Using Exercise 8.3, this shows that our carrousel monodromy around $\partial \delta$ is the composition $(-\mathrm{id}) \circ (+\mathrm{id})$, so $\zeta_{\delta^*}(t) = (1 + t)$. Finally, by Theorem 8.2.2, we get:

$$\zeta_{h_{-\frac{1}{4}}}(t) = (1 - t)^{-4}(1 - t)^4(1 + t) = 1 + t.$$

We may deduce from this that the monodromy $h_{-\frac{1}{4}}$ acts on a certain basis of absolute cycles by switching them.

The zeta function ζ_{h_0} is easier to compute since it is localizable at the Morse singularity $(0, -1)$ and its local monodromy is the identity (acting on the local cycle S^1). Therefore, by Proposition 8.2.3 we get:

$$\zeta_{h_0}(t) = (1 - t)^{-\chi(F_0)}(1 - t)\zeta_{h_{0,(0,-1)}}(t) = 1,$$

since $F_0 \overset{\mathrm{ht}}{\simeq} S^1$ and $\zeta_{h_{0,(0,-1)}}(t) = 1$, in the notations of Proposition 8.2.3.

8.3 The *s*-monodromy and boundary singularities

We consider a family $P(x, s)$ of polynomials $f_s := P(\cdot, s) : \mathbb{C}^n \to \mathbb{C}$ depending polynomially on the parameter s. We show here, following [ST5], how to study the topology of f_0, up to a certain extent, by a certain type of monodromy in the parameter space.

We have seen in §4.2 and §4.3 the phenomenon of $\mu \rightleftarrows \lambda$ exchange, which takes place whenever singularities of f_s tend to some point at infinity of a compactified fibre of f_0 (see for instance Theorem 4.2.5). In the $\mathcal{F}$-class, this amounts to the interplay between the affine singularities of the polynomial and the singularities of its compactified fibres on the hyperplane at infinity H^∞. There are finitely many one-parameter families of *boundary singularities* attached to singularities at infinity of the compactified fibres of a $\mathcal{F}$-type polynomial. Arnold has already remarked in [Ar2] that there is a close relation between the classification of meromorphic germs of type H/Z^d and the one of germs of boundary singularities [Ar1]. As in §4.2, let us consider the total space associated to the deformation $P(x, s) = f_s(x)$, which we assume of constant degree d:

$$\mathbb{Y} = \{([x : x_0], s, t) \in \mathbb{P}^n \times \mathbb{C} \times \mathbb{C} \mid \tilde{P}(x, x_0, s) - tx_0^d = 0\},$$

where $\tilde{P}$ denotes the homogenized of P in the variables x. Let $\tau : \mathbb{Y} \to \mathbb{C}$ be the projection to the t-coordinate and let $\sigma : \mathbb{Y} \to \mathbb{C}$ denote the projection to the s-coordinates. We use the notations in §4.2, $\mathbb{Y}_{s,*} := \mathbb{Y} \cap \sigma^{-1}(s)$, $\mathbb{Y}_{*,t} := \mathbb{Y} \cap \tau^{-1}(t)$ and $\mathbb{Y}_{s,t} := \mathbb{Y}_{s,*} \cap \tau^{-1}(t) = \mathbb{Y}_{*,t} \cap \sigma^{-1}(s)$. Then $\mathbb{Y}_{s,t}$ is the closure in $\mathbb{P}^n$ of the affine hypersurface $f_s^{-1}(t) \subset \mathbb{C}^n$. As usual, $\mathbb{Y}^\infty := \mathbb{Y} \cap \{x_0 = 0\} = \{P_d(x, s) = 0\} \times \mathbb{C}$ is the hyperplane at infinity of $\mathbb{Y}$. We have seen that $\mathbb{Y} \setminus \mathbb{Y}^\infty$ is nonsingular.

Definition 8.3.1 Let $\Psi := (\sigma, \tau) : \mathbb{Y} \setminus \mathbb{Y}^\infty \to \mathbb{C} \times \mathbb{C}$ be the map induced by the couple of projections. The *affine critical locus* Γ_P is the germ at $\mathbb{P}^n \times \{0\} \times \mathbb{P}$ of the closure in $\mathbb{P}^n \times \mathbb{C} \times \mathbb{P}$ of the set of points where Ψ is not a submersion. Let us call *affine discriminant* and denote by Δ_P the germ at $\{0\} \times \mathbb{P}$ of the closure in $\mathbb{C} \times \mathbb{P}$ of the image $\Psi(\Gamma_P \cap (\mathbb{C}^n \times \mathbb{C} \times \mathbb{C}))$.

It follows that Γ_P is a closed analytic set and that its affine part $\Gamma_P \cap (\mathbb{C}^n \times \mathbb{C} \times \mathbb{C})$ is just the union, over $s \in \mathbb{C}$, of the affine critical loci of the polynomials f_s. We assume from now that P is a one-parameter deformation inside the $\mathcal{F}$-class. If Γ_P is not empty, then it is a curve. An irreducible component Γ_i of the germ Γ_P can be of one of the following three types:

(1) : $\Gamma_i \cap \{\sigma = 0\} \in \mathbb{C}^n \times \{0\} \times \mathbb{C}$, which means that Γ_i is not in the neighbourhood of $\mathbb{Y}^\infty$.

(2) : $\Gamma_i \cap \{\sigma = 0\} \in \mathbb{Y}^\infty$ is a point with finite t-coordinate.
(3) : $\Gamma_i \cap \{\sigma = 0\}$ is a point with infinite t-coordinate.

Let us decompose $\Gamma_P = \Gamma_P^1 \cup \Gamma_P^2 \cup \Gamma_P^3$ according to these types, see also Figure 8.2.

Generic-at-infinity deformations of a polynomial. Let P be a one-parameter deformation of constant degree d of a polynomial f_0 such that, for all $s \neq 0$, f_s is generic-at-infinity (or $\mathcal{G}$-type), according to Definition 4.1.4. We say that these are *generic-at-infinity deformations*, for short $\mathcal{G}$-deformations. There is an equivalent way to define generic deformations of f_0, within the class of degree d polynomials, by asking that, for all $s \neq 0$, the total Milnor number $\mu(s)$ of f_s is maximum, equal to $(d-1)^n$ (see Proposition 4.1.5).

Let D be a small disk centered at the origin in the s-coordinate space $\mathbb{C}$ and let $t \in \mathbb{C}$ such that $\bar{D} \times \{t\} \cap \Delta_P = \emptyset$. There is a geometric monodromy of the fibre $F_{s,t}$ over the circle $\partial \bar{D} \times \{t\}$, which extends to a diffeomorphism of $F_{D,t} \to F_{D,t}$ isotopic to the identity.* We call it *s-monodromy*. The induced monodromy on $H_*(F_{s,t}, \mathbb{Z})$ is also called s-monodromy.

The s-monodromy clearly depends on the position of t, i.e. depends over which branch of Δ_P the small circle $\partial \bar{D} \times \{t\}$ is looping around. Figure 8.2 shows that there are three types of s-monodromies, corresponding to the three types of loops: h_{gen} for a generic value $t \notin \text{Atyp} f_0$, h_{aty} for an atypical value of t and in both cases $\Delta_P \cap D^* \times \{t\} = \emptyset$, and h_{inf} for t near to infinity such that $D^* \times \{t\}$ intersects the image of Γ_P^3.

Assuming that f_0 is $\mathcal{F}$-type and that the deformation P is generic-at-infinity, the fibres of the restriction to $D \times \{t\}$ of the map $\bar{\Psi} := (\sigma, \tau) : \mathbb{Y} \to \mathbb{C} \times \mathbb{C}$

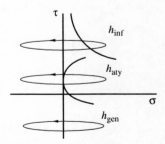

Figure 8.2. The three types of monodromy

---

* We have denoted by $F_{D,t}$ the space $(\mathbb{C}^n \times \mathbb{C}) \cap \sigma^{-1}(D) \cap \tau^{-1}(t)$.

have no singular points at the hyperplane at infinity H^∞, except for the fibre $\mathbb{Y}_{0,t}$. Then let $p_i \in \mathbb{Y}_{0,t}^\infty$ be such a singularity and let $B_i \subset \mathbb{Y}$ be a Milnor ball at p_i. We assume that D is small enough such that (B_i, D) is Milnor data for any i. Then, by excision, we have the following direct sum decomposition of the relative homology:

$$H_\star(F_{D,t}, F_{s,t}) = \oplus_i H_\star(B_i \cap F_{D,t}, B_i \cap F_{s,t}). \tag{8.9}$$

Applying to our situation Theorem 3.2.1 and its proof, especially (3.19) and the part following it, we may conclude that the relative homology $H_*(B_i \cap F_{D,t}, B_i \cap F_{s,t})$ is concentrated in dimension n.

To the pair of function germs $(\tau, \tau_{|\mathbb{Y}^\infty})$ with isolated singularities at p_i (since we are in the $\mathcal{F}$-class), one associates the pair of Milnor fibres $(B_i \cap \mathbb{Y}_{s,t}, B_i \cap \mathbb{Y}_{s,t}^\infty)$. This is an example of boundary singularities, as defined by Arnold who classified, among others, the simple germs, see [Ar1, AGV2]. In a subsequent work, Szpirglas [Sz] studied vanishing cycles of boundary singularities, from the point of view of a certain duality between the pair of Milnor fibres and their difference. As we have already mentioned, in his more recent paper [Ar2], Arnold shows that the classification of meromorphic germs of type H/Z^d is closely related to the classification of boundary singularities (see also Example 8.3.4).

Zeta-function of the s-monodromy. Let ζ_{gen} denote the zeta-function of the generic s-monodromy h_{gen} on $H_*(F_{s,t})$. Since $F_{s,t}$ is a general fibre of a $\mathcal{F}$-type polynomial, its homology $\tilde{H}_*(F_{s,t})$ is concentrated in dimension $n - 1$ (cf. Theorem 3.2.1).

Proposition 8.3.2 Let P be a deformation of a $\mathcal{F}$-type polynomial f within the $\mathcal{F}$-class and such that $\mathbb{Y}_{D,t}$ and $\mathbb{Y}_{D,t}^\infty$ are nonsingular. Then:

$$\zeta_{\text{gen}}(t) = (1 - t)^{-\chi(F_{0,t})} \cdot \prod_i [\zeta_{B_i \cap \mathbb{Y}_{s,t}}(t) \cdot \zeta_{B_i \cap \mathbb{Y}_{s,t}^\infty}^{-1}(t)].$$

Proof The s-monodromy acts on the pair $(F_{D,t}, F_{s,t})$ and is trivial on $H_*(F_{D,t})$. Moreover, $\chi(F_{D,t}) = \chi(F_{0,t})$.

Since $B_i \cap F_{D,t} = B_i \cap \mathbb{Y}_{D,t} \setminus B_i \cap \mathbb{Y}_{D,t}^\infty$, from our hypothesis we get that $\zeta_{B_i \cap F_{D,t}}(t) = 1$. From the direct sum (8.9) on which s-monodromy acts, we then get $\zeta_{(F_{D,t}, F_{s,t})}(t) = \prod_i \zeta_{B_i \cap F_{s,t}}^{-1}(t)$.

By using the Lefschetz duality or, alternatively, the multiplicativity of the zeta-function, we get $\zeta_{B_i \cap F_{s,t}}$ in terms of the zeta-functions of the boundary singularity $(\mathbb{Y}_{0,t}, \mathbb{Y}_{0,t}^{\infty})$ at the point $(p_i, 0, t)$:

$$\zeta_{B_i \cap F_{s,t}}(\mathbf{t}) = \zeta_{B_i \cap \mathbb{Y}_{s,t}}(\mathbf{t}) \cdot \zeta_{B_i \cap \mathbb{Y}_{s,t}^{\infty}}^{-1}(\mathbf{t}).$$

Then we apply the zeta-function to exact sequence of the pair $(F_{D,t}, F_{s,t})$ and get the claimed result by collecting the above formulas. $\qquad\square$

Note that $\zeta_{B_i \cap \mathbb{Y}_{s,t}}$ and $\zeta_{B_i \cap \mathbb{Y}_{s,t}^{\infty}}$ are the zeta-functions of its corresponding boundary singularity $(B_i \cap \mathbb{Y}_{s,t}, B_i \cap \mathbb{Y}_{s,t}^{\infty})$.

Note that in case t is an atypical value of f, the proof of the above statement works and yields the same formula for ζ_{aty} (provided that the hypotheses of Proposition 8.3.2 are fulfilled).

More developments about the s-monodromy can be found in [ST5] and [GS]. We end this section by two examples.

Example 8.3.3 Let $f = x^2y + x$, the Broughton example [Br2] of a polynomial with no affine singularities and one singularity at infinity: $\mu = 0$ and $\lambda = 1$. The λ corresponds to a jump of the local type of $\bar{F}_t$ at infinity from A_2 to A_3.

Consider the deformation $f_s = x^2y + x + sy^3$. This is a $\mathcal{G}$-deformation of a $\mathcal{F}$-type polynomial. The topological bifurcation set Bif_P^{top} consists of the axis $\{\sigma = 0\}$ and the affine discriminant Δ_P, which has equation $s = \alpha t^4$. We have that $\Delta_P = \mathrm{Im}\,\Gamma_P^2$ and that Δ_P is tangent to $\{\sigma = 0\}$. This tangency implies that the inequality in Theorem 4.2.5 is strict, and this forces $\lambda(0) > 0$. For $s \neq 0$, we have: $\mu(s) = 4$ and $\lambda(s) = 0$. See also Figure 8.3, left-hand side.

Remark that the boundary singularity is of type (A_2, A_1) at the point $[0 : 1] \in \mathbb{Y}_{0,t}^{\infty}$, for $t \neq 0$, and of type (A_3, A_1) when $t = 0$. For the zeta-function of the s-monodromy, Proposition 8.3.2 can be applied since $\mathrm{Sing}\,\mathbb{Y}$ and $\mathrm{Sing}\,\mathbb{Y}^{\infty}$ are

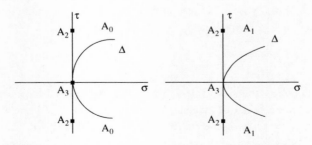

Figure 8.3. y^3-deformation and y-deformation of Broughton's example

empty. This gives the following formulas:

$$\zeta_{\text{gen}}(t) = \frac{1-t^3}{1-t} \cdot \frac{1-t^2}{1-t}, \qquad \zeta_{\text{aty}}(t) = (1-t)^{-1} \cdot \frac{1-t^4}{1-t} \cdot \frac{1-t^2}{1-t}.$$

In order to see the intermediate steps in the exchange $\mu(s) + \lambda(s) = 4 + 0 \rightarrow 0 + 1 = \mu(0) + \lambda(0)$, we consider the following deformation from [SiSm]:

$$P(x, y; s_1, s_2, s_3) = x^2y + x + s_1y + s_2y^2 + s_3y^3.$$

We get the following specializations:

$$
\begin{array}{ll}
s_3 \neq 0 & \mu + \lambda = 4 + 0 \\
s_3 = 0,\ s_2 \neq 0 & \mu + \lambda = 3 + 0 \\
s_3 = s_2 = 0,\ s_1 \neq 0 & \mu + \lambda = 2 + 0 \\
s_3 = s_2 = s_1 = 0 & \mu + \lambda = 0 + 1.
\end{array}
$$

The discriminant of the particular deformation $f_s = x^2y + x + sy$ is shown in Figure 8.3, right-hand side.

Example 8.3.4 [ST5] According to Arnold's study [Ar2], right-equivalence of fractions is very near to right-equivalence of boundary singularities (with respect to the hyperplane $Z = 0$). After [Ar2], the list of simple fractions of type $\frac{H(Y,Z)}{Z^d}$ is given in Table 8.1 (where Q_2 and Q_3 denote quadratic forms in the rest of the Y-coordinates).

We may observe that all the germs in Arnold's list have $\lambda = 0$ and that they are topologically trivial at infinity. However, deforming these singularities yields an interesting behaviour, for instance one may get nontrivial s-monodromy.

Let us consider for instance the polynomial $f = y^{k+1} + x^k$. By homogenizing and localizing in the chart $x = 1$, this corresponds to the fraction A_k in Arnold's list (Table 8.1). The hypersurface $\bar{F}_t$ is nonsingular and k-fold tangent to the hyperplane at infinity H^∞. Let us consider the following Yomdin deformation (in particular, a $\mathcal{G}$-deformation):

$$f_s = y^{k+1} + x^k + sx^{k+1}.$$

Then $\mathbb{Y}_{s,t}$ intersects transversely H^∞ for all t and all $s \neq 0$. Moreover, $\mathbb{Y}_{0,t}$ has no singularity at H^∞ and is tangent to H^∞ at the point $[1 : 0] \in H^\infty$.

We get that $\operatorname{Im}\Gamma_P^1$ is the s-axis $\{\tau = 0\}$, $\Gamma_P^2 = \emptyset$ (and therefore $\lambda(0) = 0$) and that Γ_P^3 has equation $s^k t = \alpha$ (see Figure 8.4). A slice $t = t_0$ will cut Γ_P^3 at k points, each of them being a singularity of $\mathbb{Y}_{s_i,t}$ of type A_k, for some s_i.

Table 8.1. *Arnold's list of fractions of type H/Z^d*

Type	Function	Boundary type	Conditions
A_0	Y/Z^d	(A_0, A_0)	
A_k	$(Y^{k+1} + Z + Q_2)/Z^d$	(A_0, A_k)	$k \geq 1$, $d > 1$
B_k	$(Y^2 + Z^k + Q_2)/Z^d$	(A_{k-1}, A_1)	$k \geq 2$, $d > k$
C_k	$(Y^k + YZ + Q_2)/Z^d$	(A_1, A_{k-1})	$k \geq 3$, $d > 1$
D_k	$(Y_1^2 Y_2 + Y_2^{k-1} + Z + Q_3)/Z^d$	(A_0, D_k)	$k \geq 4$, $d > 1$
E_6	$(Y_1^3 + Y_2^4 + Z + Q_3)/Z^d$	(A_0, E_6)	$d > 1$
E_7	$(Y_1^3 + Y_1 Y_2^3 + Z + Q_3)/Z^d$	(A_0, E_7)	$d > 1$
E_8	$(Y_1^3 + Y_2^5 + Z + Q_3)/Z^d$	(A_0, E_8)	$d > 1$
F_4	$(Y^3 + Z^2 + Q_2)/Z^d$	(A_2, A_2)	$d > 2$

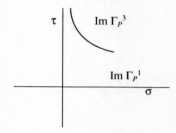

Figure 8.4. Yomdin deformation of the A_k singularity

Since at H^∞ we have a boundary singularity of type (A_0, A_k), the zeta-function of the generic s-monodromy is, according to Proposition 8.3.2 (since $\text{Sing } \mathbb{Y}$ and $\text{Sing } \mathbb{Y}^\infty$ are empty):

$$\zeta_{\text{gen}}(t) = (1 - t)^{-1 - k + k^2} \cdot \frac{1 - t^{k+1}}{1 - t}.$$

Exercises

8.1 Let $H \in \Omega''_{f,c}$. If $\Gamma(l_H, f) \cap F_c = \emptyset$, then $H_j(F_c, \mathbb{Z}) = 0$, for $j \geq n - 1$.

8.2 Let $a \in \mathbb{C}$ and $q \in \mathbb{X}_a^\infty \cap \bar{\Gamma}(l_H, f)$. Then $\hat{r}(q) = \infty$. (*Hint*: use the transversality condition of $H \in \Omega''_{f,a}$.)

8.3 Show that the complex link monodromy of the germ of a A_k-type curve singularity is $(-1)^{k-1}\text{id}$.

8.4 Let P be any deformation of a $\mathcal{F}$-type polynomial f_0 and let $\phi_d := \frac{\partial P}{\partial s}(x, 0)$.

In the notations of §4.2, prove that:

(i) $\mathbb{Y}$ is nonsingular if and only if $\Sigma_0 \cap \{\phi_d = 0\} = \emptyset$. In this case the critical set of the map $\sigma : \mathbb{Y} \to \mathbb{C}$ is $\Sigma_0 \times \mathbb{C} \subset \mathbb{Y}_{0,\mathbb{C}}$.

(ii) $\mathbb{Y}^\infty$ is nonsingular if and only if $W_0 \cap \{\phi_d = 0\} = \emptyset$. In this case the critical set of the map $\sigma_| : \mathbb{Y}^\infty \to \mathbb{C}$ is $W_0 \times \mathbb{C} \subset \mathbb{Y}_{0,\mathbb{C}}^\infty$.

8.5 Let P be a deformation within the $\mathcal{F}$-class. Then $\Gamma_P^2 \cap \{x_0 = 0\} \subset \Sigma_0 \times \{0\} \times \mathbb{C}$ and $\Gamma_P^3 \cap \{x_0 = 0\} \subset W_0 \times \{0\} \times \{\infty\}$.

8.6 Let P be a $\mathcal{G}$-deformation of a $\mathcal{F}$-type polynomial f_0. If for some $t \in \mathbb{C}$, $\mathbb{Y}_{0,t}$ is nonsingular and tangent to H^∞, then $\Gamma_P^3 \neq \emptyset$.

8.7 Let P be a deformation within the $\mathcal{F}$-class and assume that the total space $\mathbb{Y}$ is nonsingular. If for some $t \in \mathbb{C}$, $\mathbb{Y}_{0,t}$ has singularities on H^∞, then all branches of Δ_P which are images of branches of Γ_P^2 are tangent to the τ-axis.

PART III

Vanishing cycles of nongeneric pencils

9

Topology of meromorphic functions

In this chapter we show how to extend the topological study of holomorphic functions to the class of *meromorphic functions*, in the local or global settings. At the same time, polynomial functions $f : \mathbb{C}^n \to \mathbb{C}$ can be viewed as a special case of global meromorphic functions, as we explain next. Since a meromorphic function defines a pencil of hypersurfaces, our approach contains and generalizes the theory of *Lefschetz pencils*.

9.1 Singularities of meromorphic functions

Let us introduce some terminology. A *meromorphic function*, or *pencil of hypersurfaces*, on a compact complex analytic space Y, is a function $F: Y \dashrightarrow \mathbb{P}^1$ defined as the ratio of two sections P and Q of a holomorphic line bundle over Y. Let us note here that if Y is projective, then the Kodaira embedding theorem ensures the existence of global holomorphic line bundles, hence of global meromorphic functions.

Our meromorphic function $F = P/Q$ is then a holomorphic function on $Y \setminus A$, where $A := \{P = Q = 0\}$ is the *base locus* of the pencil (also called *axis*, or *indeterminacy locus*). A *germ of meromorphic function* $f : (\mathcal{Y}, y) \dashrightarrow \mathbb{P}^1$ on a space germ $(\mathcal{Y}, y)$ is the ratio $f = p/q$ of two holomorphic germs $p, q: (\mathcal{Y}, y) \dashrightarrow \mathbb{C}^1$. By definition, f is equal to $f' = p'/q'$, as germs at y, if and only if there exists a holomorphic germ u such that $u(y) \neq 0$ and such that $p = up', q = uq'$. Note that f is holomorphic on the germ at y of the complement $\mathcal{Y} \setminus \mathcal{A}$ of the axis $\mathcal{A} = \{p = q = 0\}$.

We focus on a special type of singularities of the meromorphic functions, those occurring along the indeterminacy locus. We shall see that the following definitions naturally extend those for polynomial functions that we have used in Parts I and II, and those for certain classes of regular functions on affine varieties used in [Ti4]. Our exposition relies on [ST6, Ti11, Ti8, Ti12].

From another viewpoint, Arnold [Ar2] studied germs of fractions P/Q, where P and Q are polynomials, with respect to several equivalence relations such as the right-equivalence, see also Example 8.3.4.

Nongeneric pencils. We define a new space by blowing-up along the base locus A. The idea of this construction is due to Thom and was used by Andreotti and Frankel [AF] in case of generic pencils on projective manifolds.

Definition 9.1.1 Let $G := \{(x, [s : t]) \in (Y \setminus A) \times \mathbb{P}^1 \mid F(x) = [s : t]\}$ be the graph of F, where the value ∞ is identified with $[0 : 1]$. Let $\mathbb{Y}$ denote the analytic closure of G in $Y \times \mathbb{P}^1$, namely:

$$\mathbb{Y} = \{(x, [s : t]) \in Y \times \mathbb{P}^1 \mid sP(x) - tQ(x) = 0\}.$$

In case of the germ of a meromorphic function at $(\mathcal{Y}, y)$, we similarly define the corresponding space germs, which we denote by $(G, (y, [s : t]))$, resp. $(\mathbb{Y}, (y, [s : t]))$.

This is a hypersurface in $Y \times \mathbb{P}^1$ obtained as a Nash blowing-up of Y along A. It follows that $A \times \mathbb{P}^1 \subset \mathbb{Y}$. Since G is the graph of the restriction $F_{|Y \setminus A}$, one has the analytic isomorphisms $G \simeq Y \setminus A \simeq \mathbb{Y} \setminus (A \times \mathbb{P}^1)$. The projection $\pi : \mathbb{Y} \to \mathbb{P}^1$ is a proper extension of the function $F_{|Y \setminus A}$. We may also say that the projection $\mathbb{Y} \xrightarrow{\sigma} Y$ is a blow-up of Y along the axis A, such that the meromorphic function $F : Y \dashrightarrow \mathbb{P}^1$ pulls back to a well-defined holomorphic function $\pi : \mathbb{Y} \to \mathbb{P}^1$.

$$
\begin{array}{ccc}
 & \mathbb{Y} & \\
\sigma \downarrow & & \searrow \pi \\
Y & \dashrightarrow_{F} & \mathbb{P}^1
\end{array}
\tag{9.1}
$$

We shall also consider the restriction of F (respectively that of a germ f) to $X := Y \setminus V$, where V is some compact analytic subspace of Y. Let us denote $\mathbb{X} := \mathbb{Y} \cap (X \times \mathbb{P}^1)$. The case $V = \{Q = 0\}$ is of particular interest, as we shall see in the following.

Specializations. In the general setting, a meromorphic function (as a global object or as a germ) defines a pencil of hypersurfaces on each of the spaces defined above: $\mathbb{Y}$, $\mathbb{X}$, Y, $Y \setminus A$, or $X = Y \setminus V$. The total space $\mathbb{Y}$ contains all these pencils. The members of the pencil are the fibres of the projection π to $\mathbb{P}^1$.

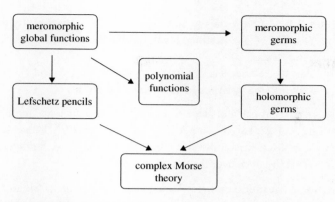

Figure 9.1. Specialization of topics

This approach covers a large field, as described in Figure 9.1. For instance, the class of *holomorphic functions* (respectively germs) represents the case when $A = \emptyset$. The class of *polynomial functions*, studied in the preceding sections, is embedded in the general context as follows: if $f : \mathbb{C}^n \to \mathbb{C}$ is a polynomial of degree d, let $\tilde{f}$ denote the homogenized of f by the new variable x_0 and let $H^\infty = \{x_0 = 0\}$ be the hyperplane at infinity. Then $\frac{\tilde{f}}{x_0^d} : \mathbb{P}^n \dashrightarrow \mathbb{P}^1$ is a meromorphic function on $Y := \mathbb{P}^n$, which coincides with f on $\mathbb{P}^n \setminus H^\infty$. Here we have $V = H^\infty$ and the axis $A := \{f_d = 0\} \subset H^\infty$. We point out, however, that this meromorphic extension of a polynomial f is far from being unique. We refer to page 158 where we explain the extension into a weighted projective space, and also to [Ti4] for more general extensions.

Finiteness of atypical locus. As in the case of polynomials (see Corollaries 1.2.13, 1.2.14), we may show that any pencil F (on Y or on $Y \setminus V$) has finitely many atypical members in either local or global settings. This finiteness result originates in Thom's work [Th2] and the arguments run as follows. We endow the space Y with a Whitney stratification $\mathcal{W}$ such that A and $V \setminus A$ are unions of strata. Then we restrict it to $Y \setminus A \simeq \mathbb{Y} \setminus (A \times \mathbb{P}^1)$ and finally we extend it to some Whitney stratification $\mathcal{S}$ of $\mathbb{Y}$, by the usual arguments (see e.g. [GWPL]).*

In case of a germ at $(\mathcal{Y}, y)$ of a meromorphic function, we consider the germ of such a Whitney stratification at $\{y\} \times \mathbb{P}^1 \subset \mathbb{Y}$.

As in §2.2, let us consider the singular locus of π with respect to $\mathcal{S}$, i.e.: $\mathrm{Sing}_{\mathcal{S}} \pi := \bigcup_{\mathcal{S}_\beta \in \mathcal{S}} \mathrm{Sing}\, \pi_{|\mathcal{S}_\beta}$. This is a closed analytic subset of $\mathbb{Y}$, due to the finiteness of the strata of $\mathcal{S}$ and to the Whitney conditions.

---

* A similar strategy has been used for the simpler version of polynomial functions in §2.2.

Proposition 9.1.2 There exists a finite set $\Lambda \subset \mathbb{P}^1$ such that the map π : $\mathbb{Y} \setminus \pi^{-1}(\Lambda) \to \mathbb{P}^1 \setminus \Lambda$ is a stratified locally trivial topological fibration (and it is C^∞ along the strata).

In particular, the restrictions $\pi_| : \mathbb{X} \setminus \pi^{-1}(\Lambda) \to \mathbb{P}^1 \setminus \Lambda$ and $F_| : Y \setminus (A \cup \pi^{-1}(\Lambda)) \to \mathbb{P}^1 \setminus \Lambda$ are stratified locally trivial fibrations.

Proof Since π is proper and since $\mathrm{Sing}_{\mathcal{S}}\pi$ is closed analytic, the image set $\Lambda = \pi(\mathrm{Sing}_{\mathcal{S}}\pi)$ is analytic and strictly included into $\mathbb{P}^1$, hence finite. By using the Thom–Mather isotopy lemma [Th2], see §A1.1, we get the stratified topological triviality of π over the complement $\mathbb{P}^1 \setminus \Lambda$.   □

Singularities at the indeterminacy locus. In this approach, the singularities of the meromorphic function F along the indeterminacy locus A are the stratified singularities of π at $(A \times \mathbb{P}^1) \cap \mathbb{Y}$. In order to define the type of singularities which are general enough, and can be reasonably handled, we shall use a *partial Thom stratification* as defined in Appendix A1.1, instead of a Whitney stratification. The former is a more general type of stratification since it does not require the Whitney (b)-regularity condition. Nevertheless, it allows us to study topological aspects, including the homotopy type, at least for isolated singularities, in both local or global contexts.[1]

Let $\mathcal{G} = \{\mathcal{G}_\alpha\}_{\alpha \in S}$ be a locally finite stratification of $\mathbb{Y}$. Let $\xi := (y, a) \in A \times \mathbb{P}^1$ be a point on a stratum $\mathcal{G}_\alpha$. We assume, without loss of generality, that $a \neq [0 : 1]$. Let $f = p/q$ be our meromorphic germ at $(\mathcal{Y}, y)$, a local representative of the germ of F at (Y, y). Then $q = 0$ is a local equation for $A \times \mathbb{P}^1$ at ξ. The *Thom regularity condition* (a_q) at $\xi \in \mathcal{G}_\alpha$ is satisfied (cf. Definition A1.1.1) if for any stratum $\mathcal{G}_\beta$ such that $\mathcal{G}_\alpha \subset \bar{\mathcal{G}}_\beta$ the relative conormal space of q on $\bar{\mathcal{G}}_\beta$ is included into the conormal of $\mathcal{G}_\alpha$, locally at ξ, i.e., $(T^*_{\mathcal{G}_\alpha})_\xi \supset (T^*_{q|\bar{\mathcal{G}}_\beta})_\xi$. We have shown in Lemma 1.2.7 that this condition is independent on q, up to multiplication by a unit.

Definition 9.1.3 Let $\mathcal{G}$ be a locally finite stratification on $\mathbb{Y}$. Assume that $Y \setminus A$ and $V \setminus A$ are unions of strata and that they coincide with the strata of some Whitney stratification $\mathcal{W}$ on $Y \setminus A$. We say that $\mathcal{G}$ is a *partial Thom stratification* ($\partial\tau$-stratification) if Thom's condition (a_q) is satisfied at any point $\xi \in A \times \mathbb{P}^1 \subset \mathbb{Y}$.

In particular, the subsets $\mathbb{X}, \mathbb{Y}\setminus\mathbb{X}, A \times \mathbb{P}^1$ and $(A \cap X) \times \mathbb{P}^1$ are unions of strata of $\mathcal{G}$. This stratification is suitable for the fibration statements, Propositions 9.1.2 and 9.1.5, and will also be used in the definition of the geometric monodromy §9.4.

We may extend a Whitney stratification on $Y \setminus A$ to a locally finite $\partial\tau$-stratification of $\mathbb{Y}$, by usual stratification theory arguments (see e.g. [GWPL]).

For instance, the Whitney stratification $\mathcal{S}$ of $\mathbb{Y}$ that we have considered in Proposition 9.1.2 is an example of $\partial\tau$-stratification, see Theorem A1.1.7 for the reason.[2]

Definition 9.1.4 Let $\mathcal{G}$ be some $\partial\tau$-stratification on $\mathbb{Y}$. We say that the following closed subset of $\mathbb{Y}$:

$$\operatorname{Sing}_{\mathcal{G}} F := \cup_{\mathcal{G}_\alpha \in \mathcal{G}} \operatorname{closure}(\operatorname{Sing} \pi_{|\mathcal{G}_\alpha})$$

is the *singular locus* of F with respect to $\mathcal{G}$. We say that F has *isolated $\mathcal{G}$-singularities* if $\dim \operatorname{Sing}_{\mathcal{G}} F \le 0$.

For the singular locus of a germ f, one modifies this definition accordingly. There is an inclusion $\operatorname{Sing}_{\mathcal{S}} \pi \subset \operatorname{Sing}_{\mathcal{S}} F$ by the above definitions and remarks.

We shall focus in the following on the new type of singularities, those along the blown-up indeterminacy locus $A \times \mathbb{P}^1$.

Local fibrations. Let us define the local fibration of a meromorphic germ. This generalizes the fibration of a holomorphic germ on a singular space (introduced by Lê D.T. [Lê3]). We shall see that it appeared in the particular cases of the localization at the singularities at infinity of a polynomial function, see e.g. §3.2.

Let $f : (Y, y) \dashrightarrow \mathbb{P}^1$ be a meromorphic function germ. In the notations of Definition 9.1.1, to every $a \in \mathbb{P}^1$, we associate the germ $\pi_{(y,a)} : (\mathbb{Y}, (y, a)) \to \mathbb{P}^1$. Considering the subset $X = Y \setminus V \subset \mathbb{Y}$, this map restricts to $X \subset \mathbb{Y}$ and we have a germ $f : (X, (y, a)) \to \mathbb{C}$, where the point (y, a) belongs to X or to the closure of X in $\mathbb{Y}$.

When the point y does not belong to the axis A, then we have the classical situation of a holomorphic germ. Actually the point (y, a) is uniquely determined by y.

In case $y \in A$, we get a different germ for each $a \in \mathbb{P}^1$. Let us see what is the local fibration at (y, a) in this case. Let $\mathcal{S}$ be a Whitney stratification of $\mathbb{Y}$ such that $\mathbb{Y} \setminus X$ is a union of strata. For all small enough radii ε, the sphere $S_\varepsilon = \partial \bar{B}_\varepsilon(y, a)$ centered at (y, a) intersects transversally all the finitely many strata of $\mathbb{Y}$ at (y, a). We use the notations: $\mathbb{Y}_W := \pi^{-1}(W)$, $\mathbb{X}_W := \mathbb{X} \cap \mathbb{Y}_W$, for any subset $W \subset \mathbb{P}^1$. We then have the following general result:

Proposition 9.1.5 For any $(y, a) \in A \times \mathbb{P}^1$ the restriction:

$$\pi : \mathbb{X}_{D^*} \cap B_\varepsilon(y, a) \to D^*. \tag{9.2}$$

is a locally trivial fibration. The restriction:

$$\pi : \mathbb{X}_D \cap B_\varepsilon(y, a) \to D \tag{9.3}$$

is a stratified trivial fibration, for all $a \in \mathbb{P}^1 \setminus \Lambda$.

Proof Lê D.T. [Lê3] showed that, if ε and the radius of D are small enough, then the projection $\pi : \mathbb{Y}_D \cap B_\varepsilon(y, a) \to D$ restricts over D^* to a stratified locally trivial fibration. The restriction to $\mathbb{X}_{D^*}$ is also locally trivial, since $\mathbb{X}_{D^*}$ is a union of strata.

For the second claim, let us remark that, if y is fixed, then *a priori* the fibration (9.2) varies with the parameter a, including the fact that the radius ε of $B_\varepsilon(y, a)$ may depend on the point a. To conclude, we use Proposition 9.1.2, which shows that π is a stratified locally trivial fibration over $\mathbb{P}^1 \setminus \Lambda$, within some neighbourhood of $\{a\} \times \mathbb{P}^1 \subset \mathbb{Y}$, and that Λ is a finite set. $\qquad \square$

This yields the following natural definition:

Definition 9.1.6 We say that the locally trivial fibration (9.2) is the *Milnor–Lê fibration** of the meromorphic function germ f *at the point* $(y, a) \in \mathbb{X}$.

Fibres of polynomial functions, revisited. A polynomial function $f : \mathbb{C}^n \to \mathbb{C}$ may be extended to a meromorphic function on a compact space, but neither the space nor the meromorphic function are unique. Let us introduce a more general extension.

The weighted projective compactification. We shall consider here the embedding of $\mathbb{C}^n \subset \mathbb{P}(w)$ into a *weighted projective space* $\mathbb{P}(w) := \mathbb{P}(w_1, \ldots, w_n, 1)$. We associate to each coordinate x_i a positive integer weight w_i, for $i \in \{1, \ldots, n\}$ and take a new variable z of weight 1. Then $\mathbb{P}(w)$ is the space of orbits under the $\mathbb{C}^*$-action on $\mathbb{C}^{n+1} \setminus \{0\}$ given by $\lambda \cdot x = (\lambda^{w_1} x_1, \cdots \lambda^{w_n} x_n, \lambda z)$. When the weights w_j are equal to 1, we get the usual projective space $\mathbb{P}^n$.

Let us write $f = f_d + f_{d-k} + f_{d-k-1} + \cdots$, where f_i is the degree i weighted-homogeneous part of f and $f_{d-k} \neq 0$. Let $\tilde{f}$ be the homogenization of f of degree d, by the variable z. We get a meromorphic function $\tilde{f}/z^d$ on $\mathbb{P}(w)$ and the Nash blown up space $\mathbb{Y} := \{s\tilde{f} - tz^d = 0\} \subset \mathbb{P}(w) \times \mathbb{P}^1$, Since $\mathbb{P}(w)$ is an orbit space of the $\mathbb{C}^*$-action on $\mathbb{C}^{n+1} \setminus \{0\}$, it has a canonical Whitney stratification by the orbit type (see e.g. [Fer] or [GWPL, p. 21]), which is moreover the coarsest Whitney stratification. We have the identification $\mathbb{C}^n = \mathbb{P}(w) \setminus \{z = 0\}$ and so the singularities of $\mathbb{P}(w)$ are contained in the part at infinity $\{z = 0\}$. We take the product stratification on $\mathbb{P}(w) \times \mathbb{C}$. This restricts to a stratification

---

* One may lookup Theorem A1.1.2.

on the subspace $\mathbb{X} \subset \mathbb{P}(w) \times \mathbb{C}$ and we then consider the coarsest Whitney stratification $\mathcal{S}$ on $\mathbb{X}$ which refines it. Then $\mathbb{Y} \setminus \{z = 0\} \simeq \mathbb{C}^n$ is a stratum of $\mathcal{S}$. Let

$$\Sigma(w) := \{\operatorname{grad} f_d = 0, f_{d-k} = 0\} \subset \mathbb{P}(w) \cap \{z = 0\}. \tag{9.4}$$

The set of *singularities of the pencil defined by f* on $\mathbb{Y}' := \mathbb{Y}_\mathbb{C}$ is by definition $\operatorname{Sing}_{\mathcal{S}} p \cap \mathbb{Y}'$. We have the following generalization of the inclusion $\mathcal{B}$-class $\subset$ $\mathcal{W}$-class. See Exercise 9.1 for an extension of this result.

Proposition 9.1.7 [LiTi] *If* $\dim \operatorname{Sing} f \leq 0$ *and if* $\dim \Sigma(w) \leq 0$, *then* $\dim \operatorname{Sing}_{\mathcal{S}} p \cap \mathbb{Y}' \leq 0$.

Proof Let $\mathbb{Y}'^\infty := \{z = 0\} \subset \mathbb{Y}'$. Since the singularities of p on $\mathbb{C}^n = \mathbb{Y}' \setminus \mathbb{Y}'^\infty$ are of dimension ≤ 0 by hypothesis, we only have to control the singularities of p at $\mathbb{Y}'^\infty = \{f_d = 0\} \times \mathbb{C} \subset (\mathbb{P}(w) \cap \{z = 0\}) \times \mathbb{C}$. We shall prove that $\mathbb{Y}'^\infty \cap \operatorname{Sing}_{\mathcal{S}} p \subset \Sigma(w) \times \mathbb{C}$.

Consider the hypersurface $\tilde{\mathbb{Y}}'$ defined by the same equation $s\tilde{f} - tz^d = 0$ but this time in $(\mathbb{C}^{n+1} \setminus \{0\}) \times \mathbb{C}$. Notice first that the subset $\tilde{\mathbb{Y}}' \cap \{z = 0\} \cap (\{f_d = 0\} \setminus \{\operatorname{grad} f_d = 0\})$ is contained in the regular part of $\tilde{\mathbb{Y}}'$. Since the $\mathbb{C}^*$-action on the factor $\mathbb{C}$ of $\mathbb{P}(w) \times \mathbb{C}$ is trivial, the strata within $\mathbb{Y}' \cap \{z = 0\} \cap (\{f_d = 0\} \setminus \{\operatorname{grad} f_d = 0\} \times \mathbb{C})$ are products by $\mathbb{C}$. It then follows that p, which is the projection to $\mathbb{C}$, is transversal to these strata.

Next, at some point $\xi = (q, 0, t_0) \in \tilde{\mathbb{Y}}' \cap \{z = 0\} \cap (\{\operatorname{grad} f_d = 0\} \setminus \{f_{d-k} = 0\})$, we claim that there exists, locally at ξ, a Whitney stratification of $\tilde{\mathbb{Y}}'$ with the property that its strata are product-spaces by the t-coordinate (i.e., if (p, t) belongs to a stratum, then (p, t') belongs to the same stratum for all $t' \in \mathbb{C}$). Indeed, the local equation of $\tilde{\mathbb{Y}}'$ is $\tilde{f}(x, z) - tz^d = 0$ and this can be written as $f_d(x) + z^k g = 0$, where $g(x, z, t) = f_{d-k}(x) + z h(x, z, t)$. Since $f_{d-k}(q) \neq 0$, we can define, locally at ξ, a new coordinate $z' = z \sqrt[k]{g}$, by choosing a kth root of g. It follows that, locally, our hypersurface is equivalent, via an analytic change of coordinates at ξ, to the product of $\{f_d(x) + (z')^k = 0\}$ by the t-coordinate. Consequently, there exists a local Whitney stratification at ξ which is a product by the t-coordinate. Notice that $\{z' = 0\}$ corresponds to $\{z = 0\}$ at ξ, hence the complement of $\{z' = 0\}$ is nonsingular too. Taking the $\mathbb{C}^*$ quotient, we obtain a Whitney stratification which is also a product by $\mathbb{C}$. It is, locally, a refinement of our coarsest stratification $\mathcal{S}$ on $\mathbb{Y}'$. Again p is transversal to such strata.

This proves that $\operatorname{Sing}_{\mathcal{S}} p \cap \mathbb{Y}' \subset \Sigma(w) \times \mathbb{C}$. Since $\dim \Sigma(w) \leq 0$, the set $\Sigma(w) \times \mathbb{C}$ is a finite union of complex lines. The map p is transversal to such a line, so singularities of p on $\Sigma(w) \times \mathbb{C}$ can occur only if $\Sigma(w) \times \mathbb{C}$ contains point-strata of $\mathcal{S}$. But there can only be finitely many such point-strata. This ends our proof. $\qquad\square$

9.2 Vanishing homology and singularities

Generalities on the vanishing homology. We define the vanishing homology attached to a global meromorphic function and relate it to the singularities along the indeterminacy locus. The vanishing homology turns out to detect the change of topology of the fibres.

Notations. Let $a_i \in \Lambda$ be an atypical value of π and take a small enough disk D_i at a_i such that $D_i \cap \Lambda = \{a_i\}$. Let us fix some point $s_i \in \partial D_i$. As before, for any subset $W \subset \mathbb{P}^1$, we use the notations $\mathbb{Y}_W$ and $\mathbb{X}_W$ but also the following (recall (9.1)): $Y_W := \sigma(\pi^{-1}(W))$, $X_W := X \cap Y_W$, $A' := A \cap X$. We have by definition $X_s = \mathbb{X}_s$ for any $s \in \mathbb{P}^1$ and, in case $A' = \emptyset$, we have $\mathbb{X}_W = X_W$ for any W. We may also prove the following:

Lemma 9.2.1 For any $W \subset \mathbb{P}^1$, the space Y_W (resp. X_W) is homotopy equivalent to the space $\mathbb{Y}_W$ (resp. $\mathbb{X}_W$) to which one attaches along $A \times W$ (resp. $A' \times W$) the product of $A \times \text{Cone}(W)$ (resp. $A' \times \text{Cone}(W)$). In particular, if W is contractible, then $Y_W \overset{\text{ht}}{\simeq} \mathbb{Y}_W$ and $X_W \overset{\text{ht}}{\simeq} \mathbb{X}_W$.

Proof The first statement follows directly from the definition of the spaces $\mathbb{Y}$ and $\mathbb{X}$. Whenever W is contractible, we have $A \times \text{Cone}(W) \overset{\text{ht}}{\simeq} A \times W$, which implies the homotopy equivalences:

$$Y_W \overset{\text{ht}}{\simeq} \mathbb{Y}_W \cup_{A \times W} (A \times \text{Cone}(W)) \overset{\text{ht}}{\simeq} \mathbb{Y}_W.$$

The same argument applies to X_W and we get $X_W \overset{\text{ht}}{\simeq} \mathbb{X}_W$.   □

Definition 9.2.2 We call *vanishing homology of $F_{|X}$ at a_i* the relative homology $H_*(X_{D_i}, X_{s_i})$.

Remark that $H_*(X_{D_i}, X_{s_i}) \simeq H_*(\mathbb{X}_{D_i}, X_{s_i})$, by Lemma 9.2.1. The vanishing homology of meromorphic functions has of course its straightforward local counterpart. It is a natural extension of the vanishing homology of local holomorphic functions. In the latter case, the total space of the Milnor fibration [Mi2] is contractible, by the local conical structure of analytic sets [BV]. For local or global meromorphic functions, the space X_{D_i} may be not contractible and the general fibre X_s inherits from its homology.

Let $s \in \mathbb{P}^1 \setminus \Lambda$ be a general value, situated on the boundary of some larger closed disk $D \subset \mathbb{P}^1$, such that $D \supset D_i$, $\forall a_i \in \Lambda$. We identify X_s to X_{s_i}, in the following explicit manner. For each i, take a path $\gamma_i \subset D$ from s to s_i, with the usual conditions: the path γ_i has no self intersections and does not intersect any other path γ_j, except at the point s. Then Proposition 9.1.2 allows identifying X_s to X_{s_i}, by parallel transport along γ_i.

The following general result tells that vanishing homologies can be 'patched' together:[3]

Proposition 9.2.3 Let $F : Y \dashrightarrow \mathbb{P}^1$ be a meromorphic function and let $X := Y \setminus V$. Then:

(a) $H_\star(X_D, X_s) = \oplus_{a_i \in \Lambda} H_\star(X_{D_i}, X_{s_i})$.

(b) The long exact sequence of the triple (X_D, X_{D_i}, X_s) decomposes into short exact sequences which split:

$$0 \to H_\star(X_{D_i}, X_{s_i}) \to H_\star(X_D, X_s) \to H_\star(X_D, X_{D_i}) \to 0. \qquad (9.5)$$

(c) There is a natural direct sum splitting:

$$H_\star(X_D, X_{D_i}) = \oplus_{a_j \in \Lambda, j \neq i} H_\star(X_{D_j}, X_s).$$

Proof By Proposition 9.1.2, the fibration $F_| : \mathbb{Y} \setminus ((V \times \mathbb{P}^1) \cup \pi^{-1}(\Lambda) \to \mathbb{P}^1 \setminus \Lambda$ is locally trivial. Its fibre over some $b \in \mathbb{P}^1 \setminus \Lambda$ is, by definition, X_b. We then get a sequence of excisions:

$$\oplus_{a_i \in \Lambda} H_*(\mathbb{X}_{D_i}, X_{s_i}) \xrightarrow{\simeq} H_\star(\mathbb{X}_D \cap \pi^{-1}(\cup_{a_i \in \Lambda} D_i \cup \gamma_i), X_s) \xrightarrow{\simeq} H_\star(\mathbb{X}_D, X_s).$$

This also shows that each inclusion $(\mathbb{X}_{D_i}, X_{s_i}) \subset (\mathbb{X}_D, X_{s_i})$ induces an injection in homology $H_\star(\mathbb{X}_{D_i}, X_{s_i}) \hookrightarrow H_\star(\mathbb{X}_D, X_s)$. The claims (a), (b), (c) follow easily from this and from the isomorphism $H_*(X_{D_i}, X_{s_i}) \simeq H_*(\mathbb{X}_{D_i}, X_{s_i})$ implied by Lemma 9.2.1. $\qquad \square$

One would like to say that vanishing homology is supported at the singular points of F but the typical problem for a meromorphic function is that part of its singularities are outside the ground space X. This explains why we need a larger space like $\mathbb{Y}$ to define singularities: with this definition, the support of vanishing cycles is clearly included into the singular locus of π on $\mathbb{Y}$. This recalls Deligne's result [De] in cohomology: the sheaf of vanishing cycles of a holomorphic function h on a nonsingular space is supported by the singular locus of h.

Isolated singularities and their vanishing cycles. We shall further investigate the relation between singularities and vanishing homology, in case of *isolated $\mathcal{G}$-singularities* (cf. Definition 9.1.4), following [Ti11, ST6].

We show that if the singularities along the indeterminacy locus are isolated, then we can localize the variation of topology of fibres. The same type of phenomenon exists in the previously known cases: holomorphic germs [Mi2] and of polynomial functions, cf. Proposition 8.1.3 (see also [Ti4, 4.3]). This has consequences on the problem of detecting the variation of topology, especially when the underlying space Y has maximal *rectified homotopical depth* (i.e. rhd $Y \geq \dim Y$, see the definition below).[4]

Definition 9.2.4 Let Z be a complex space endowed with some Whitney stratification $\mathcal{W}$ and let $\mathcal{W}^i$ denote the union of strata of dimension $\leq i$. After [HmL1], we say that rhd $_\mathcal{W} Z \geq m$ if, for any i and any point $x \in \mathcal{W}^i \setminus \mathcal{W}^{i-1}$, the homotopy groups of $(U_\alpha, U_\alpha \setminus \mathcal{W}^i)$ are trivial up to the order $m - 1 - i$, where $\{U_\alpha\}$ is some fundamental system of neighbourhoods of x.

It is shown in [HmL1] that this does not depend on the chosen Whitney stratification, and therefore we only use the notation rhd Z. A similar definition holds in homology instead of homotopy, yielding the notion of *rectified homological depth*, denoted by rHd, which is a weaker condition. Let us point out that, according to [LM], if Y is a local complete intersection of dimension n at all its points, then rhd $Y \geq n$ (and therefore rHd $Y \geq n$). Saying that the 'variation of topology is localizable', we mean that there exist small enough balls in $\mathbb{Y}$ centered at the isolated singularities such that, outside these balls, the projection π is a trivial fibration over a small enough disk centered at $a \in \mathbb{P}^1$. The proof of the following result, in the general case of meromorphic functions, has the same structure as the proof of Proposition 8.1.3 and we may safely leave it to the reader.

Proposition 9.2.5 (localization of topology)
Let F have isolated singularities with respect to some $\partial \tau$-stratification $\mathcal{G}$ at $a \in \mathbb{P}^1$ (i.e. dim $\mathbb{Y}_a \cap \mathrm{Sing}_\mathcal{G} F \leq 0$). Then the variation of topology of the fibres of π at $\mathbb{X}_a$ is localizable at those singularities.

As in the particular settings of holomorphic germs and of polynomial functions, the localization result implies that the vanishing homology is concentrated at the isolated singularities.

Corollary 9.2.6 Let F have isolated singularities with respect to $\mathcal{G}$ at $a \in \mathbb{P}^1$ and let $\mathbb{Y}_a \cap \mathrm{Sing}_\mathcal{G} F = \{p_1, \ldots, p_k\}$. For any small enough balls $B_i \subset \mathbb{Y}$ centered at p_i, and, for small enough closed disk centered at a, $s \in \partial D$, we have:

$$H_\star(X_D, X_s) \simeq \oplus_{i=1}^k H_\star(\mathbb{X}_D \cap B_i, \mathbb{X}_s \cap B_i). \tag{9.6}$$

□

Polar curves. We show that an isolated singularity with respect to the stratification $\mathcal{G}$, at some point of $A \times \mathbb{P}^1 \subset \mathbb{Y}$, is detectable by a certain local polar curve.

Definition 9.2.7 Let $\xi = (y, a) \in A \times \mathbb{P}^1$ and let $f = p/q$ a local representative of F at ξ. Then the germ at ξ of the *polar locus* is the following space:

$$\Gamma_\xi(\pi, q) := \text{closure}\{\text{Sing}_\mathcal{G}(\pi, q) \setminus (\text{Sing}_\mathcal{G}\pi \cup (A \times \mathbb{P}^1))\} \subset \mathbb{Y}.$$

An easy exercise yields the following isomorphisms:

$$\Gamma_\xi(\pi, q) \simeq \Gamma_\xi(f, q) \simeq \Gamma_\xi(p, q).$$

Let us point out that the multiplication by a unit u may change the polar locus, i.e. $\Gamma_\xi(\pi, uq)$ is in general different from $\Gamma_\xi(\pi, q)$. Nevertheless, if the space Y is 'nice enough' (for instance if $\text{rHd } Y \geq \dim Y$), then the local vanishing homology $H_*(\mathbb{X}_D \cap B_i, \mathbb{X}_s \cap B_i)$ is concentrated in one dimension only. We have the following more precise statement:

Proposition 9.2.8 Let $\xi = (y, a) \in A \times \mathbb{P}^1$ be an isolated singularity of F with respect to $\mathcal{G}$ and let $f = p/q$ be a local representative of F. Then, for any local multiplicative unit u, either the polar locus $\Gamma_\xi(\pi, uq)$ is empty or $\dim \Gamma_\xi(\pi, uq) = 1$.

If moreover Y is of pure dimension m and $\text{rHd } (Y \setminus A) \geq m$ in some neighbourhood of ξ, then the intersection multiplicity $\text{int}_\xi(\Gamma_\xi(\pi, uq), \mathbb{Y}_a)$ is independent of the unit u.

Proof Let $\mathbb{P}T_q^*$ be the projectivized relative conormal of q. We use the independence of $\mathbb{P}T_{uq}^*$ from the multiplicative unit u, cf. Lemma 1.2.7.

The proof goes like the one of Theorem 2.1.7. Since $\dim \mathbb{P}T_q^* = m + 1$, it follows that $\Gamma_\xi(\pi, q)$ is either empty or of dimension at least one. On the other hand, since ξ is an isolated $\mathcal{G}$-singularity on $A \times \mathbb{P}^1$, the intersection of the polar locus with the divisor $A \times \mathbb{P}^1$ is, locally, included in $\{\xi\}$; therefore $\Gamma_\xi(\pi, q)$ has dimension at most one. This proves that $\dim \Gamma_\xi(\pi, q) = 1$ and the same argument holds for uq instead of q. As in the proof of Theorem 2.1.7, the polar loci $\Gamma_\xi(\pi, uq)$ are simultaneously either empty or of dimension 1, for all u, and this depends only on the space of *characteristic covectors* at $A \times \mathbb{P}^1$, see Definition 1.2.9.

To prove the second statement, we assume that we are in the nontrivial case $\dim \Gamma_\xi(\pi, q) = 1$. Consider the Milnor–Lê fibration of the function π at ξ (see Theorem A1.1.2):

$$\pi_| : \mathbb{Y}_{D^*} \cap B \to D^*, \tag{9.7}$$

as explained at the end of §9.1, where B is a Milnor ball centered at ξ and D is small enough. The restriction of q to $\mathbb{Y}_s \cap B$ has a finite number of stratified isolated singularities, which are precisely the points of intersection $\mathbb{Y}_s \cap B \cap \Gamma(\pi, q)$. Since cylindrical neighbourhoods are conical, cf. [GM2, p. 165], it follows that $\mathbb{Y}_s \cap B$ is homotopy equivalent to $\mathbb{Y}_s \cap B \cap q^{-1}(\delta)$, where δ is a small disk at $0 \in \mathbb{C}$ such that $\mathbb{Y}_s \cap B \cap \Gamma(\pi, q) = q^{-1}(\delta) \cap \mathbb{Y}_s \cap B \cap \Gamma(\pi, q)$. Let us now take a very small small disk $\hat{\delta}$ centered at $0 \in \mathbb{C}$ such that $q^{-1}(\hat{\delta}) \cap \mathbb{Y}_s \cap B \cap \Gamma(\pi, q) = \emptyset$. By a stratified retraction, it follows that $q^{-1}(\hat{\delta}) \cap \mathbb{Y}_s \cap B$ is homotopy equivalent to the central fibre $q^{-1}(0) \cap \mathbb{Y}_s \cap B$. Since the hypersurface $q^{-1}(0)$ is the product $A \times \mathbb{P}^1$, the central fibre $q^{-1}(0) \cap \mathbb{Y}_s \cap B$ is contractible; hence $q^{-1}(\hat{\delta}) \cap \mathbb{Y}_s \cap B$ is contractible too.

The total space $\mathbb{Y}_s \cap B \overset{\text{ht}}{\simeq} q^{-1}(\delta) \cap \mathbb{Y}_s \cap B$ is built by attaching to the space $q^{-1}(\hat{\delta}) \cap \mathbb{Y}_s \cap B$ finitely many cells, due to the isolated singularities of the function q on $q^{-1}(\delta \setminus \hat{\delta}) \cap \mathbb{Y}_s \cap B$. By assumption rHd $(Y \setminus A) \geq m$ and Hamm–Lê's [HmL1, Theorem 3.2.1] tells us that taking a hypersurface slice lowers rHd by at most one, hence rHd $(\mathbb{Y}_s \setminus A) \geq m - 1$. It follows that each singularity of $q^{-1}(\delta \setminus \hat{\delta}) \cap \mathbb{Y}_s \cap B$ contributes with cells of dimension exactly $m-1$, and the number of cells is equal to the corresponding local Milnor number of the function q. The sum of these numbers is, by definition, the intersection multiplicity $\text{int}_\xi(\Gamma(\pi, q), \mathbb{Y}_a)$.

We have shown that:

$$\dim H_{m-1}(\mathbb{Y}_s \cap B) = \text{int}(\Gamma_\xi(\pi, q), \mathbb{Y}_a) \tag{9.8}$$

and that $\tilde{H}_i(\mathbb{Y}_s \cap B) = 0$ for $i \neq m - 1$. This proves our claim since the proof is the same when replacing q by uq. $\qquad\square$

The polar locus defines a numerical invariant which measures the number of 'vanishing cycles at this point'.

Definition 9.2.9 We say that $\text{int}(\Gamma_\xi(\pi, q), \mathbb{Y}_a)$ is the *polar number* at ξ of π with respect to q.

We can get more precise results when lowering the generality; let us consider the following condition:

(∗) $X := Y \setminus A$ has at most isolated singularities and rHd $X \geq \dim X$.

This is more general than the setting of polynomial functions on $\mathbb{C}^n$, which we have considered in Parts I and II. Under the condition (∗) we have $\text{Sing}_G F \cap (A \times \mathbb{P}^1) \subset \text{Sing} \mathbb{Y} \cap (A \times \mathbb{P}^1)$. In the notations of Corollary 9.2.6, the Lefschetz duality for polyhedra, cf. [Dold, p. 296], see also [Br2, p. 238], implies the following duality:

$$H_*(\mathbb{X}_D \cap B_i, \mathbb{X}_s \cap B_i) \simeq H^{2m-*}(\mathbb{Y}_D \cap B_i, \mathbb{Y}_s \cap B_i),$$

where $m = \dim_{p_i} \mathbb{Y}$. Since $\mathbb{Y}_D \cap B_i$ is contractible, we get:

$$H_*(\mathbb{X}_D \cap B_i, \mathbb{X}_s \cap B_i) \simeq \tilde{H}^{2m-1-*}(\mathbb{Y}_s \cap B_i). \tag{9.9}$$

Comparing (9.9) to Definition 9.2.2, Corollary 9.2.6 and (9.8) from the proof of Proposition 9.2.8, we get:

Corollary 9.2.10 Let F be a meromorphic function on Y such that the condition $(*)$ is satisfied. Let $\xi = (y, a) \in A \times \mathbb{P}^1$ be an isolated singularity of F with respect to $\mathcal{G}$. Then the number of vanishing cycles of $F_{|Y \setminus A}$ at ξ is $\lambda_\xi = \mathrm{int}(\Gamma_\xi(\pi, q), \mathbb{Y}_a)$, where $m = \dim_\xi \mathbb{Y}_a$.

Milnor numbers. Let us assume that, in addition to the condition $(*)$, the hypersurface $\mathbb{Y}_a$ has an isolated singularity at $\xi \in A \times \{a\}$. Since $(\mathbb{Y}_a, \xi)$ is a hypersurface germ in Y, it has a well-defined Milnor–Lê fibration (see Theorem A1.1.2), and, in particular, a Milnor number, which we denote by $\mu(a)$.

It follows that the family of hypersurfaces $\mathbb{Y}_s$ has isolated singularities at $A \times \mathbb{P}$, for s close to a. There is a well-defined Milnor number $\mu_i(s)$ at each hypersurface germ $(\mathbb{Y}_s, \xi_i(s))$, where $\xi_i(s) \to \xi$ as $s \to a$, $1 \le i \le k$. In case $\mathrm{Sing}\,\mathbb{Y}$ is just the point ξ, we consider that $\mu_i(s) = 0, \forall i$.

It also follows that $\mathrm{Sing}\,\mathbb{Y}$ is a curve at ξ and this curve intersects $\mathbb{Y}_s$ at the points $\{\xi_i(s)\}_{i=1}^k$ for all s close enough to a. We have the following computation of the number $\lambda_\xi = \mathrm{int}(\Gamma_\xi(\pi, q), \mathbb{Y}_a)$ of vanishing cycles at ξ:

Proposition 9.2.11 Let F be a meromorphic function on Y satisfying the condition $(*)$. Let $\dim_\xi \mathrm{Sing}\,\mathbb{Y}_a = 0$. Then:

$$\lambda_\xi = \mu(a) - \sum_{i=1}^k \mu_i(s).$$

Proof Consider the function:

$$G = p - sq : (Y \times \mathbb{C}, \xi_i(s)) \to (\mathbb{C}, 0).$$

For any fixed s, this function is a smoothing of the germ $(G = 0, \xi_i(s)) = (\mathbb{Y}_s, \xi_i(s))$. Since ξ is an isolated singularity, the polar locus at ξ of the map $(G, \pi) : Y \times \mathbb{C} \to \mathbb{C}^2$, defined as $\Gamma_\xi(G, \pi) = \mathrm{closure}\{\mathrm{Sing}\,(G, \pi) \setminus \mathrm{Sing}\,G\}$, is at most a curve. By using polydisk neighbourhoods[5] $(P_\alpha \times D_\alpha)$ at ξ in $Y \times \mathbb{C}$, we may show that $(G, \pi)^{-1}(\eta, s) \cap (P_\alpha \times D_\alpha)$, for s close enough to a, is homotopy equivalent to the Milnor fibre $(G, \pi)^{-1}(\eta, a) \cap (P_\alpha \times D_\alpha)$ of the germ $(\mathbb{Y}_a, \xi)$. To obtain from this the space $\pi^{-1}(s) \cap (P_\alpha \times D_\alpha)$, we have to attach a number of m cells, where $m := \dim_\xi Y$. Part of these cells come from the singular points $\xi_i(s) \in \text{Sing}\, G \cap \pi^{-1}(s)$ and, by definition, their total number is $\sum_{i=1}^k \mu_i(s)$. The other part of the cells come from the intersection with $\Gamma_\xi(G, \pi)$ and their number is $r = \text{int}(\Gamma_\xi(G, \pi), \pi^{-1}(0))$. We claim that r is equal to $\dim H_{m-1}(\mathbb{Y}_s \cap B)$, which is equal to the number λ_ξ of vanishing cycles at ξ, by (9.8). This follows from the fact that $\pi^{-1}(s) \cap (P_\alpha \times D_\alpha)$ is contractible, since it is the Milnor fibre of a linear function t on a smooth space. We therefore have the following equality: $\mu(a) = r + \sum_{i=1}^k \mu_i(s)$. $\square$

We refer to Corollary 9.3.3 for the counting of the total number of vanishing cycles. Let us give an examples of meromorphic functions on $\mathbb{P}^2$; other examples may be found in [Ti11, ST6].

Example 9.2.12 [ST6] Let

$$F: Y = \mathbb{P}^2 \dashrightarrow \mathbb{P}^1, \quad F = \frac{x(z^{a+b} + x^a y^b)}{y^p z^q},$$

where $a + b + 1 = p + q$ and $a, b, p, q \geq 1$. For some $s \in \mathbb{C}$, the space $\mathbb{Y}_s$ is given by:

$$x(z^{a+b} + x^a y^b) = s y^p z^q \tag{9.10}$$

$\text{Sing}\, \mathbb{Y} \cap (Y \times \mathbb{C})$ consists of three lines: $\{[1:0:0], [0:1:0], [0:0:1]\} \times \mathbb{C}$. We are under the assumptions of Proposition 9.2.11 and we inspect each of these three families of germs with isolated singularity to see where the Milnor number jumps.

Along $[1:0:0] \times \mathbb{C}$, in chart $x = 1$, there are no jumps, since the germs have uniform Brieskorn type $(b, a + b)$. Along $[0:0:1]$, in chart $z = 1$, there are no jumps, since the type is constant A_0, for all s. Along $[0:1:0]$, in chart $y = 1$, for $s \neq 0$, the Brieskorn type is $(a + 1, q)$, with $\mu(s) = a(q - 1)$. If $s = 0$, then we have $x^{a+1} + xz^{a+b} = 0$ with $\mu(0) = a^2 + ab + b$.

There is only one jump, at $\xi = ([0:1:0], 0)$. By Proposition 9.2.11 we get $\lambda_\xi = a^2 + ab + b - a(q - 1) = b + ap$.

9.3 Fibres and affine hypersurface complements

We still denote by A the axis of some meromorphic function $F : Y \dashrightarrow \mathbb{P}^1$ and $X := Y \setminus V$. We recall that for any subset $W \subset \mathbb{P}^1$, we use the notations $\mathbb{Y}_W$, $\mathbb{X}_W$ and $Y_W := \sigma(p^{-1}(W))$, $X_W := X \cap Y_W$, $A' := A \cap X$. In case of isolated singularities with respect to the partial Thom stratification $\mathcal{G}$ (for short: $\mathcal{G}$-singularities), we have the following result on the relative homotopy type:

Theorem 9.3.1 [ST6, Ti11] *Let $D \subset \mathbb{P}^1$ be some open disk and let $\Lambda \cap D = \{a_i\}_{i=1}^p$ be the set of atypical values of F within D. Let $s \in D \setminus \Lambda$ be some typical value of F. For all $1 \le i \le p$, let F have isolated $\mathcal{G}$-singularities at $\mathbb{Y}_{a_i}$, let Y be of pure dimension m and let $\mathrm{rHd}\,(X \setminus A) \ge m$. If either of the two following conditions is fulfilled:*

(a) *X is a Stein space,*

(b) *$V = A$ and X has at most isolated singularities,*

then X_D is obtained from X_s by attaching cells of real dimension m. In particular, the quotient space X_D/X_s is homotopy equivalent to a bouquet of spheres $\vee S^m$.

Proof We have the homotopy equivalence $X_D/X_s \overset{\mathrm{ht}}{\simeq} \mathbb{X}_D/\mathbb{X}_s$ by Lemma 9.2.1. We first prove that $\tilde{H}_\star(\mathbb{X}_D/\mathbb{X}_s, \mathbb{Z})$ is concentrated in dimension m. By Proposition 9.2.5, the variation of topology of the fibres of $F_{|\mathbb{X}_D}$ is localizable at the points $\mathrm{Sing}_{\mathcal{G}} F \cap \mathbb{Y}_D$. We have to take into account all the possible positions of such a singular point $\xi = (y, a)$; namely, on X, on $V \setminus A$ or on $A \times \mathbb{P}^1 \subset \mathbb{Y}$. It will turn out that, in all the cases, the pair $(\mathbb{X}_{D_a} \cap B_\xi, X_s \cap B_\xi)$ is $(m-1)$-connected, where $B_\xi \subset \mathbb{Y}$ is a small enough ball at ξ, D_a is a small enough closed disk at a and $s \in \partial D_a$.

In case $\xi \in X$, this is just Milnor's classical result for a holomorphic function with isolated singularity [Mi2]. In the two remaining cases, the claim follows from a result due to Hamm and Lê [HmL1, Corollary 4.2.2], in the slightly more general version for partial Thom stratifications instead of Whitney stratifications (see also Corollary A1.1.5, for a special case). This result needs the condition on the rectified homological depth.

By the above proved connectivity of the pair $(\mathbb{X}_{D_a} \cap B_\xi, X_s \cap B_\xi)$ and the splitting of vanishing homology into local contributions Corollary 9.2.6, we get that the homology of $(\mathbb{X}_D, X_s)$ is zero below dimension m. We also have the annulation above m, due to the following reasons. Under the assumption (a) the space X_D, respectively X_s, is Stein of dimension m, resp. $m - 1$, since Hamm [Hm3] showed that the homology of an n-dimensional Stein space is trivial above the level n.

In case (b), we may apply the equivalences (9.8) and (9.9), using that the cohomology $\tilde{H}^*(\mathbb{Y}_s \cap B_\xi)$ is concentrated in dimension $m - 1$, by the proof of Proposition 9.2.8.

Let us remark that $\mathbb{X}_D/X_s$ is a CW-complex (since it is analytic, therefore triangulable). It is simply connected if $m \geq 2$, since the pairs $(\mathbb{X}_{D_a} \cap B_\xi, X_s \cap B_\xi)$ are $(m - 1)$-connected. Then we can map a bouquet of m-spheres into $\mathbb{X}_D/X_s$ such that this map is an isomorphism in homology. This implies, by Whitehead's theorem (cf. [Sp, 7.5.9]), that the map induces an isomorphism of homotopy groups. For CW-complexes, weak homotopy equivalence coincides with homotopy equivalence. This proves our bouquet statement.   □

Let us point out that, in case (b) of Theorem 9.3.1, we also have the local bouquet result for a meromorphic function germ:

$$(X_{D_a} \cap B_\xi)/(X_s \cap B_\xi) \overset{ht}{\simeq} \vee S^m.$$

When assuming high connectivity of the space X, we get the following immediate consequence, cf. [Ti11]:

Corollary 9.3.2 Under the hypotheses of Theorem 9.3.1, if in addition X is Stein and X_D is $(m - 1)$-connected, then $X_s \overset{ht}{\simeq} \bigvee S^{m-1}$.

Particular cases of this corollary appeared previously in several circumstances: Milnor's bouquet result [Mi2] on holomorphic germs with isolated singularity; bouquet results for generic fibres of polynomial maps with isolated singularities in the affine [Br1, Br2] and with isolated singularities at infinity [ST2], [Ti4]. Corollary 9.3.2 also relates to the bouquet theorem 3.2.1 for polynomial function: it concerns a more general setting but the hypothesis about the type of singularities is less general.

We can easily derive a formula for the total number of vanishing cycles in case of isolated $\mathcal{G}$-singularities. Let us denote by λ_a the sum of the polar numbers at the singularities on $(A \times \mathbb{P}) \cap \mathbb{Y}_a$ and by μ_a the sum of the Milnor numbers of the singularities on $\mathbb{Y}_a \setminus (A \times \{a\})$.

Corollary 9.3.3 Under the hypotheses of Theorem 9.3.1, we have:

$$\dim H_{m-1}(X_{D_a}, X_s) = \mu_a + \lambda_a,$$

$$\dim H_{m-1}(X_D, X_s) = \sum_{a \in D} \mu_a + \sum_{a \in D} \lambda_a.$$

□

Example 9.3.4 [ST6] Let $Y \subset \mathbb{P}^3$ be the nonsingular hypersurface given by $h = x^2 + z^2 + yw = 0$. Consider the meromorphic function $F = y/x$ and $X := Y \setminus \{x = 0\}$. The pencil defined by F has its axis $A = \{x = y = 0\}$ tangent to $h = 0$ at $[0 : 0 : 0 : 1]$. The general fibre F is contractible, and the unique special fibre F_0 is homotopy equivalent to $\mathbb{C} \sqcup \mathbb{C}$. Moreover X is homotopy equivalent to S^2. We get $\mu_0 = 0$ and $\lambda_0 = 1$, due to the jump $A_0 \to A_1$ at $t = 0$ (see also Exercise 9.5). We remark that all the connected components of the fibres are contractible, whereas the global vanishing homology is generated by a relative 2-cycle.

Example 9.3.5 [ST6] In Example 9.2.12, we take $X = \mathbb{P}^2 \setminus \{yz = 0\}$. Then the fibre X_0 is a disjoint union of $c + 1$ copies of $\mathbb{C}^*$, where $c = \gcd(a, b)$, therefore $\chi(X_0) = 0$. For $s \neq 0$, by a branched covering argument, we show that $\chi(X_s) = -(b + ap)$. The vanishing homology is concentrated in dimension 2, by Theorem 9.3.1. When taking $D = \mathbb{C}$, we get the Betti number $b_2(X, X_s) = \chi(X, X_s) = \chi(X) - \chi(X_s) = (b + ap)$. We have seen at 9.2.12 that the sum $\sum_{a \in \mathbb{C}} \lambda_a$ consists of a single term $\lambda_\xi = b + ap$. We can easily see that there is no other singularity, hence $\sum_{a \in \mathbb{C}} \mu_a = 0$. By applying Corollary 9.2.10, we observe that the second equality in Corollary 9.3.3 is verified.

Complements of fibres. We may draw some applications of the above results to the topology of *complements of affine hypersurfaces* $V \subset \mathbb{C}^n$. This is a topic which goes back to Zariski and van Kampen [vK], who described a general procedure to compute the fundamental group of the complement to an algebraic curve in $\mathbb{P}^2$ by slicing with linear pencils. Zariski showed that π_1 depends on the type and position of the singularities of the curve. More recently, Libgober [Li2] proved more general results on the first nontrivial higher homotopy group, say π_{n-k-1}. It follows by the classical Lefschetz Hyperplane Theorem that $\pi_{n-k-1}(\mathbb{C}^n \setminus V) = \pi_{n-k-1}(H_k \cap \mathbb{C}^n \setminus V)$, for a general linear subspace H_k of codimension k, so the problem of finding π_{n-k-1} reduces to the case of the complement of a hypersurface V with isolated singularities, in the appropriate meaning of 'isolated'.[6] Our presentation follows closely [LiTi]; we refer to the forthcoming §9.4 and §11 for further results on higher homotopy groups of the complements.

The affine hypersurface $V \subset \mathbb{C}^n$ may be a general fibre of some polynomial f, or an atypical one. In case V is a general fibre of f, we have the following general result on the homotopy type of the complement:

Proposition 9.3.6 [LiTi] Let $f : \mathbb{C}^n \to \mathbb{C}$ be any polynomial and let V be a general fibre of f (i.e. above a typical value $\mathbb{C} \setminus \text{Atyp} f$). Then $\mathbb{C}^n \setminus V \overset{ht}{\simeq} S^1 \vee S(V)$, where $S(V)$ denotes the suspension over V.

In particular, if $V \overset{ht}{\simeq} \bigvee_\lambda S^{n-1}$, then $\mathbb{C}^n \setminus V \overset{ht}{\simeq} S^1 \vee \bigvee_\lambda S^n$.

Proof Let $V = f^{-1}(c)$, where $c \notin \text{Atyp} f$, and take a small enough closed disk $D \subset \mathbb{C}$ centered at c. Take a system of simple nonintersecting paths γ_i from some point $\alpha \in \partial D$ to small enough disks D_i centered at each atypical value $b_i \in \text{Atyp} f$. Remark that $\cup_i (\gamma_i \cup D_i)$ is a deformation retract of $\mathbb{C}$, and, by using the local triviality of f, we have $\mathbb{C}^n \overset{\text{ht}}{\simeq} f^{-1}(\cup_i(\gamma_i \cup D_i))$.

The polynomial function f is a trivial fibration over D, so in particular over ∂D. Since $\mathbb{C}^n \setminus V \overset{\text{ht}}{\simeq} f^{-1}(\cup_i(\gamma_i \cup D_i)) \cup f^{-1}(\partial D)$, it follows that $\mathbb{C}^n \setminus V$ is obtained from $f^{-1}(\partial D) \overset{\text{ht}}{\simeq} \partial D \times V$ by attaching the space $f^{-1}(\cup_i(\gamma_i \cup D_i))$ over $f^{-1}(\alpha) \overset{\text{ht}}{\simeq} V$. This is like the attaching of a cone to $\partial D \times V$ over $\{\alpha\} \times V$. $\square$

Proposition 9.3.6 holds in particular if f has isolated singularities in $\mathbb{C}^n$ and $\dim \Sigma(w) \le 0$, by Proposition 9.1.7 (cf. the notations in [LiTi]).

Example 9.3.7 $V := f^{-1}(0)$, where $f : \mathbb{C}^4 \to \mathbb{C}, f = x_1^4 x_2^4 + (x_1+x_2)^6 + x_3^5 + x_4^4 + x_1^2$. Note that $\bar{V} \subset \mathbb{P}^4$ fails to be transversal to H^∞ along a two-dimensional set. Nevertheless, we may observe that $f = g(x_1, x_2) + h(x_3, x_4)$ is a sum of two polynomials in separate variables. For g, we have that $\dim \text{Sing} g \le 0$ and that there are no singularities at infinity, since $\Sigma = \emptyset$, by (9.4) and by Proposition 9.1.7 for all the weights $= 1, d = 8, k = 2$. We get by Corollary 9.3.2 that the general fibre of g is, homotopically, a bouquet $\vee S^1$. On the other hand, the polynomial h is weighted homogeneous with a unique singularity at the origin, hence its general fibre is a bouquet $\vee S^3$.

By a Thom–Sebastiani type result [Ne3], the general fibre of $f = g + h$ has the homotopy type of a bouquet $\bigvee S^3$. Now, by Proposition 9.3.6, for a general fibre V of f, the complement $\mathbb{C}^4 \setminus V$ is homotopy equivalent to $S^1 \vee \bigvee S^4$.

In case the hypersurface V is an atypical fibre of a polynomial f, we have the following result:

Proposition 9.3.8 [LiTi] Let $V = f^{-1}(0)$. If the general fibre of the polynomial function $f : \mathbb{C}^n \to \mathbb{C}$ is s-connected, $s \ge 2$, then $\pi_i(\mathbb{C}^n \setminus V) = 0$, for $1 < i \le s$, and $\pi_1(\mathbb{C}^n \setminus V) = \mathbb{Z}$.

In particular, if f has isolated singularities at infinity in the sense of Definition 9.1.4, then $\pi_i(\mathbb{C}^n \setminus V) = 0$ for $1 < i \le n - 2$.

Proof We use the notations D_i, γ_i as in the proof of Proposition 9.3.6 and take $b_1 := 0$, so that $V = f^{-1}(b_1)$. We claim that the space $T_1 := f^{-1}(\cup_{i \neq 1} \gamma_i \cup D_i)$ is homotopy equivalent to a general fibre $F := f^{-1}(\alpha)$ to which one attaches cells of dimension $\ge s + 2$, in other words that the pair (T_1, F) is $(s + 1)$-connected.

Since F is s-connected by hypothesis, we have that $(\mathbb{C}^n, F)$ is $(s + 1)$-connected. Then, by excision in homology, we have that $H_j(\mathbb{C}^n, F) = H_j(T_1, F) \oplus H_j(f^{-1}(D_1), F)$, for any j. Hence $H_j(T_1, F) = 0$ for $j \leq s+1$. By Blakers–Massey theorem [BM], the homotopy excision works within a certain range. Namely, since F is s-connected, we get that the inclusion:

$$\pi_j(T_1, F) \oplus \pi_j(f^{-1}(D_1), F) \to \pi_j(\mathbb{C}^n, F)$$

is an isomorphism for $j \leq s - 1$ and an epimorphism for $j = s$. This shows that (T_1, F) is $(s - 1)$-connected and in particular simply connected, since $s \geq 2$. We may furthermore apply the relative Hurewicz isomorphism theorem and get that $\pi_j(T_1, F)$ is trivial for $j \leq s + 1$. (We also get the isomorphism $\pi_{s+2}(T_1, F) \simeq H_{s+2}(T_1, F)$.) By Switzer's result [Sw, Proposition 6.13], it follows that T_1 is homotopy equivalent to the space F to which one attaches cells of dimension $\geq s + 2$. The claimed property is proved.

Next, we have that $\pi_j(f^{-1}(\partial D_1)) = 0$ for $1 < j \leq s$ and that $\pi_1(f^{-1}(\partial D_1)) = \mathbb{Z}$, due to the homotopy exact sequence of the fibration $f_| : f^{-1}(\partial D_1) \to \partial D_1$ and the s-connectivity of the fibre. (Note that this holds even for $s = 1$.)

Finally, $\mathbb{C}^n \setminus V$ is obtained from $f^{-1}(\partial D_1)$ by attaching the space T_1 over a general fibre F, which, we have proved above, means attaching only cells of dimension $\geq s + 2$. It follows that $\pi_j(\mathbb{C}^n \setminus V) = 0$ for $1 < j \leq s$ and that $\pi_1(\mathbb{C}^n \setminus V) = \mathbb{Z}$.

The second statement follows from the first one. Indeed, if f has isolated singularities at infinity, then the general fibre of f is $(n - 2)$-connected, by Corollary 9.3.2. □

In case of higher-dimensional singular locus, we prove a connectivity result for hypersurface complements, in the setting of the embedding $\mathbb{C}^n \subset \mathbb{P}(w)$. We refer to page 158 for the natural Whitney stratification $\mathcal{S}$ that we use in this case.

Theorem 9.3.9 [LiTi] *Let $f : \mathbb{C}^n \to \mathbb{C}$ be a polynomial function. Suppose that* $\dim \mathrm{Sing}\,_{\mathcal{S}} p \cap \mathbb{Y}_{\mathbb{C}} \leq k$, *where $\mathbb{Y}$ is the total space relative to the embedding* $\mathbb{C}^n \subset \mathbb{P}(w)$, *for some system of weights w. Then:*

(a) *for a general fibre F of f, $\pi_i(\mathbb{C}^n \setminus F) = 0$, for $2 \leq i \leq n - k - 1$.*
(b) *for an atypical fibre V of f, $\pi_i(\mathbb{C}^n \setminus V) = 0$, for $2 \leq i \leq n - k - 2$.*

Proof If we prove that the general fibre F is $(n - k - 2)$-connected, then (a) follows by Proposition 9.3.6 and (b) follows by Proposition 9.3.8. Let $\Psi :$ $\mathbb{C}^n \to \mathbb{C}^n$ be the finite map given by $(x_1, \ldots, x_n) \mapsto (x_1^{m_1}, \ldots, x_n^{m_n})$, where $m_i = N/w_i$ and N is a common multiple of all w_i, $1 \leq i \leq n$. This induces $\bar{\Psi} : \mathbb{P}(w_1, \ldots, w_n, 1) \to \mathbb{P}^n$, where $z \mapsto z$, and therefore a finite map $\mathbb{X} \subset$

$\mathbb{P}(w) \times \mathbb{C} \overset{\hat{\Psi}}{\to} \mathbb{P}^n \times \mathbb{C}$, where $\mathbb{X} := \mathbb{Y}_{\mathbb{C}} \setminus \{z = 0\}$. We shall prove our statement by reduction to the space $\mathbb{P}^n$, via the finite map $\hat{\Psi}$.

We denote by H_s the affine hyperplane $\{l_H = s\} \subset \mathbb{C}^n$, where $H \in \check{\mathbb{P}}^{n-1}$ is a hyperplane defined by a linear form $l_H : \mathbb{C}^n \to \mathbb{C}$. We consider the restriction of f to $\mathcal{H}_s := \Psi^{-1}(H_s)$. We denote by $\mathcal{S}_{\mathcal{H}_s}$ the coarsest Whitney stratification of the space $\mathbb{H}_s := \mathbb{X} \cap (\bar{\mathcal{H}}_s \times \mathbb{C})$, where $\bar{\mathcal{H}}_s$ denotes the degree N weighted projective hypersurface $\{l_H \circ \Psi(x) - sz^N = 0\}$.

By eventually refining the stratification $\mathcal{S}$, we may assume without loss of generality that the restriction of $\hat{\Psi}$ to each stratum is an unramified covering. We apply the classical method of slicing by generic hyperplanes to the images by $\hat{\Psi}$ of the strata and to the projection to $\mathbb{C}$. Then we may pull back the transversality results by $\hat{\Psi}^{-1}$.

By applying the proof of Theorem 7.2.6, see also [Ti2, Lemma 5.4], it follows that there exists a Zariski-open set $\Omega \subset \check{\mathbb{P}}^{n-1}$ and a finite set $R \subset \mathbb{C}$ such that, if $H \in \Omega$ and $s \in \mathbb{C} \setminus R$ and if $\dim \text{Sing}_{\mathcal{S}} p \geq 1$, then $\dim \text{Sing}_{\mathcal{S}_{\mathcal{H}_s}} (p_{|\mathcal{H}_s}) \leq \dim \text{Sing}_{\mathcal{S}} p - 1$.

Slicing a general fibre F of f by a "generic" hypersurface $\mathcal{H}_s$ gives a general fibre of the restriction $f_{|\mathcal{H}_s}$. Moreover, by the classical Lefschetz Hyperplane Theorem, the pair $(F, F \cap \mathcal{H}_s)$ is $\dim F - 1$ connected.*

By using at each step the pull-back by $\hat{\Psi}$, we may pursue this slicing procedure until the singularities of the restriction of f become zero-dimensional. More precisely, there exist k generic hyperplanes $H^i \in \check{\mathbb{P}}^{n-1}$ and generic $s_1, \ldots, s_k \in \mathbb{C}$ such that $A := \mathcal{H}^1_{s_1} \cap \cdots \cap \mathcal{H}^k_{s_k}$ is the global Milnor fibre of a weighted homogeneous affine complete intersection with *isolated singularity* at the origin and such that the restriction $f_{|} : A \to \mathbb{C}$ has isolated singularities in the affine and at infinity, in the sense of Definition 9.1.4. It is well known that the local and global Milnor fibres are diffeomorphic [AGV2], and on the other hand the local Milnor fibre of an isolated complete intersection singularity is homotopy equivalent to a bouquet $\vee S^{n-k}$, by Hamm's result [Hm1], see also [Ti2].

We may apply Theorem 9.3.1 to the pencil $f_{|} : A \to \mathbb{C}$ with isolated singularities at infinity, and get that A is obtained from a general fibre of $f_{|A}$ by attaching a finite number of cells of dimension $n - k = \dim A$. Since A is a Stein space homotopy equivalent to a bouquet of spheres of dimension $n - k$, it follows by Corollary 9.3.2 that the general fibre is homotopy equivalent to a bouquet of $n - k - 1$ spheres. By tracing back the vanishing of the homotopy groups in the slicing sequence, we get that the general fibre F is at least $(n-k-2)$-connected.

$\square$

---

* See also the non-generic Lefschetz Hyperplane Theorem in §10.

Remark 9.3.10 The proof of Theorem 9.3.9 also shows that the general fibre of f is $(n - k - 2)$-connected and that any atypical fibre of f is at least $(n - k - 3)$-connected.

For the later claim, we have to modify the arguments in the proofs of the above results, as follows. We slice as in the proof of Theorem 9.3.9, but instead of following a generic fibre F, we work with the atypical V. We get that the restriction $f_1 : A \to \mathbb{C}$ has isolated singularities in the affine and at infinity and that $V \cap A$ is an atypical fibre of $f_{|A}$. Then take a big enough ball B such that $V \cap A \simeq B \cap (V \cap A)$. By a polar curve argument, for the general nearby fibre $F \cap A$, we have that the pair $(F \cap A, B \cap (F \cap A))$ is $(n - k - 2)$-connected. On the other hand, $B \cap (V \cap A)$ is homotopy equivalent to $B \cap (F \cap A)$ to which one attaches $(n - k)$-cells corresponding to the isolated singularities of $f_{|A}$ on $V \cap A$. It then follows that $V \cap A$ is at least $(n - k - 2)$-connected.

This bound for the connectivity of the atypical fibres appears to be sharp, as for instance in Broughton's example in two variables $f = x + x^2 y$, where $V := f^{-1}(0)$ is not zero-connected.

Let us point out that the assumption of Theorem 9.3.9 holds in particular if $\dim(\Sigma(w) \cup \mathrm{Sing} f) \leq k$, by Proposition 9.1.7.[7]

9.4 Monodromy of meromorphic functions

Let $F : Y \dashrightarrow \mathbb{P}^1$ be a meromorphic function, let A be its axis and let $\Lambda \subset \mathbb{P}^1$ denote the set of atypical values of F. Then Λ is included in the critical set of the stratified map $\pi : \mathbb{Y} \to \mathbb{P}^1$, cf. Proposition 9.1.2, where we consider here a partial Thom stratification $\mathcal{G}$ as in Definition 9.1.3. If D is some closed disk such that $\partial D \cap \Lambda \neq \emptyset$, then we have a locally trivial stratified fibration $p_{|} : \mathbb{Y}_{\partial D} \to \partial D$ above the circle ∂D. Let $a_i \in \Lambda \cap D$. The geometric monodromy h_i at a_i is induced by a counterclockwise loop around a small circle ∂D_i centered at a_i. This is constructed by using a stratified vector field on $\mathbb{Y}_{\partial D_i}$, which lifts the oriented unitary tangent vector field of the circle ∂D_i. The monodromy h_i restricts to a geometric monodromy on $\mathbb{X}_{\partial D_i}$ and yields a characteristic homeomorphism $h_i : X_{s_i} \to X_{s_i}$, where $s_i \in \partial D_i$. The action h_i on $\mathbb{X}_{\partial D_i}$ is isotopic to the indentity and can be extended (and denoted also by h_i) to $\mathbb{X}_{D_i}$ and to $\mathbb{X}_D$ such that the extensions are isotopic to the identity. The geometric monodromy h_i is not uniquely defined but induces a well-defined representation:

$$\rho_i : \pi_1(\partial D_i, s_i) \to \mathrm{Iso}(\mathbb{X}_D, \mathbb{X}_{\partial D_i}, \mathbb{X}_{s_i}),$$

where Iso(., ., .) denotes the group of relative isotopy classes of stratified homeomorphisms, which are C^∞ along each stratum. Here we identify the fibres $\mathbb{X}_{s_i}$
to $\mathbb{X}_s$ for some fixed point $s \in \partial D$ by using a system of paths $\gamma_i \subset D$ from
$s \in \partial D$ to $s_i \in \partial D_i$, as explained in §9.2. Then let T_i denote the action induced
by $\hat{\rho}_i$ on the homology with integer coefficients.

The monodromy is not well defined on the spaces such as X_D, X_{D_i}, $X_{\partial D_i}$
(except for the fibres X_{s_i} since $X_{s_i} = \mathbb{X}_{s_i}$ by definition). However, the monodromy has a well-defined action on the relative homology since we have the
following isomorphisms:

$$H_*(\mathbb{X}_{D_i}, X_{s_i}) \simeq H_*(X_{D_i}, X_{s_i}), \quad H_*(\mathbb{X}_D, X_s) \simeq H_*(X_D, X_s)$$

induced by the corresponding homotopy equivalences of pairs, which are easily
deduced from Lemma 9.2.1 (see also Definition 9.2.2 and Proposition 9.2.3).
We shall use these identifications of the relative homologies in the following.

Let us also identify $H_*(X_D, X_s)$ to $\oplus_{a_i \in D \cap \Lambda} H_*(X_{D_i}, X_{s_i})$ as in Proposition 9.2.3. This identification depends on the chosen system of paths $\gamma_i \subset D$
from $s \in \partial D$ to $s_i \in \partial D_i$, as explained in §9.2. Under this identification, we
may prove the following *global Picard formula*.[8]

Proposition 9.4.1 For every $\omega \in H_\star(X_D, X_s)$, there is $\psi_i(\omega) \in H_\star(X_{D_i}, X_{s_i})$
such that $T_i(\omega) = \omega + \psi_i(\omega)$.

Proof We may identify the map: $T_i - \text{id} : H_{q+1}(\mathbb{X}_D, X_s) \to H_{q+1}(\mathbb{X}_D, X_s)$ to
the composed map:

$$H_{q+1}(\mathbb{X}_D, X_s) \xrightarrow{\partial} H_q(X_s) \xrightarrow{w} H_{q+1}(\mathbb{X}_{\partial D_i}, X_s) \xrightarrow{i_*} H_{q+1}(\mathbb{X}_D, X_s), \quad (9.11)$$

where w denotes the *Wang map*, which is an isomorphism by the Künneth
formula. The last morphism in (9.11) factors as follows:

$$H_{q+1}(\mathbb{X}_{\partial D_i}, X_{s_i}) \xrightarrow{i_*} H_{q+1}(\mathbb{X}_D, X_s)$$
$$\searrow \qquad \qquad \nearrow$$
$$H_{q+1}(\mathbb{X}_{D_i}, X_s)$$

where all the three arrows are induced by inclusions. Using the above identifications of relative homology groups, it follows that the submodule of
"anti-invariant cycles" $\text{Im}(T_i - \text{id} : H_*(X_D, X_s) \to H_*(X_D, X_s))$ is contained
in the direct summand $H_*(X_{D_i}, X_{s_i})$ of $H_*(X_D, X_s)$. $\square$

Note 9.4.2 Proposition 9.4.1 is a Picard type formula, since Picard showed it
at the end of the nineteenth century, for algebraic functions of two variables

with simple singularities. Lefschetz proved later the well-known formula for a loop around a quadratic singularity, in which case $\psi_i(\omega)$ is, up to sign, equal to $c\Delta$, where Δ is the quadratic vanishing cycle and c is the intersection number (ω, Δ). This became the basis of what we now call *Picard–Lefschetz theory*, which is regarded as the counterpart of the *Morse theory* for complex spaces, see e.g. [AGV2, Eb, Vas1, Vas2].

We have the following basic consequence, in full generality.[9] Assume that the considered paths in D, say $\gamma_1, \ldots, \gamma_l$, are counterclockwise ordered. Denote by $T_{\partial D}$ the monodromy around the circle ∂D.

Corollary 9.4.3 Assume that the direct sum decomposition of $H_*(X_D, X_s)$ is fixed. Then $T_{\partial D}$ determines T_i, $\forall i \in \{1, \ldots, l\}$.

In particular, $T_{\partial D}$ is trivial if and only if T_i is trivial, $\forall i \in \{1, \ldots, l\}$.

Proof The monodromy $T_{\partial D}$ is the *Coxeter element*, i.e. the composition $T_{\partial D} := T_l \circ \cdots \circ T_1$. We may then write $T_{\partial D} - \mathrm{id} = (T_l \circ \cdots \circ T_1 - T_{l-1} \circ \cdots \circ T_1) + \ldots + (T_1 - \mathrm{id})$. This determines each of the monodromies T_i, by using the Picard formula (Proposition 9.4.1) together with the fixed direct sum decomposition of $H_*(X_D, X_s)$. $\square$

Remarks 9.4.4 The statement and the proof of Proposition 9.4.1 may be dualised in a standard way from homology to cohomology. Using dualization, we obtain statements about invariant cocycles $\ker(T^i - \mathrm{id} : H^*(X_D, X_s) \to H^*(X_D, X_s))$, instead of anti-invariant cycles. A special case is that of a polynomial function $F : \mathbb{C}^n \to \mathbb{C}$, for which $X = \mathbb{C}^n$. In this setting, invariant cocycles have been studied by Neumann and Norbury [NN1]. One may extend their study to the general setting of meromorphic functions.

Zeta function of the monodromy. We have seen in Part II several formulas for the zeta function in case of polynomials on $\mathbb{C}^n$. We discuss here some issues in case of meromorphic functions. We only consider global meromorphic functions. Nevertheless, following our general remark, all results translate easily to meromorphic germs. So let $T_{\partial D}$ be the monodromy around some disk D as above. We call *zeta function of* $T_{\partial D}$ the following rational function in variable t:

$$\zeta_{(X_D, X_s)}(t) = \prod_{j \geq 0} \det[\mathrm{id} - tT_{\partial D} : H_j(X_D, X_s) \to H_j(X_D, X_s)]^{(-1)^{j+1}}.$$

We consider here the case $D = D_a$ and the zeta function of the monodromy around the value $a \in \Lambda$. Let us first assume that the meromorphic function F has isolated $\mathcal{G}$-singularities. By the direct sum splitting into local data (Corollary 9.2.6) and since the monodromy acts on each local Milnor fibration, we get:

$$\zeta_{(X_{D_a}, X_s)}(t) = \prod_{i=1}^{k} \zeta_{(X_{D_a} \cap B_i, X_s \cap B_i)}(t),$$

where $\{p_1, \ldots, p_k\} = \mathbb{Y}_a \cap \mathrm{Sing}_{\mathcal{G}} F$ and B_i is a small Milnor ball centered at p_i. We easily deduce the zeta function of the monodromy $T_{\partial D_a}$ acting on the homology of the general fibre X_s, since the monodromy acts on X_{D_a} as the identity and since $\chi(X_{D_a}) = \chi(X_a)$:

$$\zeta_{X_s}(t) = (1 - t)^{-\chi(X_a)} \prod_{i=1}^{k} \zeta_{(X_{D_a} \cap B_i, X_s \cap B_i)}^{-1}(t).$$

Another way of producing zeta function formulas is by following A'Campo's method [A'C2]. Namely, let us take any meromorphic function F on a nonsingular Y, and consider $X = Y \setminus A$. There exists a proper holomorphic modification $\phi \colon \tilde{Y} \to Y$, which is bi-holomorphic over $X \setminus \cup_{a \in \lambda} X_a$. The pull-back $\tilde{F} = F \circ \phi$ has a general fibre $\tilde{X}_s$, which is isomorphic to X_s. The monodromy over ∂D_a lifts to $\tilde{Y}$ and its action on $\tilde{X}_s$ is also the same, therefore $\zeta_{\tilde{X}_s}(t) = \zeta_{X_s}(t)$. Then we can write down a formula for the zeta function around the value $a \in \mathbb{P}^1$ in terms of the exceptional divisor and of the axis A. By expressing the result as an integral with respect to the Euler characteristic (see Viro's paper [Vi] for this technique), we can get rid of the data from the resolution.

Proposition 9.4.5 [GLM1] Let Y be nonsingular and let $X = Y \setminus A$. Then:

$$\zeta_{X_s}(t) = \int_{A \times \{a\} \cup X_a} \zeta_p(t) d\chi,$$

where ζ_p denotes the local zeta function at the point $p \in \mathbb{Y}$.   □

Further formulae for the zeta function and some of their consequences may be found in the papers by Gusein-Zade, Luengo and Melle [GLM1, GLM2, GLM4].

Monodromy and Alexander polynomials of hypersurface complements. Let us show, following [LiTi], how the monodromy of certain nonisolated hypersurface singularities is related to the first nontrivial homotopy group of the hypersurface complement. Let $\mathbb{C}^n \subset Y$ be some compactification into an n-dimensional complex variety Y (e.g. $Y = \mathbb{P}(w)$). In the framework of $\mathcal{G}$-singularities (Definition 9.1.4), let us assume that $\dim \mathrm{Sing}_{\mathcal{G}} p \le k$ and that $V = f^{-1}(0)$ is an atypical fibre. Then the generic fibre F of f is $(n - k - 1)$-connected, by Theorem 9.3.9.

Let $L_k = \mathcal{H}_1 \cap \cdots \cap \mathcal{H}_k$ denote the intersection of k generic affine hypersurfaces. By applying the Zariski–Lefschetz theorem by Hamm and Lê [HmL1] (alternatively, our Lefschetz hypersurface theorem 10.1.4 for a generic pencil, without singularities in the axis, then using recursion as explained at page 190), we get:

$$\pi_j(L_k \setminus L_k \cap V) \simeq \pi_j(\mathbb{C}^n \setminus V), \tag{9.12}$$

for $j \le n - k - 1$. By slicing with generic hyperplanes, the dimension of the set of $\mathcal{G}$-singularities drops by one at each step (easy exercise), until reaching the dimension zero, so the polynomial $f_{|L_k}$ has only isolated $\mathcal{G}$-singularities at infinity relative to the extension $L_k \subset Y$.

From Theorem 10.1.4 we then get $\pi_i(L_k \setminus L_k \cap V) = 0$, for $i < n - k - 1$. We therefore say that $\pi_{n-k-1}(L_k \setminus L_k \cap V)$ is the *first possible nontrivial homotopy group*. We may then identify $\pi_{n-k-1}(\mathbb{C}^n \setminus V)$ with the homology group of the universal cover $H_{n-k-1}(\widetilde{\mathbb{C}^n \setminus V})$. Next, the fundamental group $\mathbb{Z} \simeq \pi_1(L_k \setminus L_k \cap V) \simeq \pi_1(\mathbb{C}^n \setminus V)$ (by Proposition 9.3.8) acts on the higher homotopy groups like the change of the base point (see e.g. Bredon [Bre] for higher homotopy groups). This is equivalent to the action of $\mathbb{Z}$ as the group of deck transformations on $H_{n-k-1}(\widetilde{\mathbb{C}^n \setminus V})$. This action induces on $\pi_{n-k-1}(\mathbb{C}^n \setminus V)$ the structure of a module over the group ring $\mathbb{Z}[t, t^{-1}]$ of the fundamental group. This has the following structure as $\mathbb{Q}[t, t^{-1}]$ module:

$$\pi_{n-k-1}(\mathbb{C}^n \setminus V) \otimes \mathbb{Q} = \bigoplus_i \mathbb{Q}[t, t^{-1}]/(\lambda_i) \oplus \mathbb{Q}[t, t^{-1}]^\kappa, \tag{9.13}$$

where λ_i are polynomials in $\mathbb{Q}[t, t^{-1}]$, defined up to units. Whenever $\kappa = 0$, the product $\prod_i \lambda_i$ is called the *order of* $\pi_{n-k-1}(\mathbb{C}^n \setminus V) \otimes \mathbb{Q}$. We denote it by $\Delta(\mathbb{C}^n \setminus V)$ or by $\Delta(L_k \setminus V \cap L_k)$. In case $n - k = 2$, this is nothing else than the *Alexander polynomial* of the curve $L_k \cap V$ in L_k. In case $\kappa \ne 0$, we say that order is zero.

Lemma 9.4.6 [LiTi] Let $f : \mathbb{C}^n \to \mathbb{C}$ be a polynomial with $\mathrm{Sing}_{\mathcal{G}} p \le k$ and such that $V = f^{-1}(0)$ is the only atypical fibre.* Then

$$\pi_{n-k-1}(L_k \setminus V \cap L_k) \simeq H_{n-k-1}(F, \mathbb{Z})$$

as $\mathbb{Z}[t, t^{-1}]$-modules. In particular:

$$\Delta(L_k \setminus V \cap L_k) = \det[h_0 - t \cdot \mathrm{id} : H_{n-k-1}(F, \mathbb{Q}) \to H_{n-k-1}(F, \mathbb{Q})],$$

where h_0 is the monodromy of the general fibre F of f around the value 0.

---

* see Proposition 2.3.3 for a partial classification of polynomials in 2 variables with one atypical value.

Proof The infinite cyclic cover $\widetilde{\mathbb{C}^n \setminus V}$ is homotopy equivalent to the general fibre F of f, since Atyp$f \subset \{0\}$. Together with (9.12), this proves the first claim. We next have the following identifications:

$$
\begin{aligned}
\Delta(L_k \setminus L_k \cap V) &= \text{order of } \pi_{n-k-1}(L_k \setminus L_k \cap V) \otimes \mathbb{Q} \\
&= \text{order of } \pi_{n-k-1}(\mathbb{C}^n \setminus V) \otimes \mathbb{Q} \\
&= \text{characteristic polynomial of the deck transform on } H_{n-k-1}(\widetilde{\mathbb{C}^n \setminus V}) \\
&= \text{characteristic polynomial of the monodromy on } H_{n-k-1}(F, \mathbb{Q}).
\end{aligned}
$$

$\square$

This lemma can be used to obtain results for the homology of Milnor fibres of polynomials with one-dimensional $\mathcal{G}$-singularities, as follows:

Corollary 9.4.7 [LiTi] Under the assumptions of Lemma 9.4.6, let $k = 1$ and $n > 3$. Then the multiplicity of the factor $t - 1$ in the order $\Delta(L_k \setminus L_k \cap V)$ is equal to the dimension of $H_{n-2}(\mathbb{C}^n \setminus V, \mathbb{Q})$. In particular rank $H_{n-2}(F, \mathbb{Q}) \geq$ rank $H_{n-2}(\mathbb{C}^n \setminus V, \mathbb{Q})$ and the equality holds if and only if $\Delta(\mathbb{C}^n \setminus V)$ has 1 as unique root and $k_i = 1$ for any i. If $n = 3$ and V is irreducible, then rank $H_{n-2}(F, \mathbb{Q}) = 0$ if and only if $\Delta(\mathbb{C}^n \setminus V) = 1$.

Proof Let us consider the decomposition:

$$
\begin{aligned}
H_{n-2}(F, \mathbb{Q}) &= \\
H_{n-2}(\widetilde{\mathbb{C}^n \setminus V}, \mathbb{Q}) &= \bigoplus_{i=1}^{l'} \mathbb{Q}[t, t^{-1}]/(t-1)^{k_i} \bigoplus \oplus_{j=1}^{l''} \mathbb{Q}[t, t^{-1}]/(\lambda_j),
\end{aligned}
$$

where λ_j does not have 1 as root. We have the Leray spectral sequence $H_p(\mathbb{Z}, H_q(\widetilde{\mathbb{C}^n \setminus V}), \mathbb{Q}) \Rightarrow H_{p+q}(\mathbb{C}^n \setminus V, \mathbb{Q})$ and $H_{n-3}(\mathbb{C}^n \setminus V, \mathbb{Q}) = 0$. We derive that $H_{n-2}(\widetilde{\mathbb{C}^n \setminus V}, \mathbb{Q})^{\text{inv}} = H_{n-2}(\mathbb{C}^n \setminus V, \mathbb{Q})$, where the invariant part means taking the kernel of the multiplication by $t - 1$ on $H_{n-2}(\widetilde{\mathbb{C}^n \setminus V}, \mathbb{Q})$. The result follows for $n > 3$. If $n = 3$, then the same spectral sequence shows that the number of cyclic summands is equal to rank $H_1(\mathbb{C}^3 \setminus V, \mathbb{Z}) - 1$ and the rank of the latter homology group can be identified with the number of irreducible components of V.

$\square$

Example 9.4.8 (Libgober) Let P_a and P_b be homogeneous polynomials in 3 variables of degrees b, resp. a. Then let us consider the polynomial:

$$
f = P_a(x, y, z)^b + P_b(x, y, z)^a. \tag{9.14}
$$

Since f is homogeneous, Atyp$f = \{0\}$. If P_a and P_b are generic enough, then the singular locus of f is the union of lines corresponding to the points

$\{P_a = P_b = 0\} \subset \mathbb{P}^2$, and of course $\text{Sing}_{\mathcal{S}}p_{|\mathbb{X}}$ is just the closure of $\text{Sing} f$ in $\mathbb{X}$. The Milnor fibre of f at 0, denoted by M_f, identifies to the cover of degree ab of the complement of the projective curve $f = 0 \subset \mathbb{P}^2$. By [Li1], the computation of the Alexander polynomial of this curve yields $H_1(M_f, \mathbb{Z}) = \mathbb{Z}^{(a-1)(b-1)}$. Furthermore, by 9.4.6, the characteristic polynomial of this singularity is the Alexander polynomial of the affine curve with isolated singularities $P_a(x, y, 1)^b + P_b(x, y, 1)^a = 0$, since the slice $\{z = 1\}$ is generic. This is $((t^{a+b} - 1)(t - 1)/(t^a - 1)(t^b - 1))$. We may observe that the monodromy on the first nonvanishing homology group H_1 is not trivial.

We consider now the embedding $\mathbb{C}^n \subset \mathbb{P}^n$ and the proper extension $p : \mathbb{X} \to \mathbb{C}$. We still assume that $V = f^{-1}(0)$ is the single atypical fibre of f. We have $\text{Sing} V \subset \text{Sing} f \subset \text{Sing}_{\mathcal{S}} p$, where $\mathcal{S}$ is the Whitney stratification considered at § 2. Let H be a generic hyperplane in $\mathbb{C}^n$. Let $\mathcal{S}'$ denote the Whitney stratification induced by $\mathcal{S}$ on $(\bar{H} \times \mathbb{C}) \cap \mathbb{X}$. The extension p' to $(\bar{H} \times \mathbb{C}) \cap \mathbb{X}$ of the restriction $f_{|H}$ and its singular locus $\text{Sing}_{\mathcal{S}'} p'$ are well defined. There exist generic enough hyperplanes H such that $\bar{V} \cap \text{Sing}_{\mathcal{S}'} p' = \bar{H} \cap \bar{V} \cap \text{Sing}_{\mathcal{S}} p$.

Suppose $\dim \text{Sing} V = 1$. For each irreducible component Σ_i of $\text{Sing} V$, we denote by Δ_i the characteristic polynomial of the monodromy of the transversal local hypersurface singularity $V \cap H$. This is known as the *horizontal monodromy* corresponding to Σ_i, ever since it was named so by Steenbrink [Stb].

Let us assume that $\dim(\mathbb{X}^\infty \cap \bar{V} \cap \text{Sing}_{\mathcal{S}} p) = 1$ and denote by Σ_j^∞ some one-dimensional irreducible component of $\mathbb{X}^\infty \cap \bar{V} \cap \text{Sing}_{\mathcal{S}} p$. Then let Δ_j denote the characteristic polynomial of the monodromy of the isolated singularity at infinity of the polynomial $f_{|H}$ corresponding to some singularity out of the discrete set $\bar{H} \cap \Sigma_j^\infty$.

Let us take a large enough sphere $S_R^{2n-1} \subset \mathbb{C}^n$ and, after [Li2, 4.5–4.7], denote by Δ^∞ the order of $\pi_{n-2}(S^\infty \cap H \setminus (S^\infty \cap V \cap H))$. With these notations, we can prove the following result.[*]

Corollary 9.4.9 Let $f : \mathbb{C}^n \to \mathbb{C}$ be a polynomial with $\text{Atyp} f = \{0\}$ and $\dim \text{Sing}_{\mathcal{S}} p = 1$. Let $V = f^{-1}(0)$ and let F denote a generic fibre of f. If none of the roots of Δ_is and of Δ_js distinct from 1 is a root of Δ^∞ and $k_i = 1$ for the cyclic summands in $H_{n-1}(F, \mathbb{Q})$ corresponding to 1 then $H_{n-1}(F, \mathbb{Q}) = H_{n-1}(\mathbb{C}^{n+1} \setminus V, \mathbb{Q})$.

---

[*] slightly extending [LiTi, Corollary 4.7].

Proof Note that $f_{|H}$ does not have a single atypical value, in general, even if f has only one. Due to the genericity of H, the polynomial $f_{|H}$ has isolated $\mathcal{S}$-singularities at infinity. The divisibility theorem [Li2, 4.3] proved by Libgober can be extended to the case of isolated $\mathcal{S}$-singularities at infinity (and even to $\mathcal{G}$-singularities at infinity). The conclusion from [Li2, 4.3] holds, namely that $\Delta(\mathbb{C}^{n+1} \setminus V) = \Delta(H \setminus V \cap H)$ divides the product of $\prod_i \Delta_i \cdot (1 - t)^\kappa$, for some nonnegative integer κ.

To finish the proof, we have to apply (a slight extension of) [Li2, Theorem 4.5] to the restriction $f_{|H}$, at its isolated $\mathcal{S}$-singularities at infinity. It yields that $\Delta(H \setminus V \cap H)$ divides Δ^∞ of $f_{|H}$. Now our claim follows from Corollary 9.4.7.

$\square$

Example 9.4.10 [LiTi] Let $f : \mathbb{C}^3 \to \mathbb{C}$ be homogeneous or obtained via an automorphism applied to a homogeneous polynomial, and having transversal A_1- or A_2-singularities along each of the strata of its singular locus. If the degree of f is not divisible by 6, then the rank of $H_1(F, \mathbb{C})$ is equal to the number of irreducible components of the atypical fibre minus one. Indeed, the roots of the characteristic polynomial of the monodromy for an A_2 singularity are roots of unity of degree 6. Hence, if d is not divisible by 6, then the only root of the characteristic polynomial is 1. Therefore, by Corollary 9.4.7, the rank of $H_1(F, \mathbb{Z})$ is the equal to the rank of $H_1(\mathbb{C}^3 \setminus f^{-1}(0))$ minus one and the rank of the last homology group is equal to the number of irreducible components of $f^{-1}(0)$.

Exercises

9.1 Let $f : \mathbb{C}^n \to \mathbb{C}$ be a polynomial. If dim Sing$f \le s$ and if dim $\Sigma(w) \le s$, then dim Sing$_{\mathcal{S}}p \cap \mathbb{Y}' \le s$.

9.2 If dim $\Sigma(w) \le s$, then dim Sing$_{\mathcal{S}}p \cap \mathbb{Y}' \le s+1$. In particular, dim Sing$f \le s + 1$.

9.3 Show that in §9.2 the $\mathcal{G}$-singularities can be replaced by t-singularities, which is a more general type of singularities.

9.4 Let $V = \{Q = 0\}$ and let us assume that $X := Y \setminus V$ is homologically contractible. Using Proposition 9.2.3(a), show the direct sum decomposition:

$$\tilde{H}_*(X_s) = \oplus_{a_i \in \Lambda} H_{*+1}(X_{D_i}, X_{s_i}).$$

Deduce from this the particular case of a polynomial function $f : \mathbb{C}^n \to \mathbb{C}$, for which $X = \mathbb{C}^n$.

9.5 Show that the jump $A_0 \to A_1$, which occurs in Example 9.3.4, cannot occur in case of polynomial functions at infinity.

9.6 Let $f : \mathbb{C}^n \to \mathbb{C}$ be a polynomial function and let T_∞ be the monodromy around a large circle, which surrounds Atyp$f$. Show that T_∞ is trivial if and only if the *monodromy group* of f is trivial. (The monodromy group of f is spanned by the algebraic monodromies around the atypical values of f.)

9.7 Let $f : \mathbb{C}^n \to \mathbb{C}$ be a polynomial with unique atypical fibre $V = f^{-1}(0)$, such that dim Sing$f = 1$ and such that dim $\mathbb{X}^\infty \cap \mathrm{Sing}_{\mathcal{S}}p = 0$.

If none of the roots of Δ_is and Δ_js distinct from 1 is a root of unity of degree $d = \deg f$, and $k_i = 1$ for the cyclic summands corresponding to the root 1, then show that $H_{n-2}(F, \mathbb{Q}) = H_{n-2}(\mathbb{C}^n \setminus V, \mathbb{Q})$.

10

Slicing by pencils of hypersurfaces

Our aim is to extend the method of slicing by pencils to certain classes of nongeneric pencils. Nongeneric pencils may occur naturally in certain situations, for instance a distinguished class of nongeneric pencils is the class of polynomial functions $f : \mathbb{C}^n \to \mathbb{C}$.*

The setting of this chapter is general: a complex analytic space $X = Y \setminus V$ with arbitrary singularities, where Y is some compact complex space and V is a closed analytic subspace, and a meromorphic function on Y. Considering pencils of hypersurfaces instead of pencils of hyperplanes, although not more general in itself, has the advantage to enfold the local theory of hypersurface singularities initiated by Milnor [Mi2], see Figure 9.1.

We have introduced in §9.1 the notion of singularity of a meromorphic function and we have seen that such singularities may occur in the axis. This means that the axis A of the pencil might not be anymore in general position in Y, as it is supposed to be in the classical Lefschetz theory. Pencils which allow *isolated singularities* in the axis turn out to be a natural class of "admissible" pencils, extending the class of generic Lefschetz pencils. Indeed, isolated singularities of functions on singular spaces are manageable enough objects, as we have already seen.

In the literature, examples of nongeneric pencils occurred sporadically; more recently they got into light, e.g. [KPS, Oka3], precisely because we can use nongeneric pencils towards more efficient computations. For instance, M. Oka uses special pencils tangent to flex points of projective curves in order to compute the fundamental group of the complement, see e.g. [Oka2, Oka3]. Nongenericity also proved to be a useful concept for treating the topology of polynomial functions §10.3 and of complements of the hypersurfaces §9.3, §9.4.

---

* As we have seen in the preceding chapter, the singularities at infinity $\mathrm{Sing}^{\infty} f$ occur as singularities in the axis of the meromorphic function defined by f.

The central results in this chapter are the Lefschetz hyperplane theorems, which we prove under weak assumptions and for nongeneric pencils. The presentation is based on [Ti8, Ti9, Ti12, LiTi].

10.1 Relative connectivity of pencils

Our base space $X := Y \setminus V$, where Y is a compact complex analytic space and V is a closed complex subspace, can be for instance any quasi-projective variety. Let also X be irreducible of dimension $n \geq 2$.

We recall that by *pencil* we mean a meromorphic function, i.e. the ratio of two sections f and g of a holomorphic line bundle $L \to Y$. This defines a holomorphic function $h := f/g$ on the complement $Y \setminus A$ of the indeterminacy locus $A := \{f = g = 0\}$, which we call *axis of the pencil*, cf. §9.1. A pencil is called *generic* with respect to X when its axis A is general (i.e. stratified transversal to some Whitney stratification of the pair (Y, V)) and when the holomorphic map $h = f/g : Y \setminus A \to \mathbb{P}^1$ has only stratified Morse singularities. These singularities are finitely many, by the compactness of Y, but part of those might be outside X. Moreover, instead of only double points, we may consider pencils with any kind of stratified singularities, provided they are *isolated* (see [HmL1], [GM2]).

Pencils with singularities in the axis. By *nongeneric pencil* we mean a pencil having singularities in the axis A. The meaning of "singularities" that we consider here is *singularities with respect to a Whitney stratification*. Before defining this more precisely, let us examine an example of a pencil on a projective hypersurface where the axis is not in general position.

Example 10.1.1 Let $\hat{f} = x^2 y + xz^2$ and $\hat{g} = z^3$, in coordinates x, y, z, w, define a pencil $\hat{f}/\hat{g}$ on $\mathbb{P}^3$. We restrict it to the nonsingular surface $Y \subset \mathbb{P}^3$ defined by $yw + x^2 - z^2 = 0$. The axis $\hat{A} := \{\hat{f} = \hat{g} = 0\} \subset \mathbb{P}^3$ consists of two lines $\hat{A}_1 \cup \hat{A}_2$. One of them, namely $\hat{A}_1 := \{x = z = 0\}$, is a double line and therefore it is natural to consider that every member of the pencil $\hat{f}/\hat{g}$ is singular along this line. Now the axis $A := \hat{A} \cap Y$ of the pencil on Y consists of two points and both are on $\hat{A}_1$. The pencil on Y is a nongeneric pencil, since having singularities in the axis.

We work with the Nash blow-up of Y along A, cf. Definition 9.1.1:

$$\mathbb{Y} := \text{closure}\{(y, [s : t]) \in Y \times \mathbb{P}^1 \mid sf(y) - tg(y) = 0\}.$$

Let us set $\mathbb{X} := \mathbb{Y} \cap (X \times \mathbb{P}^1)$, the projection $p : \mathbb{Y} \to \mathbb{P}^1$, its restriction $p_{|\mathbb{X}} : \mathbb{X} \to \mathbb{P}^1$ and the projection to the first factor $\sigma : \mathbb{Y} \to Y$. The restriction of p to $\mathbb{Y} \setminus (A \times \mathbb{P}^1)$ can be identified with $h = f/g$.

As in §9.1, we fix a Whitney stratification $\mathcal{W}$ on Y such that V is a union of strata. The restriction of $\mathcal{W}$ to the open set $Y \setminus A$ induces a Whitney stratification on $\mathbb{Y} \setminus (A \times \mathbb{P}^1)$, via the identification $Y \setminus A \simeq \mathbb{Y} \setminus (A \times \mathbb{P}^1)$. We then denote by $\mathcal{S}$ the *coarsest Whitney stratification* on $\mathbb{Y}$ that coincides over $\mathbb{Y} \setminus (A \times \mathbb{P}^1)$ with the one induced by $\mathcal{W}$ on $Y \setminus A$. This stratification exists within a neighbourhood of $A \times \mathbb{P}^1$, by usual arguments (see e.g. [GWPL]), hence such stratification is well defined on $\mathbb{Y}$. We call it the *canonical stratification* of $\mathbb{Y}$ generated by the stratification $\mathcal{W}$ of Y. The canonical stratification of $\mathbb{X}$ will be the restriction of $\mathcal{S}$ to $\mathbb{X}$.

According to Definition 2.2.1, the *singular locus* of p with respect to $\mathcal{S}$ is the following closed analytic subset of $\mathbb{Y}$:

$$\operatorname{Sing}_{\mathcal{S}} p := \bigcup_{\mathcal{S}_\beta \in \mathcal{S}} \operatorname{Sing} p_{|\mathcal{S}_\beta}.$$

The *critical values* of p with respect to $\mathcal{S}$ are the points in the image $p(\operatorname{Sing}_{\mathcal{S}} p)$. By Proposition 9.1.2, both $p : \mathbb{Y} \to \mathbb{P}^1$ and $p_{|\mathbb{X}} : \mathbb{X} \to \mathbb{P}^1$ are *stratified locally trivial fibrations* over the complement $\mathbb{P}^1 \setminus \Lambda$ of the finite set $\Lambda := p(\operatorname{Sing}_{\mathcal{S}} p)$. The problem we have to deal with is what happens in the pencil when we encounter such a critical value.

Definition 10.1.2 We say that the pencil defined by the meromorphic function $h = f/g$ is a *(nongeneric) pencil with isolated singularities* if $\dim \operatorname{Sing}_{\mathcal{S}} p \le 0$. We shall say that X has the structure of a *Lefschetz fibration with isolated singularities* if there exists a (possibly nongeneric) pencil on X with isolated singularities.

The case of projective varieties. In case where Y is projective, the condition $\dim \operatorname{Sing}_{\mathcal{S}} p \le 0$ in the above definition is equivalent to the condition: *the singularities of the function p at the blown-up axis $A \times \mathbb{P}^1$ are at most isolated.* Indeed, the singularities of p outside the axis are also isolated, by the following reason. Suppose that p has nonisolated singularities and take some component $\mathcal{C}$ of the singular locus $\operatorname{Sing}_{\mathcal{S}} p$, with $\dim \mathcal{C} \ge 1$. Since $\mathcal{C}$ is necessarily contained in a single fibre of p, call it $p^{-1}(b)$, it has to intersect the axis $A \times \{b\}$, since it is a hypersurface in $p^{-1}(b)$. Then, at all points of $\mathcal{C} \cap (A \times \{b\})$, we have $\dim \operatorname{Sing}_{\mathcal{S}} p \ge 1$. This gives a contradiction.

The assumption $\dim \operatorname{Sing}_{\mathcal{S}} p \le 0$ is satisfied for instance in the following particular but significant case.

Proposition 10.1.3 Let $Y \subset \mathbb{P}^N$ be a projective variety endowed with some Whitney stratification $\mathcal{W}$ and let $\hat{h} = \hat{f}/\hat{g}$ define a pencil of hypersurfaces in $\mathbb{P}^N$ with axis $\hat{A}$. Let B denote the set of points on $\hat{A} \cap Y$ where some hypersurface of the pencil is singular or where $\hat{A}$ is not transversal to $\mathcal{W}$. If dim $B \leq 0$ and the singular locus of $h : Y \setminus A \to \mathbb{P}^1$ with respect to $\mathcal{W}$ consists of at most isolated points, then dim Sing $_{S}p \leq 0$ and Sing $_{S}p \subset B \times \mathbb{P}^1$.

Proof On $\mathbb{Y} \setminus (A \times \mathbb{P}^1)$, p is just h and its singularities are isolated by hypothesis. The notation A stays for $\hat{A} \cap Y$, as usual.

Next, let us remark that $\mathbb{Y} = \mathbb{H} \cap (Y \times \mathbb{P}^1)$, where $\mathbb{H} = \{x \in \mathbb{P}^N, [s : t] \in \mathbb{P}^1 \mid s\hat{f}(x) - t\hat{g}(x) = 0\}$. Since $\hat{f}$ and $\hat{g}$ are homogeneous of the same degree, the singularities of $\mathbb{H}$ are contained into the set $\Sigma_{\hat{A}} := (\hat{A} \times \mathbb{P}^1) \cap \{s\partial\hat{f} - t\partial\hat{g} = 0\}$, which is at most a collection of lines, by hypothesis. We endow $\mathbb{H}$ with the coarsest Whitney stratification. It follows that $\mathbb{H}$ is transversal to the strata $\mathcal{W}_i \times \mathbb{P}^1$ of $Y \times \mathbb{P}^1$, except eventually along $B \times \mathbb{P}^1$. By using that a transversal intersection of Whitney stratified sets is Whitney (see A1.1), it follows that the canonical stratification $\mathcal{S}$ of $\mathbb{Y}$ restricted to $(A \setminus B) \times \mathbb{P}^1$ contains only strata which are products by $\mathbb{P}^1$. Hence the projection p is transversal to these strata. Finally, we might have nontransversality only along the projective lines $B \times \mathbb{P}^1$. The stratification $\mathcal{S}$ may distinguish at most a finite number of point strata on $B \times \mathbb{P}^1$. Then p is still transversal to the complement of these points in $B \times \mathbb{P}^1$. $\square$

Lefschetz Hyperplane Theorem via nongeneric pencils. The classical Lefschetz Hyperplane Theorem (abbreviated LHT) asserts that, if $Y \subset \mathbb{P}^N$ is a projective variety and $H \subset \mathbb{P}^N$ a hyperplane such that $Y \setminus H$ is nonsingular of dimension $\geq n$ (more generally: if rhd $Y \setminus H \geq n$), then the pair $(Y, Y \cap H)$ is $(n-1)$-connected, i.e. $\pi_k(Y, Y \cap H) = 0$ for all $k \leq n - 1$.

Lefschetz' original proof ([Lef], see also [Lam1]) uses a generic pencil of hyperplanes to scan the space. Several generalizations to quasi-projective spaces with singularities, such as by Goresky and MacPherson [GM2], Hamm and Lê [HL1-3], use the Morse theory (a method first employed by Bott [Bo3], Andreotti and Frankel [AF]).[1]

Our strategy is based of course on the use of pencils. So let us first remark that the hyperplane H in the above statement of the LHT, although not assumed generic, may nevertheless be viewed as *a member of a generic pencil*: it is sufficient to choose a generic axis $A \subset H$, in the sense that A intersects transversely all the Whitney strata of Y. This is always possible in the projective space. Then Proposition 10.1.3 tells that the pencil containing H and having A as axis is a pencil without singularities in its axis. It is therefore a generic pencil, which may have only isolated singularities outside the axis (but maybe more complicated than only A_1-singularities). Then one may apply the pencil method and reach the same conclusion.

In our more general setting, this viewpoint leads to the following extended Lefschetz slicing principle. The presentation is based on [Ti9].

Theorem 10.1.4   *(Nongeneric Lefschetz Hypersurface Theorem)* [Ti9]
Let $X = Y \setminus V$, where Y and $V \subset Y$ are compact complex analytic spaces. Let f, g be sections of a holomorphic line bundle over Y, defining a pencil $h = f/g$ with at most isolated singularities in the axis (Definition 10.1.2) and let X_c denote a generic member of the pencil (i.e. $c \notin \Lambda$). Let $\mathrm{rhd}\, X \geq n$, where $n \geq 2$.

If one of the following two conditions is fulfilled:

(a) *$A \not\subset V$ and the pair $(X_c, A \cap X_c)$ is $(n-2)$-connected,*
(b) *$A \subset V$ and $V \supset \{g = 0\}$,*

then the pair (X, X_c) is $(n-1)$-connected.

We shall give several consequences, among which Proposition 10.1.11 states that, if X_c is not generic, then, in the conclusion of Theorem 10.1.4, we may replace X_c by a small "tube" neighbourhood X_D of X_c.

We shall also derive connectivity estimations for the fibres of polynomial functions and for complements of affine hypersurfaces.

We deduce for instance the following attaching result, by applying Switzer's [Sw, Proposition 6.13] (see also the arguments in the proof of Theorem 9.3.1):

Corollary 10.1.5   Under the hypotheses of Theorem 10.1.4, the space X is homotopy equivalent to X_c to which one attaches cells of dimension $\geq n$.

If in addition X is a Stein space of dimension n, then the attaching cells are of dimension precisely n. If moreover X is $n - 1$ connected, then the general hyperplane section X_c has the homotopy type of a bouquet of spheres $\vee S^{n-1}$.

$\square$

Proof of Theorem 10.1.4.   The proof is based on homotopy excision and on local Lefschetz type results. We shall see that the situation (b) is simpler to treat. We start by several preliminaries.

Let $\Lambda = \mathrm{Sing}\,_S p = \{b_1, \ldots, b_r\} \subset \mathbb{P}^1$ be the singular values of p. Let $K \subset \mathbb{P}^1$ be a small closed disk with $K \cap \Lambda = \emptyset$ and let $\mathcal{D}$ denote the closure of its complement in $\mathbb{P}^1$. We denote by $S := K \cap \mathcal{D}$ the common boundary, which is a circle, and choose a point $c \in S$. Then take simple, nonintersecting paths $\gamma_i \subset \mathcal{D} \setminus \cup_i D_i$ from c to $c_i \in \partial \bar{D}_i$. The configuration $\cup_i (\bar{D}_i \cup \gamma_i)$ is a deformation retract of $\mathcal{D}$. We shall also identify all fibres X_{c_i} to the fibre X_c, by parallel transport along the paths γ_i. We have denoted $A' := A \cap X = A \cap X_c$.

We have the homotopy equivalence $(X, X_c) \overset{ht}{\simeq} (X, X_K)$ and we tend to apply *homotopy excision* (Blakers–Massey theorem [BM], in the version of [BGr, Corollary 16.27]) to the pair (X, X_K), where $X = X_K \cup X_{\mathcal{D}}$.

Proposition 10.1.6 (The case $A \not\subset V$.) We assume that $A' \neq \emptyset$.

(a) If (X_c, A') is m-connected, $m \geq 0$, then:
   (i) (X_S, X_c) is at least $m + 1$ connected.
   (ii) The excision morphism $\pi_j(X_{\mathcal{D}}, X_S) \to \pi_j(X, X_K)$ is an isomorphism for $j \leq m + 1$ and an epimorphism for $j = m + 2$.
(b) If, for any $i \in \{1, \ldots, r\}$, the pair $(\mathbb{X}_{D_i}, \mathbb{X}_{c_i})$ is s-connected, then $(X_{\mathcal{D}}, X_c)$ is s-connected too.

Proof (a)(i). Note first that X_S is homotopy equivalent to the subset $\mathbb{X}_S \cup A' \times K$ of $\mathbb{X}_K$. Let I and J be two arcs of angles less than 2π but more than π which cover the circle S. We have the homotopy equivalence

$$(X_S, X_c) \overset{ht}{\simeq} (\mathbb{X}_I \cup (A' \times K) \cup \mathbb{X}_J \cup (A' \times K)), \mathbb{X}_J \cup (A' \times K)).$$

According to the homotopy excision principle (Blakers–Massey theorem), if we assume that the pairs $(\mathbb{X}_I \cup (A' \times K), \mathbb{X}_{\partial I} \cup (A' \times K))$ and $(\mathbb{X}_J \cup (A' \times K), \mathbb{X}_{\partial J} \cup (A' \times K))$ are $(m + 1)$-connected, then the following morphism:

$$\pi_j(\mathbb{X}_I \cup (A' \times K), \mathbb{X}_{\partial I} \cup (A' \times K)) \to \pi_j(\mathbb{X}_I \cup (A' \times K) \cup \mathbb{X}_J \cup (A' \times K),$$
$$\mathbb{X}_J \cup (A' \times K))$$

is an isomorphism for $j \leq 2m + 1$. This is enough to conclude that (X_S, X_c) is $(m + 1)$-connected.

We still have to prove our assumption in the homotopy excision. This goes as follows, using the local triviality of the map $p_{|X}$ over $\mathbb{P}^1 \setminus \Lambda$. The pair $(\mathbb{X}_I \cup (A' \times K), \mathbb{X}_{\partial I} \cup (A' \times K))$ is homotopy equivalent to $(\mathbb{X}_c \times I, \mathbb{X}_c \times \partial I \cup A' \times I)$. The latter is precisely the product of pairs $(\mathbb{X}_c, A') \times (I, \partial I)$. Since $(\mathbb{X}_c, A')$ is m-connected by hypothesis, it follows that the product is $(m + 1)$-connected. This follows from the fact that the spaces are CW-complexes and by using Switzer's [Sw, Proposition 6.13].

(ii). We apply homotopy excision to the pair (X, X_K), where $X = X_K \cup X_{\mathcal{D}}$. The pair $(X_{\mathcal{D}}, X_S)$ is at least zero-connected and we need the connectivity level of (X_K, X_S). By considering the triple (X_K, X_S, X_c), where $X_c \hookrightarrow X_K$ is a homotopy equivalence, we get, for any $i \geq 0$, the isomorphism:

$$\pi_{i+1}(X_K, X_S) \simeq \pi_i(X_S, X_c). \tag{10.1}$$

From this we draw that (X_K, X_S) is $(m+2)$-connected since the pair (X_S, X_c) is $(m+2)$-connected by (i).

Now the homotopy excision applies to $(X_\mathcal{D}, X_S) \hookrightarrow (X, X_K)$ and yields the desired conclusion.

(b). By Switzer's result [Sw, 6.13], the s-connectivity of the CW-relative complex $(\mathbb{X}_{D_i}, \mathbb{X}_{c_i})$ implies that, up to homotopy equivalence, $\mathbb{X}_{D_i}$ is obtained from $\mathbb{X}_{c_i}$ by attaching cells of dimension $\geq s+1$. Since $\mathbb{X}_\mathcal{D} \overset{\text{ht}}{\simeq} \cup_i \mathbb{X}_{D_i \cup Y_i}$ (via a deformation retract) and since $\mathbb{X}_{c_i} \overset{\text{ht}}{\simeq} \mathbb{X}_c$, it follows that $\mathbb{X}_\mathcal{D}$ is obtained from $\mathbb{X}_c$ by attaching cells of dimension $\geq s+1$. This shows that the pair $(\mathbb{X}_\mathcal{D}, \mathbb{X}_c)$ is s-connected, and our proof abuts since, by Lemma 9.2.1, the inclusion of pairs $(X_\mathcal{D}, X_c) \hookrightarrow (\mathbb{X}_\mathcal{D}, \mathbb{X}_c)$ is a homotopy equivalence. $\square$

Proposition 10.1.7 (The case $V \supset \{g = 0\}$.) We assume that $V \supset \{g = 0\}$. If the pair $(\mathbb{X}_{D_i}, \mathbb{X}_{c_i})$ is s-connected, for any $i \in \{1, \ldots, r\}$, then (X, X_c) is s-connected.

Proof The meromorphic function $h_{|X} = f/g$, which defines our pencil, is in this case a well-defined holomorphic function $X \to \mathbb{C}$. Since $\mathbb{C}$ retracts to $\mathcal{D}$, the spaces K and S do not occur and we simply have $(X, X_c) \overset{\text{ht}}{\simeq} (X_\mathcal{D}, X_c)$. Since $A' = \emptyset$, we get $\mathbb{X}_{D_i} = X_{D_i}$ and the claim follows by the proof of Proposition 10.1.6(b). $\square$

We have shown up to now that the connectivity of (X, X_c) depends on the connectivity of each $(\mathbb{X}_{D_i}, \mathbb{X}_{c_i})$. So we further study $(\mathbb{X}_{D_i}, \mathbb{X}_{c_i})$; we fix the index j and write simply $(\mathbb{X}_D, \mathbb{X}_c)$. We have assumed that p has only isolated singularities over the center b of D; then let $\mathbb{Y}_b \cap \operatorname{Sing}_\mathcal{S} p := \{a_1, \ldots, a_l\}$.

Let us consider small enough local Milnor–Lê balls $B_j \subset \mathbb{Y}$ at each isolated singularity a_j of $\mathbb{Y}_b$.

Proposition 10.1.8 If $\operatorname{rhd} X \geq s+1$, then, for any $j \in \{1, \ldots, l\}$, the pair $(B_j \cap \mathbb{X}_D, B_j \cap \mathbb{X}_c)$ is at least s-connected.

Proof We first prove that $\operatorname{rhd} \mathbb{X} \geq s+1$. Since on the space $X \times \mathbb{P}^1$ we have the product stratification $\mathcal{W} \times \mathbb{P}^1$, the condition $\operatorname{rhd}_\mathcal{W} X \geq s+1$ implies $\operatorname{rhd}_{\mathcal{W} \times \mathbb{P}^1} X \times \mathbb{P}^1 \geq s+2$. Since our space $\mathbb{X}$ is a hypersurface in $X \times \mathbb{P}^1$ we have $\operatorname{rhd} \mathbb{X} \geq s+1$, by [HmL1, Theorem 3.2.1].

The rectified homotopical depth of $\mathbb{X}$ gives a certain level of connectivity of the complex links of the strata of the stratification $\mathcal{S}$, according to [GM2] and [HmL1]. We may then relate the connectivity of these complex links to the connectivity of the Milnor–Lê data $(B_j \cap \mathbb{X}_D, B_j \cap \mathbb{X}_c)$. This is more special data, especially when the singularity is not in $\mathbb{X}_D$ but on its 'boundary' $\mathbb{Y} \cap (V \times \mathbb{P}^1)$.

Such a relation among the connectivities is the *local Lefschetz theorem* of Hamm and Lê [HmL1, Theorem 4.2.1 and Cor. 4.2.2]. This local result can be applied to the function $p_| : \mathbb{Y}_D \to D$ with isolated singularities and for the space $\mathbb{X}_D = \mathbb{Y}_D \setminus (V \times \mathbb{P}^1)$. It tells precisely that, since rhd $\mathbb{X} \geq s + 1$, the pair $(B_j \cap \mathbb{X}_D, B_j \cap \mathbb{X}_c)$ is at least s-connected. $\square$

Corollary 10.1.9 If rhd $X \geq s+1$, then the pair $(\mathbb{X}_D, X_c)$ is at least s-connected.

Proof For D small enough the restriction $p_{|\mathbb{X}} : \mathbb{X}_D \setminus \cup_{j=1}^l B_j \to D$ is a trivial stratified fibration, thus the inclusion of pairs $(\mathbb{X}_D, X_c \cup \mathbb{X}_D \setminus \cup_j B_j) \xrightarrow{\simeq} (\mathbb{X}_D, X_c)$ is a homotopy equivalence.

We may apply Switzer's result for CW-complexes [Sw, 6.13] to Proposition 10.1.8 and get that the space $B_j \cap \mathbb{X}_D$ is obtained from $B_j \cap X_c$ by attaching cells of dimensions $\geq s + 1$. It follows that $\mathbb{X}_D$ is obtained from $X_c \cup \mathbb{X}_D \setminus \cup_j B_j$ by attaching $B_j \cap \mathbb{X}_D$ over $B_j \cap X_c$, for each j, and we have seen that this corresponds to attaching cells of dimensions $\geq s + 1$. Our claim is proved. $\square$

End of the proof of Theorem 10.1.4. By Corollary 10.1.9 for $s = n - 1$, the pairs $(\mathbb{X}_{D_i}, \mathbb{X}_{c_i})$ are $(n - 1)$-connected and by applying Proposition 10.1.6(b) we get that $(X_{\mathcal{D}}, X_c)$ is $(n - 1)$-connected too.

Using the long exact sequence of the triple $(X_{\mathcal{D}}, X_S, X_c)$ and that (X_S, X_c) is $(n - 1)$-connected by Proposition 10.1.6(a)(i), it follows that the following morphism, induced by the inclusion:

$$\pi_j(X_{\mathcal{D}}, X_c) \to \pi_j(X_{\mathcal{D}}, X_S)$$

is an isomorphism for $j \leq n - 1$ and an epimorphism for $j = n$. The same conclusion holds for:

$$\pi_j(X_{\mathcal{D}}, X_S) \to \pi_j(X, X_K)$$

and this follows from Proposition 10.1.6(a)(ii). Since X_K retracts in a fibrewise way to X_c, the pair (X, X_K) is homotopy equivalent to (X, X_c). This ends the proof of Theorem 10.1.4(a).

The claim (b) is an immediate consequence of Propositions 10.1.7 and 10.1.8. $\square$

Note 10.1.10 The proof of Theorem 10.1.4(b) only uses the weaker condition that dim Sing $_S p' \leq 0$, where p' is the restriction of p to $\mathbb{Y}_C$.

Further remarks on the relative connectivity. We start with an observation, which goes back to Grothendieck, Hamm, Goresky–MacPherson, Hamm–Lê: one may replace the hypersurface slice X_c in the conclusion of Theorem 10.1.4 by a tubular neighbourhood of some member of the pencil, even if this is a 'very bad' one. Let $D_b \subset \mathbb{P}^1$ denote a small enough disk centered at a such that $D_b \cap p(\mathrm{Sing}\,_{\mathcal{S}}p) \subset \{b\}$.

Proposition 10.1.11 In Theorem 10.1.4, let us replace 'pencil with at most isolated singularities in the axis' by 'pencil with at most isolated singularities except at one fibre X_b, that is dim$(\mathrm{Sing}\,_{\mathcal{S}}p \cap X_a) \leq 0$, for all $a \neq b$'.

Then (X, X_{D_b}) is $(n-1)$ connected.

Proof We revisit the proof of Theorem 10.1.4, with the same notations; in particular X_c still denotes a generic member of the pencil. Our disk D_b will be one of the small disks $D_i \subset \mathcal{D}$, and we take $c \in \partial D_b$. We consider simple, nonintersecting paths $\gamma_i \subset \mathcal{D} \setminus \cup_i D_i$ from c to $c_i \in \partial \bar{D}_i$ for all c_i such that $D_i \neq D_b$. The configuration $\cup_i(D_i \cup \gamma_i)$ is a deformation retract of $\mathcal{D}$.

It follows from Proposition 10.1.6(a)(i) and from (10.1) that (X_K, X_S) is n-connected. By excising $X_{\mathcal{D}}$ from $(X, X_{\mathcal{D}})$, we get that $(X, X_{\mathcal{D}})$ is n-connected. Next, via the homotopy exact sequence of the triple $(X, X_{\mathcal{D}}, X_{D_b})$, this shows that the morphism induced by inclusion $\pi_i(X_{\mathcal{D}}, X_{D_b}) \to \pi_i(X, X_{D_b})$ is an isomorphism for $i \leq n-1$.

We then focus to the pair $(X_{\mathcal{D}}, X_{D_b})$. Since $\mathcal{D}$ and D_b are contractible, Lemma 9.2.1 implies the homotopy equivalence $(X_{\mathcal{D}}, X_{D_b}) \overset{\mathrm{ht}}{\simeq} (\mathbb{X}_{\mathcal{D}}, \mathbb{X}_{D_b})$.

By Proposition 10.1.8 and the observations before it, the pairs $(\mathbb{X}_{D_i}, \mathbb{X}_{c_i})$ are $(n-1)$-connected. A deformation retraction using the local triviality of the map $p_{|\mathbb{X}}$ yields the homotopy equivalence $\mathbb{X}_{\mathcal{D}} \overset{\mathrm{ht}}{\simeq} \mathbb{X}_{D_b} \cup_{D_i \neq D_b} \mathbb{X}_{D_i \cup \gamma_i}$

By the argument in the proof of Proposition 10.1.6(b), $\mathbb{X}_{\mathcal{D}}$ is obtained from $\mathbb{X}_{D_b}$ by attaching cells of dimension $\geq n$. Consequently, $(\mathbb{X}_{\mathcal{D}}, \mathbb{X}_{D_\delta})$ is $(n-1)$-connected, and this ends the proof of our claim.   □

Remark 10.1.12 In the conclusion of the above Proposition 10.1.11, the tube X_{D_b} may be replaced by X_b whenever the inclusion $X_b \subset X_{D_b}$ is a homotopy equivalence. This occurs for instance in the two following situations:

(a) X is compact,
(b) X_b has no singularities in the axis, i.e. $(A \times \{b\}) \cap \mathrm{Sing}\,_{\mathcal{S}}p = \emptyset$.

Applying the nongeneric LHT 10.1.4 by recursion. The conditions assumed in Theorem 10.1.4 are recursive, let us see why.

Firstly, the rectified homotopical depth condition, cf. Definition 9.2.4, passes to slices. Namely, by [HmL1, Theorem 3.2.1], rhd $X \geq n$ implies rhd $X_c \geq n-1$. Secondly, the condition on the connectivity of the pair $(X_c, A \cap X_c)$ was already used by Lamotke [Lam1]) and the idea behind it is to apply induction. Indeed, suppose that X_c is defined by the equation $f' = 0$. Then $A \cap X_c$ may be viewed as an eventually nongeneric fibre of some *second pencil* $h' = f'/g'$, on the space X_c as total space, and having as axis $A_1 := \{f' = g' = 0\}$. Let us denote by $(X_c)_\beta$ a generic member of this pencil. We now assume by the induction hypothesis that the pair $((X_c)_\beta, A_1 \cap (X_c)_\beta)$ is $(n-3)$-connected. If Theorem 10.1.4 can be applied, it yields that $(X_c, (X_c)_\beta)$ is $(n-2)$-connected. According to Proposition 10.1.11 and Remark 10.1.12, if X is compact or if $A \cap X_c$ has no singularities in the axis A_1, then this implies in turn that $(X_c, A \cap X_c)$ is $(n-2)$-connected.

10.2 Second Lefschetz Hyperplane Theorem

Reformulated in homology, the Lefschetz Hyperplane Theorem stated at §10.1 tells that, for a projective manifold Y of dimension n and some hyperplane section $Y \cap H$, the following morphism induced by the inclusion:

$$H_j(Y \cap H, \mathbb{Z}) \to H_j(Y, \mathbb{Z}) \tag{10.2}$$

is bijective for $j < n - 1$ and surjective for $j = n - 1$. The kernel of the surjection in dimension $n - 1$ is described by the *'second Lefschetz hyperplane theorem'*, whenever $Y \cap H$ is a general member of a *generic pencil*, cf. §10.1. Loosely speaking, each Morse critical point of this generic pencil produces a local vanishing cycle and those vanishing cycles generate (not freely, in general) the kernel of the map $H_{n-1}(Y \cap H, \mathbb{Z}) \to H_{n-1}(Y, \mathbb{Z})$.

We shall discuss the problem of finding this kernel in the much larger setting that we have proposed before: *nongeneric pencils*, allowing isolated singularities in the axis, as defined in §10.1. We therefore introduce the *variation maps* at each atypical value a of the pencil:

$$\mathrm{var}_a : H_*(X_c, (X_a)_{\mathrm{reg}}) \to H_*(X_c),$$

and we show that the module of *vanishing cycles* at X_a, i.e. the kernel of the surjection similar to (10.2), is generated by the images of these variation maps. Our variation maps can be viewed as global versions of the local variation maps that one defines in singularity theory, see e.g. [Lam1, Loo], [Ti4, 4.4], [NN2, §2]. It is well known that in case of nonisolated singularities, such local variation maps cannot be defined, and this is the main reason why the use of variation maps in our results would not extend to this context. Nevertheless, in case of one-dimensional singularities, Siersma [Si2] studied several other types of variation maps.[2] Our exposition is based on [Ti8, Ti12].

The setting and some notations. Let Y be a compact complex analytic space and let $V \subset Y$ be a complex analytic subspace such that $X := Y \setminus V$ is of dimension n, $n \geq 2$.

We consider a pencil defined by $h: Y \dashrightarrow \mathbb{P}^1$ with isolated singularities, as defined in 10.1.2. This is the ratio of two sections f and g of a holomorphic line bundle $L \to Y$ and defines a holomorphic function $h := f/g$ over the complement $Y \setminus A$ of the axis of the pencil $A := \{f = g = 0\}$. We have defined at 9.1.1 the total space:

$$\mathbb{Y} := \text{closure}\{(y, [s : t]) \in Y \times \mathbb{P}^1 \mid sf(y) - tg(y) = 0\},$$

which is the hypersurface in $Y \times \mathbb{P}^1$ obtained as a Nash blowing-up of Y along A. We have $\mathbb{X} := \mathbb{Y} \cap (X \times \mathbb{P}^1)$, the projection $p : \mathbb{Y} \to \mathbb{P}^1$ which extends h to a proper holomorphic function, and the projection $\sigma : \mathbb{Y} \to Y$.

We recall that, for any $M \subset \mathbb{P}^1$, we denote $\mathbb{Y}_M := p^{-1}(M)$, $\mathbb{X}_M := \mathbb{X} \cap \mathbb{Y}_M$, $Y_M := \sigma(p^{-1}(M))$ and $X_M := X \cap Y_M$. Let $\Lambda = \{a_1, \dots, a_r\}$ be the set of singular values of p with respect to the Whitney stratification $\mathcal{S}$ defined in §10.1 (i.e. issued from a Whitney stratification $\mathcal{W}$ of $Y \setminus A$, such that $\mathbb{Y} \cap (A \times \mathbb{P}^1)$ is a union of strata).

We denote by $a_{ij} \in \mathbb{Y}$ some point of $\text{Sing}_{\mathcal{S}} p \cap p^{-1}(a_i)$. We then have $\text{Sing}_{\mathcal{S}} p = \cup_{i,j} \{a_{ij}\}$. For $c \in \mathbb{P}^1 \setminus \Lambda$ we say that $\mathbb{Y}_c$, resp. $\mathbb{X}_c$, is a *general fibre* of $p : \mathbb{Y} \to \mathbb{P}^1$, resp. of $p_{|\mathbb{X}} : \mathbb{X} \to \mathbb{P}^1$. We say that Y_c, resp. X_c, is a general member of the pencil on Y, resp. on X.

At some singularity a_{ij}, in local coordinates, we take a small Milnor–Lê ball B_{ij} centered at a_{ij} and a small enough disk $D_i \subset \mathbb{P}^1$ at $a_i \in \mathbb{P}^1$, so that (B_{ij}, D_i) is Milnor data for p at a_{ij}, for indices j and i. We may and shall take the same disk D_i (provided it is small enough) for all the finitely many singularities a_{ij} in the fibre $\mathbb{Y}_{a_i}$.

Geometric monodromy. By Propositions 9.1.2, 9.1.5 and 9.2.5, the restriction $p_| : \mathbb{Y}_{D_i} \setminus \cup_j B_{ij} \to D_i$ is a trivial fibration. Actually we may construct a stratified vector field, which trivializes this fibration such that this vector field is tangent to the strata cut out by the boundaries of the balls $\mathbb{Y}_{D_i} \cap \partial \bar{B}_{ij}$.

Let us choose $c_i \in \partial \bar{D}_i$. We construct in the usual way,[*] by using a stratified vector field, a *geometric monodromy* of the fibration $p_| : \mathbb{Y}_{\partial \bar{D}_i} \to \partial \bar{D}_i$, corresponding to one counterclockwise loop around the circle $\partial \bar{D}_i$, such that this monodromy is the identity on the complement of the balls, $\mathbb{Y}_{\partial \bar{D}_i} \setminus \cup_j B_{ij}$. This restricts to a geometric monodromy on $\mathbb{X}_{\partial \bar{D}_i}$ with the same property, i.e. yields a characteristic morphism $h_i : X_{c_i} \to X_{c_i}$, which is a stratified homeomorphism and it is the identity on $X_{c_i} \setminus \cup_j B_{ij}$.

---

[*] Cf. also §9.4, and one may look up Looijenga's similar discussion in [Loo, 2.C, p. 31].

Clearly h_i is not uniquely defined with these properties, but induces a well-defined representation:

$$\rho_i : \pi_1(\partial \bar{D}_i, c_i) \to \mathrm{Iso}(X_{c_i}, X_{c_i} \setminus \cup_j B_{ij}),$$

where $\mathrm{Iso}(.,.)$ denotes the group of relative isotopy classes of stratified homeomorphisms (which are C^∞ along each stratum), similar to the representation considered in §9.4.

As shown above, the *trivial fibration*:

$$p_| : \mathbb{X}_{D_i} \setminus \cup_j B_{ij} \to D_i \qquad (10.3)$$

allows us to identify the fibre $X_{c_i} \setminus \cup_j B_{ij}$ to the fibre $X_{a_i} \setminus \cup_j B_{ij}$.

Definition 10.2.1 We denote by $X_{a_i}^*$ the fibre $X_{c_i} \setminus \cup_j B_{ij}$ of the trivial fibration (10.3). The pair of spaces $(X_{c_i}, X_{c_i} \setminus \cup_j B_{ij})$ will be denoted by $(X_{c_i}, X_{a_i}^*)$.

By the local conical structure of analytic spaces [BV], the set $B_{ij} \cap X_{a_i} \setminus \cup_j a_{ij}$ retracts to $\partial \bar{B}_{ij} \cap X_{a_i}$. Therefore $X_{a_i} \setminus \cup_j a_{ij}$ is homotopy equivalent, by retraction, to $X_{a_i} \setminus \cup_j B_{ij}$.

Definition of the variation maps. Coming back to our geometric monodromy, this induces an algebraic monodromy on any homology group:

$$\nu_i : H_q(X_{c_i}, X_{a_i}^*; \mathbb{Z}) \to H_q(X_{c_i}, X_{a_i}^*; \mathbb{Z}),$$

such that the restriction $\nu_i : H_q(X_{a_i}^*) \to H_q(X_{a_i}^*)$ is the identity. Consequently, for any relative cycle $\delta \in H_q(X_{c_i}, X_{a_i}^*; \mathbb{Z})$, the image of δ by $(\nu_i - \mathrm{id})$ is an absolute cycle. In this way, we define a *variation map*, for any $q \geq 0$:

$$\mathrm{var}_i : H_q(X_{c_i}, X_{a_i}^*; \mathbb{Z}) \to H_q(X_{c_i}; \mathbb{Z}). \qquad (10.4)$$

We consider only homology with integer coefficients; for the sake of simplicity we shall no longer indicate the ring $\mathbb{Z}$ in the following. By its definition, var_i enters naturally, as a diagonal morphism, in the following diagram:

$$
\begin{array}{ccc}
H_q(X_{c_i}) & \xrightarrow{\ \nu_i - \mathrm{id}\ } & H_q(X_{c_i}) \\[2mm]
{\scriptstyle j_*}\big\downarrow & {\scriptstyle \mathrm{var}_i}\nearrow & \big\downarrow{\scriptstyle j_*} \\[2mm]
H_q(X_{c_i}, X_{a_i}^*) & \xrightarrow{\ \nu_i - \mathrm{id}\ } & H_q(X_{c_i}, X_{a_i}^*),
\end{array}
\qquad (10.5)
$$

where j_* is induced by the inclusion.

One encounters variation maps in the literature ever since Picard and Lefschetz studied the monodromy around an isolated singularity of a 2-variables holomorphic function [PS, Lef]. Zariski used the morphism v_i − id in his well-known theorem for the fundamental group, in two variables. Generalized variation maps and Picard–Lefschetz formulas play a key role in the description of global and local fibrations of holomorphic functions at singular fibres, e.g. [Ph, Mi2, Lam1, Loo, Si2, Ga, NN2, Vas1, Vas2]. Along this tradition, our definition is a direct extension of the local variation maps (see e.g. [Lam1, Loo]) to the global setting.[3]

We shall see in the next chapter §11 how this definition can be adapted to the more delicate setting of homotopy groups.

Second Lefschetz Hyperplane Theorem for nongeneric pencils. Over time, Lefschetz hyperplane theorems have been generalized in several directions, giving rise to an extended literature. Let us only refer to Fulton's general overview [Ful], Lamotke's 'classical' modern presentation of Lefschetz theorems [Lam1] and to Goresky–MacPherson's book [GM2], which covers a lot of material. However, the description of the kernel of the surjection (10.2) for certain singular underlying spaces has been considered in a few papers and for generic pencils only.[4]

We prove and discuss here a general statement in what concerns its assumptions and which extends the Lefschetz principle to nongeneric pencils. We first define the homological depth, a classical ingredient which occurs naturally in the following theorem.

Definition 10.2.2 (Homological depth) Let $\Phi \subset \mathbb{X}$ be a discrete subset. We denote by $\mathrm{Hd}_\Phi \mathbb{X}$ the *homological depth* of $\mathbb{X}$ at Φ. We say that $\mathrm{Hd}_\Phi \mathbb{X} \geq q + 1$ if, at any point $\alpha \in \Phi$, there is an arbitrarily small neighbourhood $\mathcal{N}$ of α such that $H_i(\mathcal{N}, \mathcal{N} \setminus \{\alpha\}) = 0$, for $i \leq q$.

For a manifold M, we have $\mathrm{Hd}_\alpha M \geq \dim_\mathbb{R} M$ at any point $\alpha \in M$. The homological depth measures the defect of being a homology manifold, for the chosen coefficients. For instance, complex V-manifolds are rational homology manifolds (which means that they have a homological depth over $\mathbb{Q}$ as if they were manifolds). In case of stratified complex spaces, we have seen the notion of rectified homological depth rHd in Definition 9.2.4, which is a stronger condition since rHd $\geq k$ implies Hd $\geq k$ by definition.

Theorem 10.2.3 (Second LHT for nongeneric pencils) [Ti12]
Let $h\colon Y \dashrightarrow \mathbb{P}^1$ define a Lefschetz fibration with isolated singularities on $X = Y \setminus V$ (which means $\dim \mathrm{Sing}_{SP} \leq 0$ cf. Definition 10.1.2). Let the axis

A not be included in V. For some $k \geq 0$, suppose that the following conditions are fulfilled:

(C1) $H_q(X_c, X_c \cap A) = 0$ for $q \leq k$.
(C2) $H_q(X_c, X_{a_i}^*) = 0$ for $q \leq k$ and for all i.
(C3) $\mathrm{Hd}_{\mathbb{X} \cap \mathrm{Sing}\,_{\mathcal{S}P}}\mathbb{X} \geq k + 2$.

Then:
(a) $H_q(X, X_c) = 0$ for $q \leq k + 1$.
(b) *If* (C3) *is replaced by the following condition:*

(C3i) $\mathrm{Hd}_{\mathbb{X} \cap \mathrm{Sing}\,_{\mathcal{S}P}}\mathbb{X} \geq k + 3$,

   then the kernel of the surjection $H_{k+1}(X_c) \twoheadrightarrow H_{k+1}(X)$ is generated by the images of the variation maps var_i*, for $i = \overline{1, p}$.*

The complementary case $A \subset V$ will be discussed separately. We also analyse several special situations, such as "no singularities in the axis", i.e. $\mathrm{Sing}\,_{\mathcal{S}P} \cap (A \times \mathbb{P}^1) \cap \mathbb{X} = \emptyset$, when the Lefschetz structure of the space X turns out to be hereditary on slices.

It will be interesting in particular to see what become the conditions (C1), (C2), (C3) in such special cases, and why they are weaker than other conditions used in the literature. For instance, it is well known from [HmL3] that, if X is a *complete intersection*, then $\mathrm{rHd}\,X \geq \dim_{\mathbb{C}} X$, and this implies (see e.g. the proof of Proposition 10.2.12) that condition (C3) is satisfied in this case for $k \leq \dim_{\mathbb{C}} X - 2$.

Proof of Theorem 10.2.3. We use the same notations as in the proof of Theorem 10.1.4. We denote $A' := A \cap X_c$. Since $A \not\subset V$, we have that $A' \neq \emptyset$. Let $K \subset \mathbb{P}^1$ be a closed disk with $K \cap \Lambda = \emptyset$ and let $\mathcal{D}$ denote the closure of its complement in $\mathbb{P}^1$. We denote by $S := K \cap \mathcal{D}$ the common boundary, which is a circle, and take a point $c \in S$. Then take standard paths $\gamma_i \subset \mathcal{D} \setminus \cup_i D_i$ (nonself-intersecting, nonmutually intersecting) from c to $c_i \in \partial \bar{D}_i$. The configuration $\cup_i(\bar{D}_i \cup \gamma_i)$ is a deformation retract of $\mathcal{D}$. We shall also identify all fibres X_{c_i} to the fibre X_c, by the parallel transport along the paths γ_i.

Proposition 10.2.4 *If $H_q(X_c, A') = 0$ for $q \leq k$, then the morphism induced by inclusion:*

$$H_q(X_{\mathcal{D}}, X_c) \xrightarrow{\iota_*} H_q(X, X_c)$$

is an isomorphism for $q \leq k + 1$ and an epimorphism for $q = k + 2$.

Proof The proof uses excision in homology which, unlike the homotopy excision, requires no extra conditions. We are therefore able to get a sharper result, compared with Proposition 10.1.6(a).

We claim that, if $H_q(X_c, A') = 0$, for $q \leq k$, then $H_q(X_S, X_c) = 0$ for $q \leq k + 1$. Note first that X_S is homotopy equivalent to $\mathbb{X}_S \cup (A' \times K)$, by Lemma 9.2.1. Let I and J be two arcs which cover S. We have the homotopy equivalence $(X_S, X_c) \overset{\text{ht}}{\simeq} (\mathbb{X}_I \cup (A' \times K) \bigcup \mathbb{X}_J \cup (A' \times K), \mathbb{X}_J \cup (A' \times K))$. By excision, we get the isomorphism:

$$H_*(X_S, X_c) \simeq H_*(\mathbb{X}_I \cup (A' \times K), \mathbb{X}_{\partial I} \cup (A' \times K)).$$

Next we have the homotopy equivalences of pairs: $(\mathbb{X}_I \cup (A' \times K), \mathbb{X}_{\partial I} \cup (A' \times K)) \overset{\text{ht}}{\simeq} (\mathbb{X}_c \times I, \mathbb{X}_c \times \partial I \cup A' \times I)$ and the latter is just the product of pairs $(\mathbb{X}_c, A') \times (I, \partial I)$. Our claim follows.

By examining the exact sequence of the triple $(X_{\mathcal{D}}, X_S, X_c)$ and by using the vanishing of $H_q(X_S, X_c)$ proved above, we see that $(X_{\mathcal{D}}, X_c) \hookrightarrow (X_{\mathcal{D}}, X_S)$ gives, in homology, an isomorphism in dimensions $q \leq k + 1$ and an epimorphism in $q = k + 2$. Combining this with the excision $H_*(X_{\mathcal{D}}, X_S) \overset{\simeq}{\longrightarrow} H_*(X, X_K)$ will conclude the proof.   □

By Lemma 9.2.1, since $\mathcal{D}$ is contractible, the inclusion of pairs:

$$(X_{\mathcal{D}}, X_c) \hookrightarrow (\mathbb{X}_{\mathcal{D}}, X_c) \tag{10.6}$$

is a homotopy equivalence. We have the excision $H_*(\cup_i \mathbb{X}_{D_i}, \cup_i X_{c_i}) \overset{\simeq}{\rightarrow} H_*(\mathbb{X}_{\mathcal{D}}, X_c)$, which gives a decomposition of the homology $H_*(\mathbb{X}_{\mathcal{D}}, X_c)$ into the direct sum $\oplus_i H_*(\mathbb{X}_{D_i}, X_{c_i})$. The next lemma shows how to replace $\mathbb{X}_{D_i}$ by $\mathbb{X}_{D_i}^* := \mathbb{X}_{D_i} \setminus \text{Sing}_S p$.

Lemma 10.2.5 *If* $\text{Hd}_{\mathbb{X} \cap \text{Sing}_{S} p} \mathbb{X} \geq s + 1$, *then, for all* i, *the map induced by inclusion* $H_q(\mathbb{X}_{D_i}^*, X_{c_i}) \overset{j_*}{\rightarrow} H_q(\mathbb{X}_{D_i}, X_{c_i})$ *is an isomorphism, for* $q \leq s - 1$, *and an epimorphism, for* $q = s$.

Proof Due to the exact sequence of the triple $(\mathbb{X}_{D_i}, \mathbb{X}_{D_i}^*, X_{c_i})$, it is sufficient to prove, for all i, that $H_q(\mathbb{X}_{D_i}, \mathbb{X}_{D_i}^*) = 0$, for $q \leq s$.

Then let $B_{ij} \subset \mathbb{X}$ denote a Milnor ball centered at the singular point $a_{ij} \in \text{Sing}_S p$. Then the inclusion:

$$(\mathbb{X}_{D_i} \cap (\cup_j B_{ij}), \mathbb{X}_{D_i} \cap (\cup_j B_{ij} \setminus \{a_{ij}\})) \hookrightarrow (\mathbb{X}_{D_i}, \mathbb{X}_{D_i}^*)$$

yields an excision in homology, and the unions are disjoint. Finally, the hypothesis $\text{Hd}_{\mathbb{X} \cap \text{Sing}_{S} p} \mathbb{X} \geq s + 1$ tells that the homology of each pair $(\mathbb{X}_{D_i} \cap B_{ij}, \mathbb{X}_{D_i} \cap B_{ij} \setminus \{a_{ij}\})$ annulates up to dimension s. This ends the proof of our claim.   □

Proposition 10.2.6 If $H_q(X_{c_i}, X_{a_i}^*) = 0$, for $q \leq k$, then $H_q(\mathbb{X}_{D_i}^*, X_{c_i}) = 0$ for $q \leq k + 1$.

Proof This is the homology counterpart of Proposition 10.1.6 Let us take Milnor data (B_{ij}, D_i) at the stratified singularities a_{ij}. We shall fix the index i and therefore suppress the lower indices i.

Let $D^* = D \setminus \{a\}$. By retraction, we identify D^* to a circle and cover this circle with the union of two overlapping arcs I, J of angles less than 2π. Then $\mathbb{X}_{D^*} \overset{\mathrm{ht}}{\simeq} \mathbb{X}_I \cup \mathbb{X}_J$ and $X_c \overset{\mathrm{ht}}{\simeq} \mathbb{X}_J \simeq X_c \times J$. With these notations, we have the following isomorphisms induced by homotopy equivalences:

$$
\begin{aligned}
H_*(\mathbb{X}_D^*, X_c) \quad &\simeq \quad H_*(\mathbb{X}_{D^*} \cup (\mathbb{X}_D \setminus \cup_j B_{ij}), X_c \cup (\mathbb{X}_D \setminus \cup_j B_{ij})) \\
&\simeq \quad H_*(\mathbb{X}_I \cup \mathbb{X}_J \cup (\mathbb{X}_D \setminus \cup_j B_{ij}), \mathbb{X}_J \cup (\mathbb{X}_D \setminus \cup_j B_{ij})),
\end{aligned}
$$

and recall that $\mathbb{X}_D \setminus \cup_j B_{ij}$ is the total space of the trivial fibration (10.3) over D with fibre $X_a^* \overset{\mathrm{ht}}{\simeq} X_a \setminus \cup_j B_{ij}$. By excising $\mathbb{X}_J \cup X_a^* \times D$ from the last relative homology we get $H_*(\mathbb{X}_I, X_{\partial I} \cup X_a^* \times I)$, which in turn is isomorphic to $H_*((X_c, X_a^*) \times (I, \partial I)) \simeq H_{*-1}(X_c, X_a^*)$. Since, by hypothesis, the homology of the pair (X_c, X_a^*) annulates up to dimension k, it follows that the homology of the last product annulates up to dimension $k + 1$. $\qquad \square$

At this point, part (a) of Theorem 10.2.3 is proved. Indeed, we apply Proposition 10.2.4, we take $s = k + 1$ in Lemma 10.2.5 to have condition (C3), and we finally apply Proposition 10.2.6.

We now focus on part (b) of Theorem 10.2.3, finding the kernel of the map $H_{k+1}(X_c) \to H_{k+1}(X)$. By the long exact sequence, this is equal to the image of the boundary map $H_{k+2}(X, X_c) \overset{\partial}{\to} H_{k+1}(X_c)$, and we shall find this image. Consider the commutative diagram:

$$
\begin{array}{ccc}
H_{k+2}(X_D, X_c) & \overset{\iota_*}{\longrightarrow} & H_{k+2}(X, X_c) \\
& \searrow{\scriptstyle \partial_1} & \downarrow{\scriptstyle \partial} \\
& & H_{k+1}(X_c),
\end{array}
\qquad (10.7)
$$

where ∂ and ∂_1 are the natural boundary morphisms. Since Proposition 10.2.4 shows that ι_* is an epimorphism under condition (C1), we get:

Corollary 10.2.7 If $H_q(X_c, A') = 0$ for $q \leq k$, then we have $\operatorname{Im} \partial = \operatorname{Im} \partial_1$ in the diagram (10.7).

Using (10.6), we may identify the boundary morphism $H_{k+2}(X_{\mathcal{D}}, X_c) \xrightarrow{\partial_1} H_{k+1}(X_c)$ to the boundary morphism $H_{k+2}(\mathbb{X}_{\mathcal{D}}, X_c) \xrightarrow{\partial_1} H_{k+1}(X_c)$.

Next, the excision $H_*(\cup_i \mathbb{X}_{D_i}, \cup_i X_{c_i}) \xrightarrow{\simeq} H_*(\mathbb{X}_{\mathcal{D}}, X_c)$ allows us to identify the boundary map ∂_1 to the map ∂_2 defined as the sum of the boundary maps $\partial_i : H_{k+2}(\mathbb{X}_{D_i}, X_{c_i}) \to H_{k+1}(X_{c_i})$, where X_{c_i} is identified with X_c by parallel transport along the path γ_i.

With these identifications, we have the following commutative diagram:

$$
\begin{array}{ccc}
\oplus_i H_{k+2}(\mathbb{X}_{D_i}, X_{c_i}) & \simeq & H_{k+2}(\cup_i \mathbb{X}_{D_i}, \cup_i X_{c_i}) \\
& & \\
exc \downarrow \simeq & & \searrow^{\partial_2} \\
& & \\
H_{k+2}(\mathbb{X}_{\mathcal{D}}, X_c) & \xrightarrow{\quad \partial_1 \quad} & H_{k+1}(X_c),
\end{array}
$$

which implies that:

$$\operatorname{Im} \partial_1 = \sum_i \operatorname{Im} \partial_i. \tag{10.8}$$

Our problem is then reduced to finding the image of the boundary map $\partial_i : H_{k+2}(\mathbb{X}_{D_i}, X_{c_i}) \to H_{k+1}(X_{c_i})$, for all i. We next use condition (C3i) to further reduce it, replacing $\mathbb{X}_{D_i}$ by $\mathbb{X}_{D_i}^* := \mathbb{X}_{D_i} \setminus \operatorname{Sing}_{\mathcal{S}} p$.

Corollary 10.2.8 *If* $\operatorname{Hd}_{\mathbb{X} \cap \operatorname{Sing}_{\mathcal{S}} p} \mathbb{X} \geq k + 3$, *then, for all* i:

$$\operatorname{Im}(\partial_i : H_{k+2}(\mathbb{X}_{D_i}, X_{c_i}) \to H_{k+1}(X_{c_i}))$$
$$= \operatorname{Im}(\partial_i' : H_{k+2}(\mathbb{X}_{D_i}^*, X_{c_i}) \to H_{k+1}(X_{c_i})).$$

Proof We have that $\partial_i' = \partial_i \circ j_*$, where $j_* : H_{k+2}(\mathbb{X}_{D_i}^*, X_{c_i}) \to H_{k+2}(\mathbb{X}_{D_i}, X_{c_i})$ is induced by the inclusion. But then Lemma 10.2.5 applied for $s = k+2$ shows that j_* is surjective, and this ends the proof. $\qquad\square$

The variation map enters in the picture now:

Proposition 10.2.9 *If* $H_q(X_{c_i}, X_{a_i}^*) = 0$, *for* $q \leq k$, *then*:

$$\operatorname{Im} \partial_i' = \operatorname{Im}(\operatorname{var}_i : H_{k+1}(X_{c_i}, X_{a_i}^*) \to H_{k+1}(X_{c_i})).$$

Proof We use the second half of the proof of Proposition 10.2.6, which shows the identifications marked by "$\simeq$" in the following diagram:

$$
\begin{array}{ccc}
H_{k+2}(\mathbb{X}_{D_i}^*, X_{c_i}) & \xrightarrow{\quad \partial_i' \quad} & H_{k+1}(X_{c_i}) \\
& & \\
exc \uparrow \simeq & & \uparrow v_i - \operatorname{id} \\
& & \\
H_{k+2}(\mathbb{X}_I, X_{\partial I} \cup X_{a_i}^* \times I) & \xleftarrow{\simeq} & H_{k+1}(X_{c_i}, X_{a_i}^*) \otimes H_1(I, \partial I).
\end{array}
$$

This diagram is the identification of the boundary morphism ∂_i' via the "Wang exact sequence" to the morphism $H_{k+1}(X_{c_i}, X_{a_i}^*) \xrightarrow{v_i - \mathrm{id}} H_{k+1}(X_{c_i})$. The proof of the Wang exact sequence is given by Milnor in [Mi2, p. 67, Lemma 8.4]. Now, since the above diagram is commutative, it shows the equality $\mathrm{Im}\, \partial_i' = \mathrm{Im}\,(v_i - \mathrm{id})$. By the construction of var_i, we have $\mathrm{Im}\,(v_i - \mathrm{id}) = \mathrm{Im}\,\mathrm{var}_i$.   □

We are now able to conclude the proof of Theorem 10.2.3(b). It follows by the sequence of results: Corollary 10.2.7, equality (10.8), Corollary 10.2.8 and Proposition 10.2.9.

Some particular cases. From Theorem 10.2.3 and its proof, we may derive several versions in particular cases. We have to take into account the following facts:

Proposition 10.2.10 Let $\Sigma := X \cap \sigma(\mathrm{Sing}_S p)$ and assume that $A \cap X \neq \emptyset$.

(a) In case $\mathbb{X} \cap \mathrm{Sing}_S p = \emptyset$, the condition (C3) is empty.
(b) In case $(A \times \mathbb{P}^1) \cap \mathbb{X} \cap \mathrm{Sing}_S p = \emptyset$, one may replace condition (C3), respectively (C3i), by the following more general condition, which is a global one:

(C3)' $H_q(X, X \setminus \Sigma) = 0$, *for $q \leq k + 2$,*

respectively

(C3i)' $H_q(X, X \setminus \Sigma) = 0$, *for $q \leq k + 3$.*

(c) In case $(A \times \mathbb{P}^1) \cap \mathrm{Sing}_S p = \emptyset$, if condition (C1) is true, then (C2) is equivalent to the following:

(C2)' $H_q(X_{a_i}^*, X_{a_i}^* \cap A) = 0$, *for $q \leq k - 1$.*

Proof Since (a) is obvious, we focus on (b). By examining the proof of Theorem 10.2.3, we see that we have used the homology depth condition only to compare $\mathbb{X}_{D_i}$ to $\mathbb{X}_{D_i}^*$. We may cut off from the proof this comparison (which means Lemma 10.2.5 and Corollary 10.2.8) and start from the beginning with the space $X \setminus \Sigma$ instead of the space X. Taking into account that, under our hypothesis, $\mathbb{X}_{D_i}^* = \mathbb{X}_{D_i}^* \setminus \Sigma$, for all i, the effect of this change is that the proof yields the conclusion "$H_q(X \setminus \Sigma, X_c) = 0$, for $q \leq k+1$" and the corresponding statement for the vanishing cycles. At this final stage, condition (C3)' allows us to replace $X \setminus \Sigma$ by X.

(c) When there are no singularities in the axis, we have $A \cap X_{a_i}^* = A \cap X_c$, for any i. Then the exact sequence of the triple $(X_c, X_{a_i}^*, A \cap X_{a_i}^*)$ shows that the boundary morphism:

$$H_q(X_c, X_{a_i}^*) \to H_{q-1}(X_c, A \cap X_{a_i}^*)$$

is an isomorphism, for $q \le k$, by condition (C1). This implies our claimed equivalence.   □

Quasi-projective varieties. In case of quasi-projective varieties, we have the abundance of generic hyperplane pencils, in the sense that the axis A is transversal to the stratification. It follows from Proposition 10.1.3 that such a pencil has no singularities along the axis. We are therefore under the conditions of Proposition 10.2.10(b) and (c). Another nice aspect of quasi-projective varieties is that the Lefschetz structure is hereditary on slices: as observed in §10.1, since the axis A is chosen to be transversal hence generic, it becomes in turn a generic slice of a hyperplane slice of X, and so on.[5]

The complementary case $A \subset V$. We continue the discussion of the case $A' = \emptyset$, started in §10.1. One would be tempted to replace the condition (C1) by the condition "$H_q(X_c) = 0$, for $q \le k$", but this appears to be too restricting. Nevertheless, in case the target of $h_{|X}$ is $\mathbb{P}^1 \setminus \{b\}$, for some $b \in \mathbb{P}^1$, the situation becomes more interesting and it appears that (C1) does not play a role anymore.

Theorem 10.2.11 [Ti12] *Let* $X = Y \setminus V$ *have a structure of Lefschetz fibration with isolated singularities over* $\mathbb{C} \simeq \mathbb{P}^1 \setminus \{c\}$. *For some fixed* $k \ge 0$, *assume that* $H_q(X_c, X_{a_i}^*) = 0$ *for* $q \le k$, *where* X_c *is a general member* X_c *and* X_{a_i} *is any atypical one.*

(a) *If* $H_q(X, X \setminus \Sigma) = 0$ *for* $q \le k + 1$, *then* $H_q(X, X_c) = 0$ *for* $q \le k + 1$.
(b) *If* $H_q(X, X \setminus \Sigma) = 0$ *for* $q \le k + 2$, *then* $\ker(H_{k+1}(X_c) \twoheadrightarrow H_{k+1}(X)) = \sum_i^p \operatorname{Im} \operatorname{var}_i$.

Proof We show that the conditions of Theorem 10.2.3 are fulfilled and that its proof works with some modifications which simplify it. Since the target of our holomorphic function $h_{|X}$ is $\mathbb{C}$, we have $X_D \overset{\text{ht}}{\simeq} X$. Then the conclusions of Proposition 10.2.4 and of Corollary 10.2.7 hold with no assumption on k. Therefore (C1) does not enter any more as a condition in our proof. On the other hand, since our condition on the axis implies $(A \times \mathbb{P}^1) \cap \mathbb{X} \cap \operatorname{Sing}_{SP} = \emptyset$, Proposition 10.2.10(b) below shows that we can use the conditions (C3)' and (C3i)' instead of (C3) and (C3i) respectively. Finally condition (C2) is itself an assumption of the above theorem.   □

Comparing to the rHd condition.

Proposition 10.2.12 Theorem 10.2.3 holds if we replace the conditions (C2) and (C3), respectively (C3i), by the single condition:

(C4) $\mathrm{rHd}\, X \geq k + 2$.
respectively:
(C4i) $\mathrm{rHd}\, X \geq k + 3$.

Proof The condition $\mathrm{rHd}\, X \geq q$ implies $\mathrm{rHd}\, \mathbb{X} \geq q$, since $\mathbb{X}$ is a hypersurface in $X \times \mathbb{P}^1$ and we can apply again [HmL3, Theorem 3.2.1]. This in turn implies $\mathrm{Hd}_\alpha \mathbb{X} \geq q$, for any point $\alpha \in \mathbb{X}$, by definition.

Next, $\mathrm{rHd}\, X \geq q$ implies that the homology of the pair $(\mathbb{X}_{D_i}, X_{c_i})$ vanishes up to dimension $q - 1$, by Proposition 10.1.8 and the observations before it (where rhd has to be simply replaced by rHd in all arguments). Up to now we have shown that (C1) + (C4) imply claim (a) of Theorem 10.2.3.

On the other hand, if we assume (C4i), then, besides the vanishing of the homology of $(\mathbb{X}_{D_i}, X_{c_i})$ up to $k + 2$ (shown just above), it follows that $H_q(\mathbb{X}_{D_i}^*, X_{c_i}) = 0$ for $q \leq k + 1$, by Lemma 10.2.5. Now the proof of Proposition 10.2.6 shows in fact that the annulation of the homology of $(\mathbb{X}_{D_i}^*, X_{c_i})$ up to $k + 1$ is equivalent to the annulation of the homology of the pair $(X_{c_i}, X_{a_i}^*)$ up to k, which is just condition (C2). Therefore Theorem 10.2.3(b) applies. $\qquad\square$

10.3 Vanishing cycles of polynomials, revisited

We recall from §9.1 that a polynomial function $f : \mathbb{C}^n \to \mathbb{C}$ may naturally be considered as a nongeneric pencil of hypersurfaces on the space $\mathbb{C}^n$, which is itself a particular quasi-projective variety. If $\deg f = d$, then we have the meromorphic function $h = \tilde{f}/z^d : \mathbb{P}^n \dashrightarrow \mathbb{P}^1$, where $\tilde{f}$ is the homogenized of f with respect to the new variable z and the axis of the pencil is $A = \{f_d = 0\} \subset H^\infty$. Here we have $Y = \mathbb{P}^n$, $V = H^\infty = \{z = 0\} \subset \mathbb{P}^n$, and we are in the situation managed by Theorem 10.2.11 for a pencil on $X = \mathbb{C}^n$ with $h_{|\mathbb{C}^n} = f$ and $\Sigma = \mathrm{Sing}\, f$.

We show here that the following more general setting can still be treated with the same theory: the polynomial function f has isolated singularities, but *no condition* on its singularities in the axis, which may be nonisolated. Moreover, we shall see that we have a more precise grip on the variation maps.

For a big enough radius of some ball $B \subset \mathbb{C}^n$ centered at the origin, we have:

$$X_{a_i} \cap B \overset{\mathrm{ht}}{\simeq} X_{a_i},$$

for any i, since the distance function has a finite set of critical values on the algebraic sets X_{a_i}. With arguments like in §3.1, for small Milnor balls B_{ij} at the critical points of f on X_{a_i} and for a small enough disk D_i, we prove that the map:

$$f_| : f^{-1}(D_i) \cap B \setminus \cup_j B_{ij} \to D_i \qquad (10.9)$$

is a trivial fibration, essentially since the fibres of f over D_i are transversal to ∂B and transversal to the boundaries of the Milnor balls; the claim then follows by Ehresmann's theorem.

This implies, as in §10.2, that there is a well-defined geometric monodromy representation at each $a_i \in \text{Atyp} f \subset \mathbb{C}$, $\rho_i : \pi_1(\partial \bar{D}_i, c_i) \to \text{Iso}(X_{c_i}, X_{c_i} \setminus (\complement B \bigcup \cup_j B_{ij}))$. We have seen in §10.2 that this induces a variation map:

$$\text{var}_i : H_k(X_{c_i}, X_{a_i}^*) \to H_k(X_{c_i}),$$

where in our special setting $X_{a_i}^*$ is a notation for the subset $X_{c_i} \setminus (\complement B \bigcup \cup_j B_{ij})$ of X_{c_i}. This notation is suggested by the fact that $X_{a_i} \setminus \text{Sing} f$ is homotopy equivalent to $X_{a_i} \cap B \setminus \cup_j B_{ij}$, which in turn may be identified to $X_{c_i} \setminus (\complement B \bigcup \cup_j B_{ij})$, since both are fibres of the trivial fibration (10.9).

Corollary 10.3.1 Let $f : \mathbb{C}^n \to \mathbb{C}$ be a polynomial function with isolated singularities. Then:

(a) If $H_q(X_{c_i}, X_{a_i}^*) = 0$ for $q \le k$ and for any i, then $\tilde{H}_q(X_c) = 0$ for $q \le k$.
(b) [NN2, Theorem 2.3][6] The variation map $\text{var}_i : H_\star(X_{c_i}, X_{a_i}^*) \to \tilde{H}_\star(X_{c_i})$ is injective, for any $i \ge 0$. In particular, we have $H_q(X_c) \simeq \sum_i \text{Im var}_i$ for the first integer $q \ge 1$ such that $H_q(X_c) \ne 0$.

Proof We apply Theorem 10.2.11. Of course the homology of $\mathbb{C}^n$ is trivial, but the relative homology $H_{\star+1}(\mathbb{C}^n, X_c)$ is isomorphic to $\tilde{H}_\star(X_c)$ by the long exact sequence of the pair $(\mathbb{C}^n, X_c)$. By Lemma 3.2.2, we have the general decomposition into the direct sum of vanishing cycles at each atypical fibre X_{a_i}, namely $H_{\star+1}(\mathbb{C}^n, X_c) \simeq \oplus_i H_{\star+1}(X_{D_i}, X_{c_i})$, when the system of paths γ_i is fixed.

Since the fibres of f are Stein spaces of dimension $n - 1$, their homology groups are trivial in dimensions $\ge n$. The condition (C3)$'$ is largely satisfied, since $(\mathbb{C}^n, \mathbb{C}^n \setminus \text{Sing} f)$ is $(2n - 1)$-connected. Hence part (a) follows from Theorem 10.2.11(a). For part (b), remark first that, by the above arguments, the boundary map $\partial_i : H_{\star+1}(X_{D_i}, X_{c_i}) \to \tilde{H}_\star(X_{c_i})$ is injective, for any i. Next we replace X_{D_i} by $X_{D_i}^* := X_{D_i} \setminus \text{Sing} f$, since $(X_{D_i}, X_{D_i}^*)$ is $(2n - 1)$-connected, and we deduce that the boundary morphism $\partial_i' : H_{\star+1}(X_{D_i}^*, X_{c_i}) \to \tilde{H}_\star(X_{c_i})$ is injective for all $i \geq 0$. By the diagram used in the proof of Proposition 10.2.9, we identify $H_{*+1}(X_{D_i}^*, X_{c_i})$ to $H_*(X_{c_i}, X_{a_i}^*)$ and $\ker \partial_i'$ to $\ker \mathrm{var}_i$. $\qquad\square$

The image of the 'pseudo-embedding' $\iota : X_{a_i}^* := X_{a_i} \cap (B \setminus \cup_j B_{ij}) \hookrightarrow X_{c_i}$ plays here the role of the boundary of the Milnor fibre in the local case. We may therefore call $\mathrm{Im}\,\iota_*$ the group of 'boundary cycles' at a_i. We immediately get the following consequence.[7]

Corollary 10.3.2 The invariant cycles under the monodromy at a_i are exactly the boundary cycles, i.e. the following sequence is exact:

$$H_*(X_{a_i}^*) \xrightarrow{\iota_*} H_*(X_{c_i}) \xrightarrow{v_i - \mathrm{id}} H_*(X_{c_i}).$$

Proof We have the following commutative diagram, by (10.5), where the first row is the exact sequence of the pair $(X_{c_i}, X_{a_i}^*)$:

$$
\begin{array}{ccccc}
H_*(X_{a_i}^*) & \xrightarrow{\;\iota_*\;} & H_*(X_{c_i}) & \xrightarrow{\;j_*\;} & H_*(X_{c_i}, X_{a_i}^*) \\
 & & & \searrow_{\,v_i - \mathrm{id}} & \downarrow{\mathrm{var}_i} \\
 & & & & H_*(X_{c_i}).
\end{array}
$$

We have the equality $\mathrm{Im}\,\iota_* = \ker j_*$. Since $v_i - \mathrm{id} = \mathrm{var}_i \circ j_*$, and since var_i is injective by Corollary 10.3.1, the claim follows. $\qquad\square$

This result may be considered as a counterpart, in a nonproper situation, of the well-known 'invariant cycle theorem' proved by Clemens [Cl]. The latter holds for proper holomorphic functions $g : X \to D$, in cohomology (thus 'invariant co-cycle theorem' would be more appropriate), where X is a Kähler manifold. It says that the following sequence is exact:

$$H^\star(X) \xrightarrow{\;j^*\;} H^\star(X_c) \xrightarrow{\;h - \mathrm{id}\;} H^\star(X_c),$$

where h denotes the monodromy around the center of the disk D (assumed to be the single critical value of g).

It is natural to ask if an invariant cycle result similar to Corollary 10.3.2 holds for more general classes of nonproper pencils.

11

Higher Zariski–Lefschetz theorems

We discuss in this last chapter the problem of finding natural generators for the 'vanishing homotopy' of a pencil of hypersurfaces on an analytic variety X. The solution of this problem will yield analogues of the Second Lefschetz Hyperplane Theorem for homotopy groups, which are called in the literature *Zariski–Lefschetz theorems*. At the same time, this generalizes to higher homotopy groups the *Zariski–van Kampen theorem* for the fundamental group.

It turns out that our geometric method allows us to prove such results in a general setting, which includes for instance the use of nongeneric pencils.

11.1 Homotopy variation maps

We use the same notations as before: Y is a compact complex analytic space, $V \subset Y$ is a closed analytic subspace such that $X := Y \setminus V$ is connected, of pure complex dimension $n \geq 3$. Assume that there is a pencil on X defining a *Lefschetz fibration with isolated singularities* on X, in the sense of Definition 10.1.2, involving the Whitney stratification $\mathcal{S}$ of $\mathbb{Y}$.

For $a_i \in \Lambda := \{a_1, \ldots, a_r\}$, at some singularity $a_{ij} \in \mathrm{Sing}_{\mathcal{S}} p \cap \mathbb{Y}_{a_i}$, in local coordinates, we take *Milnor data* (B_{ij}, D_i) for p at a_{ij}. Moreover, we may keep the same disk D_i for all j, provided that it is small enough.

Since the function $p : \mathbb{Y} \to \mathbb{P}^1$ has isolated stratified singularities, the fibres of p are endowed with the stratification induced by $\mathcal{S}$ except that we have to introduce the point-strata $\{a_{ij}\}$. Every fibre has therefore a natural induced Whitney stratification. For the construction of a geometric monodromy at a_i we refer to §10.2. The exposition follows closely [Ti8, Ti10].

Construction of the homotopy variation map. We assume from now on that $X_{c_i} \setminus \cup_j B_{ij}$ is path connected, for all i. For instance, if the general fibre X_c

is path-connected and if $(X_{c_i}, X_{c_i} \setminus \cup_j B_{ij})$ is 1-connected then $X_{c_i} \setminus \cup_j B_{ij}$ is path connected. (The assumptions of Theorem 11.2.1(b) will allow these considerations.)

We have to keep track of the base points of homotopy groups, so let us fix $u \in S^{q-1}$ and $v \in X_{c_i} \setminus \cup_j B_{ij}$. Let $\gamma : (D^q, S^{q-1}, u) \to (X_{c_i}, X_{c_i} \setminus \cup_j B_{ij}, v)$, for $q \geq 3$, be some continuous map, and let $[\gamma]$ be its homotopy class in $\pi_q(X_{c_i}, X_{c_i} \setminus \cup_j B_{ij}, v)$. The homotopy class $[h_i \circ \gamma]$ is a well defined element of $\pi_q(X_{c_i}, X_{c_i} \setminus \cup_j B_{ij}, v)$ and does not depend on the representative h_i in its relative isotopy class. Consider the map $\rho : (D^q, S^{q-1}, u) \to (D^q, S^{q-1}, u)$ which is the reflection into some fixed generic hyperplane through the origin, which contains u. Then $[\gamma \circ \rho]$ is the inverse $-[\gamma]$ of $[\gamma]$. We use the additive notation since the relative homotopy groups π_q are abelian for $q \geq 3$.

Consider now the map:

$$\mu_i : \pi_q(X_{c_i}, X_{c_i} \setminus \cup_j B_{ij}, v) \to \pi_q(X_{c_i}, X_{c_i} \setminus \cup_j B_{ij}, v) \qquad (11.1)$$

defined as follows: $\mu_i([\gamma]) = [(h_i \circ \gamma) * (\gamma \circ \rho)] = [h_i \circ \gamma] - [\gamma]$.

The notation "$*$" stands for the operation on maps which induces the group structure of the relative π_q. The map μ_i is well defined and it is an automorphism of $\pi_q(X_{c_i}, X_{c_i} \setminus \cup_j B_{ij}, v)$, because of the abelianity of the relative homotopy groups π_q, for $q \geq 3$. This construction does not extend to the case $q = 2$.[1] In case $q = 1$ there are the monodromy relations which enter in the well-known Zariski–van Kampen theorem.

Lemma 11.1.1 The map $v_i := (h_i \circ \gamma) * (\gamma \circ \rho) : (D^q, S^{q-1}, u) \to (X_{c_i}, X_{c_i} \setminus \cup_j B_{ij}, v)$ is homotopic, relative to S^{q-1}, to a map $(S^q, u) \to (X_{c_i}, v)$.

Proof We first observe that for the restriction of γ to S^{q-1}, which we shall call γ', we have $h_i \circ \gamma' = \gamma'$, since the geometric monodromy h_i is the identity on $\text{Im}\, \gamma'$. Without loss of generality, we may and shall suppose in the following that the geometric monodromy h_i is the identity over a small tubular neighbourhood $\mathcal{T}$ of $X_{c_i} \setminus \cup_j B_{ij}$ within X_{c_i}.

Let v_i' denote the restriction of v_i to S^{q-1}. We claim that we can shrink $\text{Im}\, v_i'$ to the point v through a homotopy within $\text{Im}\, v_i'$. The precise reason is that, by the definition of ρ, the map $\gamma' * (\gamma' \circ \rho)$ is homotopic to the constant map to the base-point v, through a deformation of the image of $\gamma' * (\gamma' \circ \rho)$ within itself. We may moreover extend this homotopy to a continuous deformation of the map v_i in a small collar neighbourhood τ of S^{q-1} in D^q, which has its image in the tubular neighbourhood $\mathcal{T}$, such that it is the identity on the interior boundary of τ (which is a smaller sphere $S_{1-\varepsilon}^{q-1} \subset D^q$). Then we extend this deformation as the identity to the rest of D^q. In this way we have constructed a homotopy between v_i and a map $(D^q, S^{q-1}, u) \to (X_{c_i}, v, v)$, which is nothing else than a map $(D^q, u) \to (X_{c_i}, v)$. This represents an element of the group $\pi_q(X_{c_i}, v)$. □

The construction in the above proof is well defined up to homotopy equivalences, hence we get:

Definition 11.1.2 The map defined in the proof of Lemma 11.1.1:

$$\text{hvar}_i : \pi_q(X_{c_i}, X_{c_i} \setminus \cup_j B_{ij}, v) \to \pi_q(X_{c_i}, v), \tag{11.2}$$

is called *the homotopy variation map* of the monodromy h_i.

Let us point out that hvar_i is a morphism of groups. Indeed, due to the abelianity we have:

$$\text{hvar}_i([\gamma] + [\delta]) = [(h_i \circ \gamma) * (h_i \circ \delta)] - [\gamma * \delta] = [h_i \circ \gamma] + [h_i \circ \delta] - [\gamma] - [\delta]$$

$$= [h_i \circ \gamma] - [\gamma] + [h_i \circ \delta] - [\delta] = \text{hvar}_i([\gamma]) + \text{hvar}_i([\delta]).$$

By its geometric definition, the homotopy variation map enters in the following commutative diagram:

$$
\begin{array}{ccc}
\pi_q(X_{c_i}, X_{c_i} \setminus \cup_j B_{ij}, v) & \xrightarrow{\ \mu_i\ } & \pi_q(X_{c_i}, X_{c_i} \setminus \cup_j B_{ij}, v) \\
& \searrow_{\text{hvar}_i} & \uparrow_{j_\sharp} \\
& & \pi_q(X_{c_i}, v),
\end{array}
\tag{11.3}
$$

where $j_\sharp$ is induced by the inclusion $(X_{c_i}, v) \to (X_{c_i}, X_{c_i} \setminus \cup_j B_{ij}, v)$.

For simplicity reasons, we use the notation in Definition 10.2.1 and we write $X_{a_i}^*$ for the fibre $X_{c_i} \setminus \cup_j B_{ij}$ of the trivial fibration (10.3). Thus the pair of spaces $(X_{c_i}, X_{c_i} \setminus \cup_j B_{ij})$ will be denoted by $(X_{c_i}, X_{a_i}^*)$.

Keeping in mind the significance of $X_{a_i}^*$, we shall use in the remainder the following notation for the variation map, instead of (11.2):

$$\text{hvar}_i : \pi_q(X_{c_i}, X_{a_i}^*, \cdot) \to \pi_q(X_{c_i}, \cdot). \tag{11.4}$$

The action of π_1. The fundamental group $\pi_1(X_c, \cdot)$ acts on the higher homotopy groups $\pi_q(X_c, \cdot)$ and on any relative homotopy group, where X_c appears at the second position. Under the hypotheses of Theorem 11.2.1(b), $X_{a_i}^*$ is path connected for all i and $\pi_1(X_{a_i}^*, \cdot) \xrightarrow{i_*} \pi_1(X_{c_i}, \cdot)$ is an isomorphism. Having this assumption, we may and shall identify the two actions of $\pi_1(X_{a_i}^*, \cdot)$ and of $\pi_1(X_{c_i}, \cdot)$ modulo the isomorphism i_*.

We therefore have the action of $\pi_1(X_{a_i}^*, \cdot)$ on those higher relative homotopy groups of pairs where $X_{a_i}^*$ is the second space of the pair, and we also have the action of $\pi_1(X_{a_i}^*, \cdot)$, via the isomorphism i_*, on $\pi_q(X_{c_i}, \cdot)$ and on the higher relative homotopy groups of pairs, where X_{c_i} is the second space of the pair.

Lemma 11.1.3 *The image of* hvar_i *is invariant under the action of* $\pi_1(X_{c_i}, \cdot)$.

Proof For some $\beta \in \pi_1(X_{a_i}^*, \cdot)$, we have $\mathrm{hvar}_i(\beta[\gamma]) = \beta[h_i \circ \gamma] - \beta[\gamma] = \beta \mathrm{hvar}_i([\gamma])$, since, by the functoriality, the action of β commutes with the morphism induced by the geometric monodromy map $h_i : X_{c_i} \to X_{c_i}$. $\qquad\square$

We shall denote in the following by $\hat{\pi}_q$ the quotient of the corresponding higher homotopy group by the action of $\pi_1(X_c, \cdot)$ or by that of $\pi_1(X_{a_i}^*, \cdot)$ eventually through the isomorphism i_*. We shall use the same notation $\hat{v}$ for the passage to the quotients of some morphism v which commutes with the action of π_1.

11.2 General Zariski–Lefschetz theorem for nongeneric pencils

Let us now state the general Zariski–van Kampen–Lefschetz type result, which uses the homotopy variation map (11.2). It is a faithfull analogue of Theorem 10.2.3 (the nongeneric Second LHT). The ingredients were defined before, except of the homotopy depth condition 'hd', which is also entirely analogous to the homology depth Hd, cf. Definition 10.2.2. The exposition follows [Ti8, Ti10].

Theorem 11.2.1 *[Ti8, Ti10] Let* $h \colon Y \dashrightarrow \mathbb{P}^1$ *define a pencil with isolated singularities such that the axis A of the pencil is not included in V and such that the general fibre X_c on $X = Y \setminus V$ is path connected. Let us consider the following three conditions:*

(i) $(X_c, X_c \cap A)$ *is k-connected;*

(ii) (X_c, X_a^*) *is k-connected, for any critical value a of the pencil;*

(iii) $\mathrm{hd}_{\overline{X} \cap \mathrm{Sing}_{SP}} \overline{X} \geq k + 2.$

Then we have the following two conclusions:

(a) *If for some $k \geq 0$ the conditions* (i), (ii) *and* (iii) *hold, then* $\pi_q(X, X_c, \cdot) = 0$ *for all $q \leq k + 1$.*

(b) *If for some* $k \geq 2$ *the conditions* (i) *and* (ii) *hold, and if condition*
   (iii) *is replaced by the following one:*
(iii') $\text{hd}_{\mathbb{X} \cap \text{Sing}_{SP}} \mathbb{X} \geq k + 3$,
   then the kernel of the surjective map $\pi_{k+1}(X_c, \cdot) \twoheadrightarrow \pi_{k+1}(X, \cdot)$ *is*
   generated by the images of the variation maps hvar_i.

The conditions in our theorem are satisfied by a large class of spaces X. For instance, the conditions (ii) and (iii), respectively (iii'), are both fulfilled as soon as X is a singular space that is a local complete intersection of dimension $n \geq k + 2$, respectively $n \geq k + 3$ (see also Definition 10.2.2 and after it).

In what concerns part (a) of Theorem 11.2.1, this is a general (nongeneric) LHT for homotopy groups, which recovers Theorem 10.1.4 with a significantly weakened hypotheses.

The focus is here on part (b) of Theorem 11.2.1, which treats the 'homotopy vanishing cycles'. Since the homology counterpart of it is Theorem 10.2.3, we concentrate here on those more delicate arguments for homotopy groups. The result does not cover the case $k = 1$ because of the technical restriction in the definition of the variation map (11.2).

Theorem 11.2.1 represents a synthetic viewpoint on Zariski–van Kampen-type results for higher homotopy groups, having in the background the pioneering work of Lefschetz [Lef], Zariski and van Kampen [vK]. It recovers several cases considered before in the literature and at the same time it extends the range of applicability of the Zariski–van Kampen–Lefschetz principle.[2]

We shall discuss in §11.3 several consequences and particular, cases. In particular, we show how to get a version of Theorem 11.2.1 in the case of complements of singular projective spaces. Some other possible applications of Theorem 11.2.1 parallel the ones for homology groups, which have been discussed in §10.

Proof of Theorem 11.2.1. For the connectivity part (a), we follow in the beginning the proof of Theorem 10.1.4.

Definition 11.2.2 One says that an inclusion of pairs of topological spaces $(N, N') \hookrightarrow (M, M')$ is a *q-equivalence* if this inclusion induces an isomorphism of relative homotopy groups for $j < q$ and a surjection for $j = q$.

Lemma 11.2.3 Let (X_c, A') be k-connected for $k \geq 0$. Then the inclusion of pairs $\iota : (X_{\mathcal{D}}, X_c) \hookrightarrow (X, X_c)$ is a $(k + 2)$-equivalence, for $k \geq 0$.

Proof This can be extracted from page 189. Under the assumption (X_c, A') is k-connected, $k \geq 0$, Proposition 10.1.6(a)(i) shows that (X_S, X_c) is $(k + 1)$-connected. From the exact sequence of the triple $(X_{\mathcal{D}}, X_S, X_c)$, we deduce that the inclusion $(X_{\mathcal{D}}, X_c) \hookrightarrow (X_{\mathcal{D}}, X_S)$ is a $(k+2)$-equivalence. Next, Proposition 10.1.6(a)(ii) shows that, by homotopy excision, the inclusion of pairs $(X_{\mathcal{D}}, X_S) \hookrightarrow (X, X_K)$ is a $(k+2)$-equivalence. Since X_K retracts in a fibrewise way to X_c, the pair (X, X_K) is homotopy equivalent to (X, X_c). $\qquad\square$

We turn our attention to the Nash blow-up $\mathbb{X}$. We have that the inclusion $(X_{\mathcal{D}}, X_c) \to (\mathbb{X}_{\mathcal{D}}, X_c)$ is a homotopy equivalence, by applying Lemma 9.2.1. Next, Proposition 10.1.6(b) shows that $(\mathbb{X}_{\mathcal{D}}, X_c)$ is $(k+1)$-connected, $k \geq 0$, provided that $(\mathbb{X}_{D_i}, X_{c_i})$ is $(k+1)$-connected for all i.

Then we replace $\mathbb{X}_{D_i}$ by $\mathbb{X}_{D_i}^* := \mathbb{X}_{D_i} \setminus \mathrm{Sing}_{\mathcal{S}} p$, using the homotopy depth assumption.

Lemma 11.2.4 If $\mathrm{hd}_{\mathbb{X} \cap \mathrm{Sing}_{\mathcal{S}} p} \mathbb{X} \geq q+1$, for $q \geq 1$, then the inclusion of pairs $(\mathbb{X}_{D_i}^*, X_{c_i}) \overset{j}{\hookrightarrow} (\mathbb{X}_{D_i}, X_{c_i})$ is a q-equivalence, for all i.

Proof In the exact sequence of the triple $(\mathbb{X}_{D_i}, \mathbb{X}_{D_i}^*, X_{c_i})$, it is sufficient to prove that $(\mathbb{X}_{D_i}, \mathbb{X}_{D_i}^*)$ is q-connected. By the homotopy depth assumption, for all j, the pair $(\mathbb{X}_{D_i} \cap B_{ij}, \mathbb{X}_{D_i} \cap B_{ij} \setminus \{a_{ij}\})$ is q-connected. Then, by Switzer's result for CW-complexes [Sw, 6.13], the space $\mathbb{X}_{D_i} \cap B_{ij}$ is obtained from $B_{ij} \setminus \{a_{ij}\}$ by attaching cells of dimensions $\geq q+1$.[*] It follows that $\mathbb{X}_{D_i}$ is obtained from $\mathbb{X}_{D_i}^*$ by replacing each $B_{ij} \setminus \{a_{ij}\}$ with B_{ij}, which amounts, as we have seen, to attaching cells of dimensions $\geq q+1$. So our claim is proved. $\qquad\square$

To conclude part (a) of the proof of Theorem 11.2.1, we need the following key result.

Proposition 11.2.5 If $(X_{c_i}, X_{a_i}^*)$ is k-connected, $k \geq 0$, then $(\mathbb{X}_{D_i}^*, X_{c_i})$ is $(k+1)$-connected.

Proof We shall suppress the lower indices i in the following.
Let $D^* := D \setminus \{a\}$. We retract D^* to the circle ∂D and cover this circle by two arcs $I \cup J$. We have the homotopy equivalences $\mathbb{X}_{D^*} \overset{ht}{\simeq} \mathbb{X}_I \cup \mathbb{X}_J$ and $X_c \overset{ht}{\simeq} \mathbb{X}_J$. We recall that $\mathbb{X}_D \setminus \cup_j B_{ij}$ is the total space of the trivial fibration $p_| : \mathbb{X}_D \setminus \cup_j B_{ij} \to D$. Its fibre is homotopy equivalent to $X_c \setminus \cup_j B_{ij}$ and has been denoted by X_a^*. Recall that $\mathbb{X}_D^* := \mathbb{X}_D \setminus \mathrm{Sing}_{\mathcal{S}} f$. Then the following inclusions are homotopy equivalences of pairs. They induce isomorphisms in all relative homotopy groups.

---

[*] Note that, if $a_{ij} \notin \mathbb{X}_{a_i}$ then the two spaces of the pair are identical and the connectivity of the pair is infinite; in this case there is nothing to attach.

$$(\mathbb{X}_D^*, X_c) \hookleftarrow (\mathbb{X}_{D^*} \cup (\mathbb{X}_D \setminus \cup_j B_{ij}), X_c) \hookrightarrow$$
$$\hookrightarrow (\mathbb{X}_{D^*} \cup (\mathbb{X}_D \setminus \cup_j B_{ij}), \mathbb{X}_J \cup (\mathbb{X}_D \setminus \cup_j B_{ij})) \hookleftarrow$$
$$\hookleftarrow (\mathbb{X}_I \cup \mathbb{X}_J \cup (\mathbb{X}_D \setminus \cup_j B_{ij}), \mathbb{X}_J \cup (\mathbb{X}_D \setminus \cup_j B_{ij})).$$

Let us consider the following homotopy excision:

$$(\mathbb{X}_I \cup \mathbb{X}_J \cup (\mathbb{X}_D \setminus \cup_j B_{ij}), \mathbb{X}_J \cup (\mathbb{X}_D \setminus \cup_j B_{ij})) \hookleftarrow (\mathbb{X}_I, \mathbb{X}_{\partial I} \cup (\mathbb{X}_I \setminus \cup_j B_{ij})). \quad (11.5)$$

The right-hand side pair is homotopy equivalent to $(\mathbb{X}_I, X_c \times \partial I \cup X_a^* \times I)$, which is the product of pairs $(X_c, X_a^*) \times (I, \partial I)$. Since (X_c, X_a^*) is k-connected by our hypothesis, this product is $(k + 1)$-connected. This follows from the fact that the spaces are CW-complexes and by using Switzer's [Sw, Proposition 6.13].

By the Blakers–Massey theorem (cf. Gray [BGr, Cor. 16.27]), it will follow that the excision (11.5) is a $(k + 1 + q)$-equivalence provided that the pair $(\mathbb{X}_J \cup (\mathbb{X}_D \setminus \cup_j B_{ij}), \mathbb{X}_{\partial I} \cup (\mathbb{X}_I \setminus \cup_j B_{ij}))$ is q-connected. If $q \geq 0$, then this proves our claim that the original pair $(\mathbb{X}_D^*, X_c)$ is $(k + 1)$-connected.

It therefore remains to evaluate the level q. Since the inclusions $\mathbb{X}_I \setminus \cup_j B_{ij} \hookrightarrow \mathbb{X}_D \setminus \cup_j B_{ij}$ and $\mathbb{X}_J \setminus \cup_j B_{ij} \hookrightarrow \mathbb{X}_D \setminus \cup_j B_{ij}$ are homotopy equivalences, it follows that the following inclusions of pairs are homotopy equivalences, and therefore induce isomorphisms in all relative homotopy groups:

$$(\mathbb{X}_J \cup (\mathbb{X}_D \setminus \cup_j B_{ij}), \mathbb{X}_{\partial I} \cup (\mathbb{X}_I \setminus \cup_j B_{ij})) \hookrightarrow$$
$$\hookrightarrow (\mathbb{X}_J \cup (\mathbb{X}_D \setminus \cup_j B_{ij}), \mathbb{X}_{\partial I} \cup (\mathbb{X}_D \setminus \cup_j B_{ij})) \hookleftarrow$$
$$\hookleftarrow (\mathbb{X}_J \cup (\mathbb{X}_J \setminus \cup_j B_{ij}), \mathbb{X}_{\partial J} \cup (\mathbb{X}_J \setminus \cup_j B_{ij})) \hookleftarrow (\mathbb{X}_J, \mathbb{X}_{\partial J} \cup (\mathbb{X}_J \setminus \cup_j B_{ij})).$$

We also have:

$$(\mathbb{X}_J, \mathbb{X}_{\partial J} \cup (\mathbb{X}_J \setminus \cup_j B_{ij})) \overset{\text{ht}}{\simeq} (\mathbb{X}_c \times J, X_c \times \partial J \cup X_a^* \times J) = (X_c, X_a^*) \times (J, \partial J).$$

Since (X_c, X_a^*) is supposed k-connected, this implies that the above product of pairs is at least $(k + 1)$-connected, so the level q is $k + 1 > 0$. $\qquad\square$

We next prove the more delicate part (b) of Theorem 11.2.1. From part (a) we know that $\pi_{k+1}(X_c, \cdot) \twoheadrightarrow \pi_{k+1}(X, \cdot)$ is a surjection. The kernel of it is invariant under the the $\pi_1(X_c, \cdot)$-action on $\pi_{k+1}(X_c, \cdot)$, by the functoriality of this action. So in order to prove (b) it suffices to show that the image of $\hat{\partial} : \hat{\pi}_{k+2}(X, X_c, \cdot) \to \hat{\pi}_{k+1}(X_c, \cdot)$ is generated by the images $\text{Im}\,(\widehat{\text{hvar}_i})$, since

Im (hvar_i) is also invariant under the π_1-action, see the end of §11.1. This will be done in the last part of the proof.

But first, let us observe that we may replace $\hat{\partial}$ by $\hat{\partial}_0$. Indeed, we has the natural commutative triangle:

$$
\begin{array}{ccc}
\pi_{k+2}(X_{\mathcal{D}}, X_c, \bullet) & \xrightarrow{\ \iota_\sharp\ } & \pi_{k+2}(X, X_c, \bullet) \\
& \searrow{\scriptstyle \partial_0} & \ \ \downarrow{\scriptstyle \partial} \\
& & \pi_{k+1}(X_c, \bullet),
\end{array}
\qquad (11.6)
$$

where ∂ and ∂_0 are boundary morphisms and ι is the inclusion of pairs $(X_{\mathcal{D}}, X_c) \hookrightarrow (X, X_c)$. The fundamental group acts on this diagram and the action is functorial. By Lemma 11.2.3, the inclusion of pairs $\iota : (X_{\mathcal{D}}, X_c) \hookrightarrow (X, X_c)$ is a $(k+2)$-equivalence, for $k \geq 0$, and taking quotients by the action of π_1 preserves this property. Consequently, we have the equality:

$$
\mathrm{Im}\,\hat{\partial} = \mathrm{Im}\,\hat{\partial}_0. \qquad (11.7)
$$

For all i, let $\iota_i : (\mathbb{X}_{D_i}, X_{c_i}) \to (\mathbb{X}_{\mathcal{D}}, X_c)$ denote the inclusion of pairs and let $\partial_i : \pi_{k+2}(\mathbb{X}_{D_i}, X_{c_i}, \bullet) \to \pi_{k+1}(X_{c_i}, \bullet)$ denote the boundary morphism in the pair $(\mathbb{X}_{D_i}, X_{c_i})$. Consider now the following commutative diagram, for $k \geq 1$:

$$
\begin{array}{ccc}
\oplus_i \pi_{k+2}(\mathbb{X}_{D_i}, X_{c_i}, \bullet) & & \\
{\scriptstyle \sum \iota_{i\sharp}} \downarrow & \searrow{\scriptstyle \sum \partial_i} & \\
\pi_{k+2}(\mathbb{X}_{\mathcal{D}}, X_c, \bullet) & \xrightarrow[\ \partial_0\ ]{} & \pi_{k+1}(X_c, \bullet),
\end{array}
\qquad (11.8)
$$

where, for all i, we identify $(X_{c_i}, \bullet)$ to $(X_c, \bullet)$ together with the base points "$\bullet$" by using the paths γ_i, as explained in the proof of Theorem 10.1.4. We use the additive notations $\sum \partial_i$ and $\sum \iota_{i\sharp}$ since for $k \geq 1$ the groups $\pi_{k+1}(X_c, \bullet)$ and $\pi_{k+2}(\mathbb{X}_{\mathcal{D}}, X_c, \bullet)$ are abelian. By the functoriality of the action of $\pi_1(X_c, \bullet)$, we get the induced "hat" diagram:

$$
\begin{array}{ccc}
\oplus_i \hat{\pi}_{k+2}(\mathbb{X}_{D_i}, X_{c_i}, \bullet) & & \\
{\scriptstyle \sum \hat{\iota}_{i\sharp}} \downarrow & \searrow{\scriptstyle \sum \hat{\partial}_i} & \\
\hat{\pi}_{k+2}(\mathbb{X}_{\mathcal{D}}, X_c, \bullet) & \xrightarrow[\ \hat{\partial}_0\ ]{} & \hat{\pi}_{k+1}(X_c, \bullet).
\end{array}
\qquad (11.9)
$$

Proposition 11.2.6 Let $(\mathbb{X}_{D_i}, X_{c_i})$ be $(k+1)$-connected, $k \geq 1$, for all i. Then $\operatorname{Im} \hat{\partial}_0 = \operatorname{Im}(\sum_i \hat{\partial}_i)$ in diagram (11.9).

Proof We use Hurewicz maps between the relative homotopy and homology groups (denoted $\mathcal{H}_i$ and $\mathcal{H}$ below). We have the following commutative diagram:

$$
\begin{array}{ccc}
\oplus_i \hat{\pi}_{k+2}(\mathbb{X}_{D_i}, X_{c_i}, \cdot) & \xrightarrow{\sum \hat{\iota}_{i\sharp}} & \hat{\pi}_{k+2}(\mathbb{X}_{\mathcal{D}}, X_c, \cdot) \\
\downarrow{\scriptstyle \oplus_i \mathcal{H}_i} & & \downarrow{\scriptstyle \mathcal{H}} \\
\oplus_i H_{k+2}(\mathbb{X}_{D_i}, X_{c_i}) & \xrightarrow{\sum \iota_{i*}} & H_{k+2}(\mathbb{X}_{\mathcal{D}}, X_c).
\end{array}
$$

The additive notations are used because the relative homotopy groups are abelian for $k \geq 1$.

Under our hypothesis and by the relative Hurewicz isomorphism theorem (see e.g. [Sp, p. 397]), we get that the Hurewicz map $\mathcal{H}_i : \hat{\pi}_{k+2}(\mathbb{X}_{D_i}, X_{c_i}, \cdot) \to H_{k+2}(\mathbb{X}_{D_i}, X_{c_i})$ is an isomorphism, for any i. The same is $\mathcal{H}$, since $(\mathbb{X}_{\mathcal{D}}, X_c)$ is $(k+1)$-connected, by Proposition 10.1.6(b). Next, $\sum \iota_{i*}$ is a homology excision, thus an isomorphism.

It follows then from the diagram that $\sum \hat{\iota}_{i\sharp}$ is an isomorphism too, which proves our claim. $\qquad\square$

To complete the proof of the theorem, we need to find the image of the map $\hat{\partial}_i : \hat{\pi}_{k+2}(\mathbb{X}_{D_i}, X_{c_i}, \cdot) \to \hat{\pi}_{k+1}(X_{c_i}, \cdot)$, $k \geq 1$, for all i.

Corollary 11.2.7 If $\operatorname{hd}_{\mathbb{X} \cap \operatorname{Sing}_{SP} \mathbb{X}} \geq k+3$, $k \geq 0$, then the following equality holds, for all i:

$$
\begin{aligned}
\operatorname{Im}(\hat{\partial}_i : \hat{\pi}_{k+2}(\mathbb{X}_{D_i}, X_{c_i}, \cdot) \to \hat{\pi}_{k+1}(X_{c_i}, \cdot)) = \\
= \operatorname{Im}(\hat{\partial}_i' : \hat{\pi}_{k+2}(\mathbb{X}_{D_i}^*, X_{c_i}, \cdot) \to \hat{\pi}_{k+1}(X_{c_i}, \cdot)).
\end{aligned}
$$

Proof By applying Lemma 11.2.4 for $q = 2$, we get that $j_\sharp : \pi_{k+2}(\mathbb{X}_{D_i}^*, X_{c_i}, \cdot) \to \pi_{k+2}(\mathbb{X}_{D_i}, X_{c_i}, \cdot)$ is a surjection. Since $\partial_i' = \partial_i \circ j_\sharp$, we get our claim after taking the quotient by the $\pi_1(X_{c_i}, \cdot)$-action. $\qquad\square$

Proposition 11.2.8 If $(X_{c_i}, X_{a_i}^*)$ is k-connected, $k \geq 2$, then:

$$
\operatorname{Im} \hat{\partial}_i' = \operatorname{Im}(\widehat{\operatorname{hvar}_i} : \hat{\pi}_{k+1}(X_{c_i}, X_{a_i}^*, \cdot) \to \hat{\pi}_{k+1}(X_{c_i}, \cdot)).
$$

Proof We work in this proof with a fixed index i; we shall write c, a, μ instead of c_i, a_i, μ_i respectively. The final purpose will be to show the commutativity of the diagram (11.13) below. The idea is to compare it with the analogous diagram for homology groups, via the Hurewicz maps. However, we cannot do this directly, since we do not have the k-connectivity of X_c from the upper-right corner of (11.13); we only have that the pair (X_c, X_a^*) is k-connected. We therefore construct a proof in two steps.

Step 1. Let us consider the following diagram, not claiming that it is commutative:

$$
\begin{array}{ccc}
\pi_{k+2}(\mathbb{X}_D^*, X_c \times I, \cdot) \simeq \pi_{k+2}(\mathbb{X}_D^*, X_c, \cdot) & \xrightarrow{\;\partial''\;} & \pi_{k+1}(X_c, X_a^*, \cdot) \\[4pt]
{\scriptstyle exc}\Big\uparrow & & \Big\uparrow{\scriptstyle \mu} \qquad (11.10)\\[4pt]
\pi_{k+2}(\mathbb{X}_I, X_c \times \partial I \cup X_a^* \times I, \cdot) & \xleftarrow{\;\simeq\;} & \pi_{k+1}(X_c, X_a^*, \cdot),
\end{array}
$$

where $\mu := \mu_i$ is the morphism defined in (11.1) and ∂'' denotes the boundary morphism of the homotopy exact sequence of the triple $(\mathbb{X}_D^*, X_c, X_a^*)$. The fundamental group $\pi_1(X_a^*, \cdot)$ acts on the groups at the right-hand side of this diagram and, via certain isomorphisms, on the groups at the left-hand side. These isomorphisms are $\pi_1(X_a^*, \cdot) \simeq \pi_1(X_c, \cdot) \simeq \pi_1(X_c \times \partial I \cup X_a^* \times I, \cdot)$ which follow from the connectivity hypothesis on (X_c, X_a^*) and from the inclusion of spaces $X_a^* \times I \subset X_c \times \partial I \cup X_a^* \times I \subset X_c \times I$. Therefore the natural diagram (11.10) passes to the quotient by the π_1-action and yields:

$$
\begin{array}{ccc}
\hat{\pi}_{k+2}(\mathbb{X}_D^*, X_c \times I, \cdot) \simeq \hat{\pi}_{k+2}(\mathbb{X}_D^*, X_c, \cdot) & \xrightarrow{\;\hat{\partial}''\;} & \hat{\pi}_{k+1}(X_c, X_a^*, \cdot) \\[4pt]
{\scriptstyle \widehat{exc}}\Big\uparrow{\scriptstyle \simeq} & & \Big\uparrow{\scriptstyle \hat{\mu}} \qquad (11.11)\\[4pt]
\hat{\pi}_{k+2}(\mathbb{X}_I, X_c \times \partial I \cup X_a^* \times I, \cdot) & \xleftarrow{\;\simeq\;} & \hat{\pi}_{k+1}(X_c, X_a^*, \cdot).
\end{array}
$$

We now claim that (11.11) is a *commutative diagram*. Let us see, there is the following diagram in homology:

$$
\begin{array}{ccc}
H_{k+2}(\mathbb{X}_D^*, X_c) & \xrightarrow{\;\partial''\;} & H_{k+1}(X_c, X_a^*) \\[4pt]
{\scriptstyle exc}\Big\uparrow{\scriptstyle \simeq} & & \Big\uparrow{\scriptstyle h_*-\mathrm{id}} \qquad (11.12)\\[4pt]
H_{k+2}(\mathbb{X}_I, X_c \times \partial I \cup X_a^* \times I) & \xleftarrow{\;\simeq\;} & H_{k+1}(X_c, X_a^*),
\end{array}
$$

which is a Wang type diagram, and therefore it is *commutative*. Milnor explained a simpler version of it in [Mi2, p. 67] and his explanation holds true in the relative homology. We have used it in the proof of Proposition 10.2.9.

Let us prove the claim. We parallel the homotopy groups diagram (11.11) by the homology groups diagram (11.12) and connect them by the Hurewicz homomorphism. We get in this way a 'cubic' diagram. Notice that the homology version of the homotopy map μ is just $h_* -$ id, by the geometric definition of μ given in (11.1). By the naturality of the Hurewicz morphism, all the maps between homotopy groups are in commuting diagrams with their homology versions (as 'faces' of the cubic diagram). Moreover, the Hurewicz theorem may be applied each time. The connectivity conditions that the four Hurewicz morphisms become isomorphisms are fullfilled: (X_c, X_a^*) is k-connected by hypothesis and $(\mathbb{X}_D^*, X_c)$ is $(k+1)$-connected by Proposition 11.2.5. This allows us to identify the homotopy diagram (11.11) to the homology one (11.12), and, since the later commutes, it is the same with the former. Our claim is therefore proved.

Step 2. We claim that the following diagram:

$$
\begin{array}{ccc}
\hat{\pi}_{k+2}(\mathbb{X}_D^*, X_c, \bullet) & \xrightarrow{\ \hat{\partial}'\ } & \hat{\pi}_{k+1}(X_c, \bullet) \\[4pt]
\widehat{exc} \big\uparrow \simeq & & \big\uparrow \widehat{hvar} \\[4pt]
\hat{\pi}_{k+2}(\mathbb{X}_I, X_c \times \partial I \cup X_a^* \times I, \bullet) & \xleftarrow{\ \simeq\ } & \hat{\pi}_{k+1}(X_c, X_a^*, \bullet),
\end{array}
\tag{11.13}
$$

which differs from (11.11) by the upper right corner only, is *commutative* too.

Let us justify why we may replace the morphisms μ and $\hat{\partial}''$ in (11.11) by the morphisms hvar and $\hat{\partial}'$ respectively. We have the following diagram:

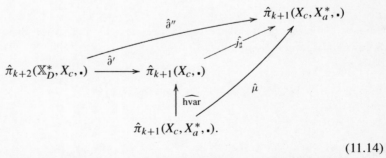

$$\tag{11.14}$$

The two triangles of this diagram are commutative by the following reasons: (1) the right-hand side triangle coincides with diagram (11.3) after having taken quotients by the action of $\pi_1(X_c, \bullet)$, and (2) the upper-left triangle is commutative by the naturality of the boundary morphism.

Let $\alpha \in \pi_{k+1}(X_c, X_a^*, \cdot)$. As explained in the construction of the homotopy variation map §11.1, an element of the form $\mu(\alpha)$ is homotopic, relative to X_a^*, to a certain element of the absolute group $\pi_{k+1}(X_c, \cdot)$, which we have denoted by $\mathrm{hvar}(\alpha)$. Furthermore, let us remark that, by the commutativity of diagram (11.11), $\hat{\mu}(\alpha) = \hat{\partial}''(\beta) \in \hat{\pi}_{k+1}(X_c, X_a^*, \cdot)$ for some $\beta \in \hat{\pi}_{k+2}(\mathbb{X}_D^*, X_c, \cdot)$. We get $\widehat{\mathrm{hvar}}(\alpha) = \hat{\partial}'(\beta)$.

Hence we can 'pull-back' the upper-right corner of the diagram (11.11), as shown in diagram (11.14). This completes the proof of the commutativity of (11.13).

The proof of our statement follows now from the commutativity of the diagram (11.13) and the fact that the bottom row and the excision from the left are both isomorphisms. $\qquad\square$

The scheme of the proof of Theorem 11.2.1(b) is as follows: apply one after the other (11.7), Proposition 11.2.6, Corollary 11.2.7 and Proposition 11.2.8.

11.3 Specializations of the Zariski–Lefschetz theorem

In the following we discuss what becomes in some particular cases Theorem 11.2.1, i.e. the generalized Zariski–Lefschetz theorem.

Aspects of the homotopy depth condition. Let us show that, if the space X satisfies the condition $\mathrm{rhd}\,\mathbb{X} \geq k + 2$ (as assumed in the statement of the Lefschetz Hyperplane Theorem 10.1.4), then this replaces the conditions (ii) and (iii).

Indeed, $\mathrm{rhd}\,X \geq k + 2$ implies $\mathrm{rhd}\,\mathbb{X} \geq k + 2$ by [HmL3, Theorem 3.2.1] applied to the hypersurface $\mathbb{X}$ of $X \times \mathbb{P}^1$. This in turn implies $\mathrm{hd}_v \mathbb{X} \geq k + 2$, for any point $v \in \mathbb{X}$, by definition. Moreover, the condition $\mathrm{rhd}\,\mathbb{X} \geq k + 2$ implies that the pair $(\mathbb{X}_{D_i}, X_{c_i})$ is $(k + 1)$-connected, in other words Lemma 11.2.4 is replaced by Corollary 10.1.9. Therefore Proposition 11.2.5 is not needed anymore and we have a shortcut to the proof of part (a) of Theorem 11.2.1 not using at all the condition (ii). The path connectivity of X_c is also not needed here; it only comes up in part (b), in the construction of the homotopy variation map.

In what concerns Part (b) of Theorem 11.2.1, the condition $\mathrm{rhd}\,X \geq k + 3$ implies not only (iii'), as shown just above, but implies also the condition (ii), which is important in the proof of Part (b). Let us prove this.

We claim that every pair $(X_{c_i} \cap B_{ij}, X_{c_i} \cap \partial B_{ij})$ is k-connected. Firstly, $X_{c_i} \cap B_{ij}$ is $(k+1)$-connected since $\mathbb{X}_D \cap B_{ij}$ is contractible by the local conical structure, and the pair $(\mathbb{X}_D \cap B_{ij}, X_{c_i} \cap B_{ij})$ is $(k + 1)$-connected by Proposition 10.1.8.

Secondly, rhd $X \geq k + 3$ implies rhd $X_{a_i} \geq k + 2$, by [HmL3, Theorem 3.2.1]. This implies $\mathrm{hd}_{a_{ij}} X_{a_i} \geq k + 2$, which shows that $X_{a_i} \cap \partial B_{ij}$ is k-connected (by the local conical structure of X_{a_i}). Since $X_{a_i} \cap \partial B_{ij}$ is homeomorphic to $X_{c_i} \cap \partial B_{ij}$, our claim is proved.

Now, by using Switzer's result for CW-complexes [Sw, 6.13] we get that the space $X_{c_i} \cap B_{ij}$ is obtained from $X_{c_i} \cap \partial B_{ij}$ by attaching cells of dimensions $\geq k + 1$.

It then follows that X_{c_i} is obtained from $X_{c_i} \setminus \cup_j B_{ij}$ by attaching $X_{c_i} \cap B_{ij}$ over $X_{c_i} \cap \partial B_{ij}$, for each j, and we have seen that this corresponds to attaching cells of dimensions $\geq k + 1$. It then follows that the pair $(X_{c_i}, X_{c_i} \setminus \cup_j B_{ij}) = (X_{c_i}, X_{a_i}^*)$ is k-connected, which is condition (ii).

Pencils on X having no singularities in the axis. Let us consider the situation $(A \times \mathbb{P}^1) \cap \mathbb{X} \cap \mathrm{Sing}_{\mathcal{S}} p = \emptyset$, which means that the pencil has no singularities along the axis on X. However, in our setting there might be singularities (which we still assume isolated, as before) along the axis on Y, and they certainly influence the topology of the pencil on X. This situation may be discussed somewhat more as follows (see also Exercise 11.1 for an interesting subcase).

We claim that the homotopy depth condition (iii), resp. (iii'), may be replaced by the following more general, global condition:

$$(X, X \setminus \Sigma) \text{ is } (k + 2)\text{-connected, resp. } (k + 3)\text{-connected,} \tag{11.15}$$

where $\Sigma := \sigma(\mathrm{Sing}_{\mathcal{S}} p) \subset Y$.[3] Indeed, we notice that we have used the homotopy depth condition only for comparing $\mathbb{X}_{D_i}$ to $\mathbb{X}_{D_i}^*$. So we may get rid of this comparison by replacing everywhere in the proof of Theorem 11.2.1(a) the space X by the space $X \setminus \Sigma$. In particular, this reduces to tautologies Lemma 11.2.4 and Corollary 11.2.7. We then get the conclusion of Theorem 11.2.1 for the inclusion $X_c \hookrightarrow X \setminus \Sigma$ instead of the inclusion $X_c \hookrightarrow X$. But, at this stage, we may just replace X by $X \setminus \Sigma$ since the condition (11.15) tells that $X \setminus \Sigma$ and X have isomorphic homotopy groups up to π_{k+1}, which suffices to our purpose.

Still in case $(A \times \mathbb{P}^1) \cap \mathbb{X} \cap \mathrm{Sing}_{\mathcal{S}} p = \emptyset$, let us observe a second fact: in Theorem 11.2.1, if the condition (i) is fulfilled, then the condition (ii) becomes equivalent to:

$$(X_{a_i}^*, A \cap X_{a_i}^*) \text{ is } (k - 1)\text{-connected.} \tag{11.16}$$

Indeed, when there are no singularities in the axis, we have $A \cap X_{a_i}^* = A \cap X_c = A'$, for any i and our claim follows from the exact sequence of the triple $(X_c, X_{a_i}^*, A')$.

The classical case: generic pencils on the projective space. From the preceding observations on pencils without singularities in the axis, we may derive the following statement for *complements* $\mathbb{P}^n \setminus V$ *of arbitrarily singular subspaces* $V \subset \mathbb{P}^n$.

Its proof will go by induction, and is based on the abundance of *generic pencils* in $\mathbb{P}^n$, in the sense that the pencils have no singularities along the axis and such that the axis A is transversal to all strata and is not included into the subspace V.

Corollary 11.3.1 Let $V \subset \mathbb{P}^n$ be a singular algebraic projective space of pure dimension, not necessarily irreducible. Let $H \subset \mathbb{P}^n$ be some hyperplane transversal to all strata of some Whitney stratification of V. Then there exist generic pencils of hyperplanes having H as a generic member and we have:

(a) $H \cap (\mathbb{P}^n \setminus V) \hookrightarrow \mathbb{P}^n \setminus V$ is a $(n + \operatorname{codim} V - 2)$-equivalence.
(b) If $\dim V \leq 2n - 5$, then the kernel of the surjection $\pi_{n + \operatorname{codim} V - 2}(H \cap (\mathbb{P}^n \setminus V), \cdot) \twoheadrightarrow \pi_{n + \operatorname{codim} V - 2}(\mathbb{P}^n \setminus V, \cdot)$ is generated by the images of the variation maps hvar_i of such a generic pencil.

Proof We may choose a hyperplane A of $H = \mathbb{P}^{n-1}$, hence $\operatorname{codim}_{\mathbb{P}^n} A = 2$, such that A is transversal to all strata of Y. The set of generic axes A is a Zariski–open subset of the set of all axes $A \subset H \subset \mathbb{P}^n$. Let us then fix in the following such an axis and consider the corresponding pencil.

The part (a) is classical and well-known and its proof goes by induction on $\dim V$. By repeated slicing with generic hyperplanes, the dimension of V goes down by 1 at each step until we reach the dimension zero. At each stage we choose a generic pencil.

Let us see that the conditions of Theorem 11.2.1 are satisfied at the induction step. We have $X := \mathbb{P}^n \setminus V$. The generic fibre X_c of our pencil is path connected at each step. Condition (iii), resp. (iii'), is empty at every step, since a generic pencil has no singularities in $\mathbb{P}^{n-j} \setminus V$.

The condition (i) in case $\dim V = 0$ (hence V consists of finitely many points) is the level of connectivity of the pair $(\mathbb{P}^n \setminus V, \mathbb{P}^{n-1})$, which is known to be equal to $2n - 2$. Indeed, we have that $(\mathbb{P}^m, \mathbb{P}^{m-1})$ is $(2m - 1)$-connected and the homotopy depth of $\mathbb{P}^m$ is $2m$.

If condition (i) is satisfied, then (ii) is equivalent to the condition "$(X_{a_i}^*, A')$ is $(k - 1)$-connected" as shown above, cf. (11.16) of §11.3. Let us notice that $X_{a_i}^* = X_{a_i}$ in our case. But then the pair (X_{a_i}, A') is of the same type as the pair in the conclusion of our statement, namely X_{a_i} is the complement of some projective space in $\mathbb{P}^{n-1}$ and A' is a generic hyperplane in $\mathbb{P}^{n-1}$, and therefore this condition too can be reproduced at each step by the conclusion of Theorem 11.2.1(a).

We also need to check the initial conditions for our induction, i.e. the start of the induction. In case dim $V = 0$, we have that $(X_{a_i}, A') = (\mathbb{P}^m \setminus V, \mathbb{P}^{m-1} \setminus V)$ and this is $(2m - 2)$-connected. Then Theorem 11.2.1(a) says that the higher pair $(\mathbb{P}^{m+1} \setminus V, \mathbb{P}^m \setminus V)$ is $(2m - 1)$-connected. This becomes the conditions (i), and (ii) in its equivalent form (11.16), for the next induction step.

In conclusion, after a number equal to dim V of steps of induction we get the connectivity $2n - \dim V - 2$, which was our claim. So far we have proved part (a).

Part (b) follows from (a) and from Theorem 11.2.1(b), since all conditions are fulfilled, modulo the fact that our homotopy variation map works for homotopy groups starting with level 3 (by its definition). This fact amounts to the condition $2n - \dim V - 2 \geq 3$. The proof is now complete. $\qquad\square$

Part (b) of this Corollary is a new result since we consider the general situation of complements of arbitrarily singular subspaces. There are three cases not covered by the condition dim $V \leq 2n - 5$, namely the case $n = 3$, dim $V = 2$, and the cases $n = 2$ and dim $V = 1$ or dim $V = 0$. The last one is trivial, since clearly the kernel we are looking for is just $\mathbb{Z}$. The second case is covered by the classical Zariski–van Kampen theorem. The only open case is therefore the case of complements of surfaces in $\mathbb{P}^3$ (see Exercise 11.2).[4]

Exercises

11.1 Prove the following complementary case of Theorem 11.2.1, which is the homotopy analogue of Theorem 10.2.11:

Let $h : Y \dashrightarrow \mathbb{P}^1$ define a pencil with isolated singularities, such that V contains a member of the pencil. Le X_c be path connected and let $(X_c, X_{a_i}^*)$ be k-connected, where $c \notin \Lambda$ is a regular value of the pencil and $a \in \Lambda$ is a critical one. Let $\Sigma := X \cap \sigma(\mathrm{Sing}_S p)$.

 (i) If $k \geq 0$ and $(X, X \setminus \Sigma)$ is $(k + 1)$-connected, then the inclusion $X_c \hookrightarrow X$ is a $(k + 1)$-equivalence.

 (ii) If $k \geq 2$ and if $(X, X \setminus \Sigma)$ is $(k + 2)$-connected, then the kernel of the surjection $\pi_{k+1}(X_c) \twoheadrightarrow \pi_{k+1}(X)$ is generated by the images of the variation maps hvar_i.

11.2 Prove that the case $k = 1$ in Theorem 11.2.1(b) can be recovered by showing that the construction of the homotopy variation map may be carried out in case $q = 2$.

Appendix 1
Stratified singularities

A1.1 Stratifications of singular spaces and fibration theorems

'Whitney stratifications' is a key device for studying singular analytic spaces, in the real or complex setting. An excellent reference to this topic is [GWPL] and an introduction, with historical remarks, may be found in [GM2]. Let us briefly recall some basics. A *stratification* of an analytic space $\mathcal{X}$ (we stick to this setting) is a partition of $\mathcal{X}$ into a locally finite collection $\mathcal{S} = \{S_i\}_i$ of connected locally closed submanifolds, such that, if $\mathcal{S}_i \cap \overline{\mathcal{S}_j} \neq \emptyset$, then $\mathcal{S}_i \subset \overline{\mathcal{S}_j}$ and in this case we write $\mathcal{S}_i < \mathcal{S}_j$. The stratification $\mathcal{S}$ is called a *Whitney stratification* if any two strata $\mathcal{S}_i < \mathcal{S}_j$ satisfy the conditions (a) and (b), as follows. Let $x_k \in \mathcal{S}_j$ be a sequence of points converging to some $y \in \mathcal{S}_i$. Suppose that in some local coordinate system the tangent planes $T_{x_k}\mathcal{S}_j$ converge to some plane T in the appropriate Grassmannian. Then:

(a) $T \supset T_y \mathcal{S}_i$.

Let moreover $y_k \in \mathcal{S}_i$ be a sequence of points converging to the same point $y \in \mathcal{S}_i$. Suppose that, in the same local coordinate system, the secant lines $l_i = \overline{x_k y_k}$ converge to some line l. Then:

(b) $T \supset l$.

It was proved that condition (b) implies (a). We shall use the following well-known facts, and send to the textbooks [GM2, GWPL] for the appropriate references:

1 Any closed analytic (or subanalytic) subset $\mathcal{X}$ of an analytic manifold admits a Whitney stratification such that its strata are analytic manifolds. In the complex analytic setting, we call this a *complex Whitney stratification*.

219

2 The transversal intersection of two Whitney stratified subspaces is a Whitney stratified space whose strata are the intersections of the strata of the two spaces (see e.g. [GWPL, p. 19]).

3 Whitney stratifications are topologically trivial along the strata. More generally, one has *Thom–Mather's first isotopy lemma*: given a proper stratified submersion $f : \mathcal{X} \to \mathbb{K}^p$, there is a stratum preserving homeomorphism $\phi : \mathcal{X} \to f^{-1}(t_0) \times \mathbb{K}^p$, which is smooth on each stratum and commutes with the projection on $\mathbb{K}^p$; in particular the fibres of f are homeomorphic by stratum preserving homeomorphisms, which are C^∞-diffeomorphisms on each stratum.

Let us introduce a stratification which is weaker than a Whitney stratification but still allows us to prove a local isotopy theorem, which is useful in the study of the topology of polynomial and meromorphic functions.

Let $g : \mathcal{X} \to \mathbb{K}$ be a $\mathbb{K}$-analytic function, where $\mathbb{K} = \mathbb{C}$ or $\mathbb{R}$. Suppose that $\mathcal{X}$ is endowed with a complex (resp. real) stratification $\mathcal{G} = \{\mathcal{G}_\alpha\}_{\alpha \in \Lambda}$, such that $g^{-1}(0)$ is a union of strata.[*]

Definition A1.1.1 We say that $\mathcal{G}$ is a $\partial\tau$-stratification (*partial Thom stratification*) relative to $g^{-1}(0)$ if, for any two strata $\mathcal{G}_\alpha < \mathcal{G}_\beta$ with $\mathcal{G}_\alpha \subset g^{-1}(0)$ and $\mathcal{G}_\beta \subset \mathcal{X} \setminus g^{-1}(0)$, the Thom (a_g)-regularity condition is satisfied at all the points of $\mathcal{G}_\alpha$.

One may look up for instance [GWPL, ch. I] for the definition of the (a_g) regularity condition, introduced by Thom. This is a relative (a)-type condition, which, in terms of conormal spaces, amounts to the following: two strata $\mathcal{G}_\alpha < \mathcal{G}_\beta$ satisfy the *Thom (a_g)-regularity condition* if and only if, at any point $\xi \in \mathcal{G}_\alpha$, the relative conormal space of g on $\bar{\mathcal{G}}_\beta$ is included into the conormal of $\mathcal{G}_\alpha$, i.e., $(T^*_{g|\bar{\mathcal{G}}_\beta})_\xi \subset (T^*_{\mathcal{G}_\alpha})_\xi$.

Local $\partial\tau$-stratifications exist since local Thom stratifications relative to a function exist, see e.g. [HmL1, Théorème 1.2.1], [Hiro], [Be] ('local' means in some neighbourhood of a point x, for any $x \in \mathcal{X}$). Nevertheless $\partial\tau$-stratifications are less demanding than Thom stratifications.[**]

Fibration theorems. The existence of Thom–Whitney stratifications and of partial Thom stratifications allows us to prove fibration theorems. A well-known one is the *Milnor–Lê's local fibration theorem*, in the complex setting:

Theorem A1.1.2 [Lê3] *Let $h : (\mathcal{X}, x) \to (\mathbb{C}, 0)$ be a holomorphic function germ. For any ball B_ε centered at x of sufficiently small radius $\varepsilon > 0$ and any $0 < \delta \ll \varepsilon$, the restriction:*

$$h_| : B_\varepsilon \cap h^{-1}(D^*_\delta) \to D^*_\delta \tag{1.1}$$

---

[*] It is well known that we may refine a given stratification such that this assumption also holds.

[**] Also called Thom–Whitney stratifications, since these are Whitney stratifications which satisfy in addition the Thom (a_f)-regularity condition relative to some function f for any couple of strata $\mathcal{G}_\alpha < \mathcal{G}_\beta$. See also Theorem A1.1.7 below.

is a locally trivial C^∞ fibration. Its fibre is called Milnor–Lê *fibre. The isotopy type of the fibration does not depend on the choice of the radii ε and δ.*

Theorem A1.1.3 (open local isotopy) [Ti4]
Let $(\mathcal{X}, x) \subset (\mathbb{K}^N, x)$ be a $\mathbb{K}$-analytic irreducible space germ, $\dim_x \mathcal{X} \geq 2$. Let $h : (\mathcal{X}, x) \to (\mathbb{K}, 0)$ be a $\mathbb{K}$-analytic map transversal to a $\partial \tau$-stratification $\mathcal{G}$ relative to $g^{-1}(0)$. Suppose moreover that $\mathcal{X} \setminus g^{-1}(0)$ is nonsingular. Then, for any small enough $\varepsilon > 0$ and $0 < \delta \ll \varepsilon$, the following restriction of h:

$$h_| : B_\varepsilon \cap h^{-1}(D_\delta) \cap (\mathcal{X} \setminus g^{-1}(0)) \to D_\delta$$

is a topologically trivial fibration.

Proof We give the proof from [Ti4] since in the same manner we may prove several fibration theorems, e.g. Lê's theorem A1.1.2.

The stratification $\mathcal{G}$ may fail to be Whitney, hence we cannot apply Thom's first isotopy lemma. Also $\mathcal{G}$ might not be (c)-regular in the sense of Bekka [Be, §2, Def. 1.1], nevertheless it is of a similar flavour.

The idea of proof is to lift by h the unit real (resp. complex) vector field $\partial / \partial t$ on D_δ to a vector field on $\mathcal{X}$, tangent to both $S_\varepsilon := \partial \overline{B_\varepsilon}$ and to all positive levels of g.

First choose $\varepsilon > 0$ such that, for all $0 < \varepsilon' \leq \varepsilon$, the sphere $S_{\varepsilon'}$ is transversal to all strata of $\mathcal{G}$ and to all positive dimensional strata of the stratification induced by $\mathcal{G}$ on $h^{-1}(0)$ (by e.g. [HmL1, Theorem 1.3.2] or [Loo, Lemma 2.2]). Our hypothesis on h implies that the levels of h are transversal to the levels of g different from 0, at any point of $B_\varepsilon \setminus g^{-1}(0)$, for ε small enough. We may therefore define a continuous vector field $\mathbf{v}$ on $B_\varepsilon \cap h^{-1}(D_\delta) \cap (\mathcal{X} \setminus g^{-1}(0))$, without zero, which is a pull-back of the unit vector field $\partial / \partial t$ on D_δ and is tangent to the levels of g. This follows for instance from the fact that $\mathcal{G}$ is, so to say, 'partially' (c)-regular, in the sense that $|g|^2$ is a control function for two strata $\mathcal{G}_\alpha < \mathcal{G}_\beta$ satisfying $\mathcal{G}_\alpha \subset g^{-1}(0)$ and $\mathcal{G}_\beta \subset \mathcal{X} \setminus g^{-1}(0)$. Therefore, we may still construct a continuous vector field $\mathbf{v}$ as done by Bekka in [Be, p. 61, point 1].

But $\mathbf{v}$ might not be tangent to the sphere S_ε, so we have to modify it in the neighbourhood of $S_\varepsilon \cap h^{-1}(D_\delta)$. We need the following:

Lemma A1.1.4 Suppose that $h : (\mathcal{X}, x) \to (\mathbb{K}, 0)$ is transversal to $\mathcal{G}$ except possibly at x. Then for some $0 < \alpha \ll \varepsilon, 0 < \delta \ll \varepsilon$ we have that $h^{-1}(h(q))$ is transversal to $S_\varepsilon \cap g^{-1}(g(q))$, for any $q \in S_\varepsilon \cap h^{-1}(D_\delta) \cap g^{-1}(D_\alpha^*)$.

Proof Our hypothesis implies that $\Gamma_\mathcal{G}(g, h) \cap g^{-1}(0) \subset \{x\}$. Indeed, if $y \in g^{-1}(0)$, $y \neq x$, then h is transversal to the level of g at any point within $\mathcal{N}_y \cap \mathcal{X} \setminus g^{-1}(0)$, where $\mathcal{N}_y$ is some small enough neighbourhood of y. It follows that $\Gamma_\mathcal{G}(g, h) \cap S_\varepsilon \cap g^{-1}(D_\alpha) = \emptyset$, for small enough ε and $0 < \alpha \ll \varepsilon$. We recall that ε was chosen small enough such that S_ε is transversal to all positive-dimensional strata of the form $h^{-1}(0) \cap \mathcal{G}_\beta$, with $\mathcal{G}_\beta \in \mathcal{G}$.

If the conclusion of this lemma is not true, then there would exist a sequence of points $p_i \in S_\varepsilon \cap g^{-1}(D_\alpha^*) \subset \mathcal{X} \setminus g^{-1}(0)$, with $p_i \to p \in S_\varepsilon \cap h^{-1}(0) \cap g^{-1}(0)$, such that the intersection of tangent spaces $T_{p_i} h^{-1}(h(p_i)) \cap T_{p_i} g^{-1}(g(p_i))$ is contained in $T_{p_i}(S_\varepsilon \cap \mathcal{X})$. Then, provided that the following limits exist (which we may assume without loss of generality), we would have:

$$\lim T_{p_i} h^{-1}(h(p_i)) \cap \lim T_{p_i} g^{-1}(g(p_i)) \subset T_p S_\varepsilon. \tag{1.2}$$

On the other hand, let $\mathcal{G}_\phi \subset g^{-1}(0)$ be the stratum containing p. We have that $\lim T_{p_i} h^{-1}(h(p_i)) \supset T_p \mathcal{G}_\phi$, since $\mathcal{G}$ is a $\partial \tau$-stratification. Now, since h is transversal to $\mathcal{G}_\phi$ and since $p \in h^{-1}(0)$, the set $h^{-1}(0) \cap \mathcal{G}_\phi$ must have positive dimension and we have assumed that S_ε is transversal to it. But this contradicts (1.2).   □

We now complete the last part of the proof of Theorem A1.1.3. From Lemma A1.1.4, it follows that there exists a continuous vector field **w** without zero on some collar of $S_\varepsilon \cap h^{-1}(D_\delta) \cap g^{-1}(D_\alpha^*) \subset \overline{B_\varepsilon} \cap h^{-1}(D_\delta) \cap g^{-1}(D_\alpha^*)$, which is a lift of $\partial / \partial t$ by h and is tangent to $S_{\varepsilon'} \cap g^{-1}(p)$, for any $p \in D_\alpha^*$ and $\varepsilon' \leq \varepsilon' \leq \varepsilon$, for some ε' close to ε.

Lastly, there exists another continuous vector field **u** without zero on $B_\varepsilon \cap h^{-1}(D_\delta) \setminus g^{-1}(D_\alpha)$, which is again a lift of $\partial / \partial t$ and is tangent to the sphere S_ε (by the choice of ε).

We then glue those three **u**, **v**, **w** by a partition of unity and get a vector field which trivializes the fibration $h_|$.   □

We stick to the complex setting for the remainder. We have the following Lefschetz type result extending Hamm and Lê's statement [HmL1, Corollary 4.2.2] from the point of view of the stratification involved in the statement.

Corollary A1.1.5 Let $(\mathcal{X}, x)$ be a complex analytic space germ and let $h, g : (\mathcal{X}, x) \to (\mathbb{C}, 0)$ be holomorphic functions such that h is transversal to a $\partial \tau$-stratification with respect to $g^{-1}(0)$, except possibly at x. Assume that $\mathcal{X} \setminus g^{-1}(0)$ is nonsingular of dimension n. Then for any ε small enough, any $0 < \delta \ll \varepsilon$ and any $\eta \in D_\delta^*$, the pair:

$$(B_\varepsilon \cap h^{-1}(D_\delta) \cap \mathcal{X} \setminus g^{-1}(0), B_\varepsilon \cap h^{-1}(\eta) \cap \mathcal{X} \setminus g^{-1}(0))$$

is $(n-1)$-connected.

Proof Under the stated conditions, the following pair:

$$(B_\varepsilon \cap h^{-1}(D_\delta) \cap \mathcal{X} \setminus g^{-1}(0), (S_\varepsilon \cap h^{-1}(D_\delta)) \cup (B_\varepsilon \cap h^{-1}(\eta) \cap \mathcal{X} \setminus g^{-1}(0)))$$

is $(n-1)$-connected, by [HmL1, Theorem 4.2.1]. We should check that the proof in [HML1], still works in our setting.

By Lemma A1.1.4, the restriction of h to $S_\varepsilon \cap h^{-1}(D_\delta) \cap \mathcal{X} \setminus g^{-1}(0)$ induces a topologically trivial fibration over D_δ, and the conclusion follows.

□

Local stratified triviality An interesting and delicate problem would be to find reasonable converses to Thom's first isotopy lemma or to Theorem A1.1.3. Let us discuss an aspect of this matter, following [BMM, Ti4].

So let $\mathcal{X} \subset \mathbb{C}^N$ be a complex analytic set endowed with a Whitney (a)-regular stratification. Let $\mathcal{G}_\alpha$ be a stratum, $\dim \mathcal{G}_\alpha > 0$, let $x \in \mathcal{G}_\alpha$ and let $h : (\mathbb{C}^N, x) \to (\mathbb{C}^p, 0)$ be a submersion transversal to $\mathcal{G}_\alpha$. Let D_η denote the open ball centered at $0 \in \mathbb{C}^p$, of radius η and let $B_\varepsilon \subset \mathbb{C}^N$ be some ball centered at x, of radius ε. For $\varepsilon > 0$ small enough, $\mathcal{G}$ induces a Whitney (a)-regular stratification on $h^{-1}(0)$; therefore $\mathcal{X} \cap (B_\varepsilon \cap h^{-1}(0))$ is stratified, its strata being the intersections of $h^{-1}(0)$ with the strata of $\mathcal{G}$. Then $D_\eta \times (\mathcal{X} \cap (B_\varepsilon \cap h^{-1}(0)))$ is endowed with the product stratification.

Definition A1.1.6 [BMM, 4.1] The stratification $\mathcal{G}$ of $\mathcal{X}$ satisfies the *local stratified triviality* property (abbreviated LST) if and only if for any point $x \in \mathcal{X}$ and any submersion $h : (\mathbb{C}^N, 0) \to (\mathbb{C}^p, 0)$ transversal to $\mathcal{G}_\alpha$ at x, where $x \in \mathcal{G}_\alpha$ and $0 < p \leq \dim \mathcal{G}_\alpha$, there are $\eta > 0$, $\varepsilon > 0$, some neighbourhood $U \subset \mathbb{C}^N$ of x and a stratified homeomorphism Φ such that the following diagram commutes:

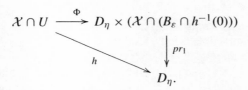

One notices that the LST property is preserved when slicing by hyperplanes transversal to the strata. Moreover, the LST property is preserved when restricting to the closure $\overline{\mathcal{G}_\beta}$ of some stratum, since Φ is a stratified homeomorphism. The LST property is verified for instance by Whitney stratifications, by Thom's first isotopy lemma.

Theorem A1.1.7 [BMM, Ti4] *Let $\mathcal{X} \subset \mathbb{C}^N$ be endowed with a Whitney (a)-regular complex stratification $\mathcal{G}$, such that $g^{-1}(0)$ is a union of strata and that $\mathcal{G}$ verifies the LST property. Then $\mathcal{G}$ is a $\partial\tau$-stratification relative to $g^{-1}(0)$.*

This is a reformulation of a result due to Briançon, Maisonobe and Merle [BMM, Theorem 4.2.1]. The original proof uses $\mathcal{D}$-modules techniques. We give bellow the geometric proof from [Ti4].

Proof Let $Y \subset g^{-1}(0)$ be the locus where the (a_g)-regularity fails and assume that $Y \neq \emptyset$. Let $\mathcal{T}$ be some complex (a_g)-regular stratification of $\mathcal{X}$, which refines $\mathcal{G}$ and such that new strata occur only within Y; we remark that this is possible by the definition of Y. Let $\mathcal{T}_0 \subset Y$ be a stratum of maximal dimension among the 'new' strata. We may assume that $\dim \mathcal{T}_0 = 0$. (If not so, then we reduce the problem to this case by slicing with a plane V with $\dim V = N - \dim \mathcal{T}_0$. Then the slice $\mathcal{X} \cap V$ inherits the same properties as the ones assumed for $\mathcal{X}$.)

Then let $y \in \mathcal{T}_0$ be an isolated point of our set Y, which is assumed to be of dimension zero, by the reduction argument. Now let $\mathcal{G}_0$ be the stratum of $\mathcal{G}$ containing y. Notice that $\dim \mathcal{G}_0 > 0$, by the definition of $\mathcal{T}$. Our hypothesis *ad absurdum* amounts to:

$$(T^*_{g|\mathcal{X}})_y \not\subset T^*_{\mathcal{G}_0}(\mathbb{C}^N). \tag{1.3}$$

Then there is some stratum $\mathcal{G}_1 \in \mathcal{G}$, $\mathcal{G}_1 \subset \mathcal{X} \setminus \{g^{-1}(0)\}$, of dimension ≥ 2, such that:

$$(T^*_{g|\overline{\mathcal{G}_1}})_y \not\subset T^*_{\mathcal{G}_0}(\mathbb{C}^N). \tag{1.4}$$

Let us pick up $\xi \in (T^*_{g|\overline{\mathcal{G}_1}})_y \setminus T^*_{\mathcal{G}_0}(\mathbb{C}^N)$ and identify ξ with a linear form l on $\mathbb{C}^N$. We then consider the germ at y of the polar locus $\Gamma_{\mathcal{G}}(l, g_{|\overline{\mathcal{G}_1}})$. By the definition of ξ, we have $\Gamma_{\mathcal{G}}(l, g_{|\overline{\mathcal{G}_1}}) \cap g^{-1}(0) = \{y\}$, and by typical 'polar curve arguments' (as in the proof of Theorem 7.1.2) we show that $\dim \Gamma_{\mathcal{G}}(l, g_{|\overline{\mathcal{G}_1}}) = 1$.

The LST property implies that the *complex links* $\mathrm{CL}(\overline{\mathcal{G}_1}, y)$ and $\mathrm{CL}(\overline{\mathcal{G}_1} \cap g^{-1}(0), y)$ are contractible (see Definitions 6.1.4 and A1.2.3). We consider the germ at y of the map $(l, g_|) : \overline{\mathcal{G}_1} \to \mathbb{C}^2$ and take a small enough in the *polydisk neighbourhood* $\mathcal{P}$ of y (see [Lê2] and also [Ti2] for the definition). Then the slice $l^{-1}(l(y) + \varepsilon) \cap \mathcal{P}$, for $\varepsilon > 0$ small enough, is just the complex link $\mathrm{CL}(\overline{\mathcal{G}_1}, y)$. This slice is obtained, up to homotopy type, by attaching a finite number of cells of dimension $= \dim_{\mathbb{C}} \mathcal{G}_1 - 1$ to $l^{-1}(l(y) + \varepsilon) \cap g^{-1}(0) \cap \mathcal{P}$, which is just the complex link $\mathrm{CL}(\overline{\mathcal{G}_1} \cap g^{-1}(0), y))$ (e.g. lookup the proof of Theorem 4.2.5). These cells correspond to the singular points of the function g on $l^{-1}(l(y) + \varepsilon) \cap \mathcal{P}$ (namely to one singular point there corresponds at least one cell), which are exactly the points where the polar curve $\Gamma_{\mathcal{G}}(l, g_{|\overline{\mathcal{G}_1}})$ intersects $l^{-1}(l(y) + \varepsilon) \cap \mathcal{P}$. Since $\Gamma_{\mathcal{G}}(l, g_{|\overline{\mathcal{G}_1}})$ is a nonempty curve, the number of these intersection points is greater than zero, hence the number of cells is also greater than zero. This contradicts the contractibility of $\mathrm{CL}(\overline{\mathcal{G}_1}, y)$ and finishes our *reductio ad absurdum* proof. $\square$

The Milnor fibre of holomorphic germs. We have seen by the Milnor–Lê theorem A1.1.2 that to any holomorphic function germ $h : (\mathcal{X}, x) \to (\mathbb{C}, 0)$ we associate a Milnor type fibration. It has been proved by Milnor [Mi2] that the Milnor fibre $M_h(x)$ has a nice structure in case of a function with isolated singularity on a nonsingular space, see § 3.2.

More generally, in case h has a *stratified isolated singularity* at x, i.e. $\dim \text{Sing}_{\mathcal{W}} h = 0$ for some Whitney stratification $\mathcal{W}$ on $\mathcal{X}$ (cf. Definition 2.2.1), we have some more information on the homotopy type of the Milnor fibre $M_h(x)$. Lê pointed out that, under certain conditions, the space $\mathcal{X}$ has 'Milnor's property' in homology (which means that the reduced homology of the Milnor fiber of h is concentrated in dimension $\dim \mathcal{X} - 1$, where $\mathcal{X}$ is purely dimensional). Then the *Milnor–Lê number* $\mu(h)$ is well defined as the rank of this homology group. Milnor's property is satisfied, cf. [Lê5], if $(\mathcal{X}, x)$ is a singular complete intersection and more generally, if rHd $(\mathcal{X}, x) \geq \dim(\mathcal{X}, x)$, where rHd $(\mathcal{X}, x)$ denotes the *rectified homological depth*[*] of $(\mathcal{X}, x_0)$.

If we are interested in the structure of the Milnor fibre up to homotopy type, then similar results hold, namely the Milnor fibre is homotopically a bouquet of spheres $\vee S^{n-1}$ as soon as $\mathcal{X}$ is irreducible, of pure dimension n and is a complete intersection or has maximal rectified homotopical depth, i.e. rhd $(\mathcal{X}, x) \geq \dim(\mathcal{X}, x)$. In the general case, we have the following result, the proof of which relies on a 'controlled attaching of cells' provided by the carrousel monodromy:

Theorem A1.1.8 [Ti2] *Let $h : (\mathcal{X}, x) \to (\mathbb{C}, 0)$ have a stratified isolated singularity with respect to a local Whitney stratification $\mathcal{W} = \{\mathcal{W}_i\}_{i=1}^q$. Then we have the following homotopy equivalence for the Milnor fibre:*

$$M_h(x) \overset{\text{ht}}{\simeq} \text{CL}(\mathcal{X}, x) \vee \vee_{i=1}^q (\bigvee S^{k_i}(\text{CL}_{\mathcal{X}}(\mathcal{W}_i))),$$

where $k_i = \dim_{\mathbb{C}} \mathcal{W}_i$, $\text{CL}_{\mathcal{X}}(\mathcal{W}_i)$ is the complex link of the stratum $\mathcal{W}_i$ and S^{k_i} means the k_i-times repeated suspension.[**] □

The number of objects in the last bouquet $\bigvee$ depends on polar invariants; one may find the details in [Ti2].

A1.2 Stratified Morse theory, characteristic classes and cycles

Stratified Morse theory. We recall here very few basics on the stratified Morse theory and send the reader to [GM1, GM2].

Let M be a complex analytic manifold and let $X \subset M$ be a complex analytic subvariety, endowed with a complex Whitney stratification $\mathcal{S}$.

---

[*] See Definition 9.2.4.

[**] By convention, the suspension over the empty set is the 0-sphere S^0, i.e. two points.

Definition A1.2.1 Let $N \subset M$ be the germ of a closed complex submanifold. We say that N is a *normal slice* to some stratum S of X, at some point $x \in S$, if N is transversal to S and $N \cap S = \{x\}$.

A covector $(x, \xi) \in T_S^* M$ is called *normally nondegenerate* if $(x, \xi) \notin \overline{T_{S'}^* M}$, for all strata $S' \neq S$.

Definition A1.2.2 Let $x \in S$ and let $g : (M, x) \to (\mathbb{K}, 0)$, where $\mathbb{K} = \mathbb{R}$ or $\mathbb{C}$, be a C^∞ function germ, such that dg_x is normally nondegenerate with respect to S. We say that x is a *stratified Morse critical point* of g with respect to the stratification $\mathcal{S}$ if x is a classical Morse critical point (real or complex) of the restriction $g_{|S}$ to the submanifold S.

The key notion of the complex stratified Morse theory is the complex link [GM2, p. 161, Def. 2.2]. Let $S \in \mathcal{S}$, $x \in S$ and let $h : (M, x) \to (\mathbb{C}, 0)$ be a holomorphic function germ such that $dh_x \in T_S^* M$ is normally nondegenerate. Let N be a normal slice to S at x. The local Milnor–Lê fibration (Theorem A1.1.2) of the restriction $h_{|X \cap N}$ has an isolated stratified critical point at x.

Definition A1.2.3 We call *complex link* of the stratum S the Milnor fibre:

$$\mathrm{CL}_X(S) := X \cap N \cap B_\delta(x) \cap \{h = u\} \quad \text{for} \quad 0 < |u| \ll \delta \ll 1.$$

of the Milnor–Lê fibration of h. We call *complex link of X at x* the complex link $\mathrm{CL}_X(\{x\})$ of the point-stratum $\{x\}$, also denoted by $\mathrm{CL}(X, x)$.

The pair of spaces:

$$\mathrm{NMD}(S) := (X \cap N \cap B_\delta(x), X \cap N \cap B_\delta(x) \cap \{h = u\}) \quad \text{for} \quad 0 < |u| \ll \delta \ll 1,$$

is called the *normal Morse datum* of S.

Goresky and MacPherson [GM2, p.163, Theorem 2.3] showed that the stratified homeomorphism type of the complex link $\mathrm{CL}_X(S)$ as well as the normal Morse datum $\mathrm{NMD}(S)$ are independent of all choices (in particular, of the choice of the point $x \in S$) and thus they are indeed invariants of the stratum S in X.

Definition A1.2.4 Let $\alpha : X \to \mathbb{Z}$ be a constructible function with respect to the stratification $\mathcal{S}$ of X. Then its *normal Morse index* along S:

$$\eta(S, \alpha) := \chi(\mathrm{NMD}(S), \alpha) = \chi(X \cap N \cap B_\delta(x), \alpha) - \chi(\mathrm{CL}_X(S), \alpha) \quad (1.5)$$

is a well-defined invariant of the stratum S.[*]

---

[*] We refer the reader to Schürmann's book [Sch] for all details and results on constructible functions and constructible sheaves.

In particular, when $\dim(S) = \dim(X)$, we have $\mathrm{CL}_X(S) = \emptyset$ and we get:

$$\eta(S, \alpha) = \chi(\mathrm{NMD}(S), \alpha) = \alpha(x). \tag{1.6}$$

Chern classes and characteristic cycles. We give a very brief account on several notions used in §6.2 and §6.3 and refer to [Sch, ScTi] for the full details. Let $F(S)$ denote the group of constructible functions with respect to the stratification S. Let $L(S)$ be the group of *conic Lagrangian cycles* generated by the conormal spaces $T_{\bar{S}}^*M := \overline{T_S^*M}$ to the closures of strata S of our stratification.

Definition A1.2.5 The *characteristic cycle map* is defined as follows:

$$cc : F(S) \to L(S); \quad cc(\alpha) = \sum_{S \in \mathcal{S}} (-1)^{\dim S} \eta(S, \alpha) \cdot T_{\bar{S}}^*M,$$

where the sum is locally finite.

The map cc is an isomorphism:

$$cc : F(S) \xrightarrow{\sim} L(S), \tag{1.7}$$

the injectivity follows by (1.6), and the surjectivity by induction on $\dim(X)$. Denoting by $L(X, M)$ be the group of all conic Lagrangian cycles, for all Whitney stratifications on X, it is not difficult to see that the characteristic cycle $cc(\alpha) \in L(X, M)$ is independent of the choice of the stratification $\mathcal{S}$.

The following index theorem is one of the main applications of stratified Morse theory to the theory of characteristic cycles. A proof in terms of real Morse theory for constructible sheaves can be found in [Sch, pp. 290–292].

Theorem A1.2.6 *Let* $f : M \to [a, d[\subset \mathbb{R}$ *be a smooth function, where* $d \in \mathbb{R} \cup \infty$, *such that* $f_{|X}$ *is proper and the stratified critical locus* $\mathrm{Sing} f$ *is compact with image* $f(\mathrm{Sing} f) \subset [a, b]$. *Then, for any* $\alpha \in F(S)$, *the global intersection index* $\sharp([cc(\alpha)] \cup [df(M)])$ *is equal to the Euler characteristic* $\chi(X \cap \{f \leq r\}, \alpha)$, *for all* $r \in]b, d[$. $\qquad\square$

Let us briefly recall the main ingredients in MacPherson's definition of his (dual) Chern classes of a constructible function and the relation to the theory of characteristic cycles.* These enter in the following key diagram used in §6.3:

$$
\begin{array}{ccccc}
F(X) & \xleftarrow[\sim]{\check{\mathrm{Eu}}} & Z(X) & \xrightarrow{\check{c}_*^{M\;a}} & H_*(X) \\
& \underset{cc}{\searrow} & {\scriptstyle cn}\Big\downarrow{\scriptstyle\sim} & \underset{c^*(T^*M_{|X})\cap s_*}{\nearrow} & \\
& & L(X, M) & &
\end{array}
\tag{1.8}
$$

---

* Some other useful references for this topic are [BDK, Dub2, Gi, Ken, Sab].

The notations $F(X)$ and $Z(X)$ stay for the groups of constructible functions and cycles respectively, in the complex analytic or algebraic context. We denote by $H_*(X)$ either the Borel–Moore homology group in even degrees $H_{2*}^{BM}(X, \mathbb{Z})$ or the Chow group $CH_*(X)$.

It follows from (1.7) that the morphism $cc : F(X) \xrightarrow{\sim} L(X, M)$ from the above diagram is an isomorphism. The vertical map cn is the correspondence $Z \mapsto T_Z^* M$. By definition, $L(X, M)$ is the group of cycles generated by the conormal spaces, so cn is an isomorphism. We shall explain the other notations and give an idea of the proof of the commutativity of the diagram (1.8).

MacPherson's Chern class transformation. Some properties of the local Euler obstruction Eu_X have been recalled in §6.2; in particular, Eu_X is a constructible function. The maps $\check{\mathrm{Eu}}$ and $\check{c}_*^{Ma}$ from diagram (1.8) were defined in §6.3. The *dual MacPherson Chern class transformation*:

$$\check{c}_*^M := \check{c}_*^{Ma} \circ \check{\mathrm{Eu}}^{-1} : F(X) \to H_*(X), \tag{1.9}$$

also defined in §6.3, agrees up to sign with MacPherson's original definition of his Chern class transformation c_*^M:

$$\check{c}_i^M(\alpha) = (-1)^i \cdot c_i^M(\alpha) \in H_i(X).$$

The dual Euler obstruction $\check{\mathrm{Eu}}_Z$ and the dual Chern–Mather classes $\check{c}_*^{Ma}(Z)$ were originally defined in terms of the Nash blow-up $\hat{Z}$ of Z. The handling of this objects in §6.3 needs their interpretation in the language of the conormal space $T_Z^* M := \overline{T_{Z_{\mathrm{reg}}}^* M}$ of Z in M.

For $\check{c}_*^{Ma}(Z)$ we have the following interpretation in terms of the Segre classes:

$$\check{c}_*^{Ma}(Z) = c^*(T^* M_{|Z}) \cap s_*(T_Z^* M), \tag{1.10}$$

the details and the appropriate references may be found in [ScTi]. This obviously yields the commutativity of the right triangle in (1.8).

The description of the *dual Euler obstruction* $\check{\mathrm{Eu}}_Z$ of Z in terms of the conormal space $T_Z^* M$ and 1-forms may be found in [Sch, §5.2]. We have the following micro-local description of the dual Euler obstruction as a local intersection number of conormal cycles at $dr_z \in T^* M$:

$$\check{\mathrm{Eu}}_Z(z) = \sharp_{dr_z}([T_Z^* M] \cup [dr(M)]), \tag{1.11}$$

The Euler obstruction $\check{\mathrm{Eu}}_{\bar{S}}$ is constructible with respect to the stratification $\mathcal{S}$, and it turns out that its characteristic cycle verifies the following equality (see [Sch1, ScTi]):

$$cc(\check{\mathrm{Eu}}_{\bar{S}}) = T_{\bar{S}}^* M. \tag{1.12}$$

This proves the commutativity of the left triangle in (1.8).

Appendix 2

Hints to some exercises

1.2. Use the Tarski–Seidenberg theorem.

1.4. Use the Lefschetz number of the local monodromy at some critical point.

1.5. Use the curve selection lemma.

1.6. Use the function ρ from Example 1.2.3. Let $E_r := \{x \in \mathbb{C}^n \mid \rho(x) < r\}$ for some $r > 0$. Then the local Milnor fibre of f at $0 \in \mathbb{C}^n$ (i.e. $f^{-1}(c) \cap E_\varepsilon$, for some small enough ε and $0 < |c| \ll \varepsilon$) is diffeomorphic to the global fibre $f^{-1}(c)$, since $f^{-1}(c)$ is transversal to $\partial \overline{E_r}$, $\forall r \geq \varepsilon$. See also §3.1.

1.7. Let $\mathbf{v}(x)$ be the projection of the vector field $\overline{\mathrm{grad} f(x)}$ to the (real or complex) tangent hyperplane of the level of ρ, at the point $x \in N$. This is the following nowhere zero vector field $\mathbf{v}(x) := \overline{\mathrm{grad} f(x)} - \frac{\langle \overline{\mathrm{grad} f(x)}, \mathrm{grad}\, \rho \rangle}{\|\mathrm{grad}\, \rho\|^2} \mathrm{grad}\, \rho$, where $\langle a, b \rangle = \sum_{i=1}^n a_i \bar{b}_i$. By renormalization, we get the (real or complex) vector field $\mathbf{w}(x) := \frac{\mathbf{v}(x)}{\langle \mathbf{v}(x), \mathrm{grad} f(x)\rangle}$, which is nowhere zero on N and it is a lift of $\partial/\partial t$ since $\langle \mathbf{w}(x), \mathrm{grad} f(x)\rangle = 1$.

1.8. Use the curve selection lemma and the definition of ρ_E regularity.

2.1. $\mathcal{B}$-class. We have $\overline{X_t} = \{\tilde{f} - t x_0^d = 0\} \subset \mathbb{P}^n$. Also $\mathrm{Sing} f \cap X_t = \mathrm{Sing} X_t$ and $\overline{\mathrm{Sing} X_t} \subset \mathrm{Sing} \overline{X_t}$ and therefore $\overline{\mathrm{Sing} X_t} \cap H^\infty \subset (\mathrm{Sing} \overline{X_t}) \cap H^\infty$.

Now $\mathrm{Sing} \overline{X_t} = \{\frac{\partial \tilde{f}}{\partial x_i} = 0 \mid i = 0, \ldots, n\}$ and so $(\mathrm{Sing} \overline{X_t}) \cap H^\infty = \{\frac{\partial f_d}{\partial x_i} = 0, f_{d-1} = 0 \mid i = 1, \ldots, n\} = \Sigma$.

3.4 One of the equalities was already proved in that section. The second equality goes with a similar proof.

7.1. If the singularity at p of l is on a stratum of positive dimension of X_c, then the complex link of the restriction $l_{|X_c}$ is a nontrivial bouquet of spheres.

7.5. Use also Corollary 1.2.6 and Proposition 1.2.12.

7.6. Consider the germ of the map $(F, s) : \mathbb{C}^{n+2} \to \mathbb{C}^2$. See also the proof of Proposition 3.3.6.

9.1. The stratified singularities of the restriction of p to $\Sigma(w) \times \mathbb{C}$ can only occur on a finite number of fibres of p. Since $\dim \Sigma(w) \leq s$, it follows that $\dim \mathbb{Y}'^\infty \cap \mathrm{Sing}_S p \leq s$.

9.2. The proof of Proposition 9.1.7 shows that, if $\dim \Sigma \leq s$, then $\dim \mathbb{X}^{\infty} \cap \operatorname{Sing}_{\mathcal{S}} p \leq s$. This implies that $\dim \operatorname{Sing}_{\mathcal{S}} p \leq s + 1$. In turn, this yields $\dim \operatorname{Sing} f \leq s + 1$, since $\mathbb{C}^{n+1}$ is a stratum of the stratification $\mathcal{S}$ of $\mathbb{X}$.

Notes

Chapter 1

1 The proof of the finiteness of the set of atypical values has been sketched by Thom [Th2] and uses stratifications. A complete proof along these lines can be deduced from Verdier's study on Bertini–Sard theorems [Ve]. Another proof using the resolution of singularities was given by Pham [Ph, Appendix].

2 This proper extension of f has been systematically used in the study of polynomial functions [Br1, Ph, Pa1, ST2] etc. We may use other proper extensions of f, by embedding $\mathbb{K}^n$ into weighted projective spaces or toric varieties, see [A'CO, Ti4, Za, LiTi].

3 For complex polynomial functions, the transversality to big spheres (i.e. the ρ_E-regularity) was used in [Br2, p. 229] and later in [NZ], where it is called *M-tameness*.

4 In case $n = 2$, a similar procedure was used by Hà H.V. and Lê D.T. [HàL].

5 The relative conormal has been introduced by Teissier, and later used by Henry, Merle and Sabbah, see [Te3], [HMS]. A more refined concept is the *characteristic cycle*, introduced by Kashiwara.

6 In the complex setting, a Whitney stratification at infinity (i.e. $\mathbb{C}^n$ is a stratum and $\mathbb{X}_{\mathbb{C}}^\infty$ is a union of strata) has automatically the Thom (a_{x_0})-property. This was proved by Briançon, Maisonobe and Merle [BMM]. A geometric proof is given in [Ti4, Theorem 2.9], see Appendix A1.1.

7 Known as the 'Malgrange condition', since used by Malgrange [Mal]), it was first formulated by Pham [Ph, 2.1] in the complex case. Our definition includes a localized version at infinity and works also in the real case, cf. [Ti7].

8 One finds in [Pa2] several interesting inequalities and all the details of the last part of the proof, which is due to Parusiński.

Chapter 2

1 Local-at-infinity polar curves were used in many papers [Pa1, ST2, Ti4, ST4, ST5].

2 Affine polar curves in several variables were introduced in [Ti4] and turn out to be very useful, as we show in Part II.

3 Broughton [Br2] considered for the first time $\mathcal{B}$-type polynomials and studied the topology of their general fibres. The $\mathcal{W}$-class of polynomials was introduced in [ST2], while $\mathcal{F}$-type polynomials were considered in [Li2, ST5, ST8].

231

4 The equivalence (a) ⇔ (b) of Theorem 2.2.5 was proved in particular cases by: Lê and Hà [HàL] for curves, Parusiński [Pa1] for $\mathcal{B}$-type polynomials, and Siersma and Tibăr [ST2] for $\mathcal{W}$-type polynomials.

5 In two complex variables, the equivalence (a) ⇔ (b) of Theorem 2.2.5 has been proved by Lê and Hà [HàL]. In his survey, Durfee [Du2] gives another equivalence in terms of a resolution at infinity. For this viewpoint, one may also consult [LêWe1, LêWe, ACD].

6 One of the simple proofs of the line embedding theorem (Abhyankar–Moh–Suzuki theorem) was given by Rudolph [Rud] and is based on knot theory exclusively. Together with the Jacobian Conjecture formulation at 2.3.4, this gives an almost completely topological viewpoint over the Jacobian problem in $\mathbb{C}^2$.

7 One may consult [CP] for a similar criterion, which presents an effectivity viewpoint.

Chapter 3

1 Links at infinity were studied notably by Neumann [Neu], see also Eisenbud and Neumann's monograph mainly for the local setting [EN]. Local links are intensely studied since the pioneering work by Milnor, Brieskorn, Hirzebruch, Mumford etc.

2 Milnor also shows that his two fibrations are equivalent.

3 The condition 'trivial φ-controlled local relative homology at infinity at t_0' may be interpreted as absence of vanishing cycles at infinity at t_0. More precisely, this means that the support supp $\Phi_{\tau - t_0}(i_* \mathbb{C}_{\mathbb{X}})$ of the sheaf of vanishing cycles of $\tau - t_0$ does not intersect $\mathbb{X}_{t_0}^\infty$. (Here $i : \mathbb{X} \subset \mathbb{P} \times \mathbb{C}$ denotes the inclusion.) In this terms, Parusiński has proved in [Pa2] a result related to Proposition 3.1.9.

4 The bouquet theorem 3.2.1 is a global version of the local bouquet theorems by J. Milnor's [Mi2, Theorem 6.5] and by Lê D.T. [Lê5, Theorem 5.1] in the singular case. In the global setting, bouquet theorems were proved to different degrees of generality by Broughton [Br2], Siersma and Tibăr [ST2], Tibăr [Ti4] and Parusiński [Pa2]. Theorem 3.2.1 extends the one for polynomials with 'good behaviour at infinity': tame [Br2], quasi-tame [Ne2], M-tame [NZ]. Recall that 'tame' implies 'quasi-tame', which implies in its turn 'M-tame' [Ne1], [NZ].

Under the additional hypothesis dim $\Sigma = 0$, the (weaker) homology counterpart of Theorem 3.2.1 was proved (with a different proof) by Broughton [Br2, Theorem 5.2]. In this case he also got Corollary 3.3.3(a).

5 The total Milnor number μ_f has an algebraic interpretation as the dimension of the quotient algebra $\mathbb{C}[x_1, \ldots, x_n] / \left(\frac{\partial f}{\partial x_1}, \ldots, \frac{\partial f}{\partial x_n} \right)$, see [Mi2].

Chapter 4

1 This formula was observed by Dimca and Parusiński and may be found for instance in [Di1].

2 There are more other aspects of this 'opposite behaviour' among the local and the global affine settings, for certain quantifiers (e.g. Hodge numbers).

3 This behaviour was been noticed in several examples by Siersma and Smeltink [SiSm].

4 We have worked in [BT] with analytic braids, which have a similar definition.

Chapter 5

1 Couples of space-function germs (X_s, g_s), varying with the parameter s, have been been investigated, from the point of view of the stability or variation of the Milnor fibre, in [JT].

2 Bodin proved in [Bo3] the topological equivalence in case there are no singularities at infinity and in case $n = 2$ with some additional hypotheses, by using results due to L. Fourrier [Fo], which involve resolution of singularities.

3 One may ask if the formal homotopy class of the boundary is a topological invariant, which is *a priori* a stronger condition than being an invariant of topologically trivial deformations. The answer is *yes* in the particular case when this boundary is diffeomorphic to the standard sphere S^{2n-1}, by Morita's work [Mo]. Morita formulae express the formal homotopy class (as an element of $\pi_{2n-2}(SO_{2n-1}/U_{n-1})$) in terms of the Milnor number μ and of the signature of the Milnor fibre, which are both topological invariants. However this proof does not work for other types of links, in particular since it uses the fact that the restriction of the tangent bundle of $\mathbb{R}^{2n}$ to the unit sphere S^{2n-1} admits a canonical trivialization.

4 This case was previously proved by Hà and Zaharia [HZ].

Chapter 6

1 Global polar invariants for affine hypersurfaces have been first used in [Ti6] in order to characterize the equisingularity at infinity of families of affine hypersurfaces.

2 Massey's approach [Mas] is based on a reformulation of MacPherson's theory in terms of characteristic cycles of constructible sheaves or holonomic D-modules. One may compare for example with [BDK, Dub2, Gi, Sab].

3 The idea of proving properness of projections from the global polar set by using genericity conditions at infinity can be traced back to Hamm[Hm3].

4 Such a formula for $\chi(X)$ has been first proved for hypersurfaces with isolated singularities in [Ti6] and then for general X in [Ti14]. See Exercise 6.2.

Chapter 7

1 A probably very incomplete list of authors who studied polar curves looks as follows: Greuel, Giusti, Henry, Langevin, Casas, Ephraim, Steenbrink, Zucker, Michel, Spivakovsky, Siersma, Weber, Gabrielov, Płoski, Gwozdzewicz, Némethi, Lipman, Oka, Massey, Maugendre, Gusein-Zade, Garcia Barroso, Snoussi, Kuo, Parusiński, Corral, Lenarcik, Campillo, Olivares, Ueda, Ebeling, Bondil, Masternak, Wall.

2 We have defined in [Ti6, §3] polar intersection multiplicities for such a family.

3 In this context, the nongeneric multiplicity $\mu_p(\bar{Y}) + \mu_p(\bar{Y} \cap H^\infty)$ was used in [Ti6, 3.7].

4 A survey of this topic can be found in Griffiths' paper [Gri].

5 The *exchange principle* was originally used in the framework of total absolute curvature of knots and embedded real manifolds, by Milnor [Mi1], Chern and Lashof [ChL], Kuiper [Ku].

Chapter 8

1 The construction of a model of the typical fibre and of geometric monodromies along loops in $\mathbb{C} \setminus \text{Atyp}f$ acting on this model appeared in [Ti6].
2 For the carrousel construction, one may also consult [Ti1, Ti2, Ti13].
3 To compute ζ_i^{rel} and $\zeta_{\delta*}$ we may use the method given in [Ti1] based on a Mayer–Vietoris argument for an annular decomposition of the carrousel disk. This works reasonably well for a carrousel with only first-order smaller carrousels (i.e. when the branches of the germs of $\bar{\Delta}$ have only one Puiseux pair), but computations become very quickly too complicated when the complexity of the $\bar{\Delta}$ increases.
4 For the Lefschetz number in case of functions on singular spaces, one may consult [Ti1].

Chapter 9

1 Instead of endowing $\mathbb{Y}$ with a stratified structure, another strategy for studying the topology of the meromorphic function F would be to further blow up $\mathbb{Y}$ in diagram (9.1), such that the pull-back of $\{P = 0\} \cup \{Q = 0\}$ becomes a divisor with normal crossings. We may then use the data provided by this divisor in order to get information. In this spirit, some results were found in the polynomial case, in two variables, by Fourrier [Fo] and Lê and Weber [LêWe]; computation of the zeta functions of the monodromy has been done for polynomials and particular meromorphic germs (namely for Y nonsingular and $X = Y \setminus A$) by Gusein-Zade, Melle and Luengo [GLM2, GLM4].
2 We may also construct a canonical (minimal) $\partial \tau$-stratification; we send to [Ti4] for the details.
3 This type of result was observed before in different particular situations, see e.g. [Br2, §5], [Si1, ST2, NN1, DN1].
4 The notion of rectified homotopical depth was introduced by Hamm and Lê [HmL1] in order to explain Grothendieck's predictions that the *homotopical depth* (see Definition 10.2.2 for the homology version) was the cornerstone for the Lefschetz type theorems on singular spaces [G].
5 Polydisk neighbourhoods were first used by Lê [Lê3].
6 Libgober [Li2, §1] considers the case when V has at most isolated singularities and $\bar{V}$ is transversal to the hyperplane at infinity in $\mathbb{P}^n$.
7 This result supersedes the connectivity estimation proved in [DP] and other more particular results previously proved by Dimca and Kato.
8 In the particular case of polynomial functions, the Picard type formula appeared in [NN2] and [DN2]. It was independently noticed, in the general setting of meromorphic functions, in the arXiv preprint of [ST6]).
9 This result was remarked in [DN1] for polynomial functions and in [ST6] for meromorphic functions, for nonsingular Y and for $X = Y \setminus A$.

Chapter 10

1 For the bibliography up to 1988, one may look up [GM2]. It appears that in some of these generalizations (e.g. in [GM2, Thm. 1.2, p.199]), under the respective hypotheses, *generic pencils* do exist and their use yield alternative proofs.

2 Our approach starts in the spirit of the Lefschetz method [Lef], as presented by Thom in his Princeton talk in 1957 and by Andreotti and Frankel in their paper [AF]. This vein has been exploited in relatively few papers ever since; we may mention the interesting ones by Lamotke [Lam1], Chéniot [Ch1, Ch2] and Eyral [Ey]. The use in the statement of our Theorem 10.2.3 of the comparison between the general element of the pencil and the axis comes from Lamotke [Lam1] and occurs in Chéniot's statements in [Ch1, Ch2]. Our setting being far more general, we follow a different strategy and use in a crucial way specific geometric constructions and results of stratified singularity theory.

3 Chéniot [Ch2] works with a different kind of variation map.

4 For *generic pencils of hyperplanes* on quasi-projective manifolds, by Chéniot [Ch2], and on complements in $\mathbb{P}^n$ of hypersurfaces with isolated singularities and for higher homotopy groups, by Libgober [Li2]. The extension of Theorem 10.2.3 to homotopy groups is investigated in [Ti8, Ti10] and some subsequent papers by Chéniot, Eyral and Libgober, and will be discussed in Chapter 11.

5 The fact that the Lefschetz structure on quasi-projective varieties is hereditary on slices was observed before by several authors, e.g. Chéniot [Ch1].

   For *generic pencils of hyperplanes*, condition $(C2)'$ has been used by Chéniot [Ch1, Ch2] and Eyral [Ey], respectively condition $(C3)'$ has been used by C. Eyral in proving a version of the first LHT (one should compare to [Ey, Proof of Theorem 2.5]). Therefore, via Proposition 10.2.10 and the observations in case of quasi-projective varieties, Theorem 10.2.3 recovers the results in the cited articles.

6 This is Neumann–Norbury's result [NN2, Theorem 2.3] modulo an identification, by some excision, of our variation map to the local variation maps used in [NN2].

7 This may also be deduced, modulo some identifications, from the more general result [NN1, Theorem 1.4].

Chapter 11

1 It is an interesting challenge to cary over the construction of the variation map in case $q = 2$.

2 A tremendous effort has been done to enlarge the Lefschetz slicing principle. There is a long list of articles around this topic, such as [AF,Ch1,Ch2,Ful,FL,GM2, HmL1,HmL3,Lam1,Lef,Li2,Loo,Mi2]. Results of Zariski–van Kampen type have been proved by Libgober [Li2] for generic Lefschetz pencils of hyperplanes and in the particular case when X is the complement in $\mathbb{P}^n$ or $\mathbb{C}^n$ of a hypersurface with isolated singularities. In recent papers after 2002, Chéniot and Libgober [CL] and Chéniot and Eyral [CE] treat with different backgrounds some particular cases of Theorem 11.2.1–the first version of which appeared in [Ti8].

3 Under those conditions, Part (a) of Theorem 11.2.1 specializes to a connectivity statement, which recovers Eyral's main result [Ey], where such a condition was used for *generic pencils*.

4 In the particular case of a hypersurface $V \subset \mathbb{C}^n$ with at most isolated singularities and transversal to the hyperplane at infinity, a related result, but with different background, has been proved by Libgober [Li2], including at the π_2 level. See also Exercises 11.1 and 11.2.

References

[AM] S. S. Abhyankar, T. T. Moh, Embeddings of the line in the plane, *J. Reine Angew. Math.* **276** (1975), 148–166.

[A'C1] N. A'Campo, Le nombre de Lefschetz d'une monodromie, *Indag. Math.* **35** (1973), 113–118.

[A'C2] N. A'Campo, La fonction zêta d'une monodromie, *Comment. Math. Helv.* **50** (1975), 233–248.

[A'CO] N. A'Campo, M. Oka, Geometry of plane curves via Tschirnhausen resolution tower, *Osaka J. Math.* **33** (1996), 1003–1033.

[AF] A. Andreotti, T. Frankel, The Lefschetz theorem on hyperplane sections, *Ann. of Math.* **69** (1959), 713–717.

[AF] A. Andreotti, T. Frankel, The second Lefschetz theorem on hyperplane sections, in *Global Analysis, Papers in Honor of K. Kodaira*, Princeton Univ. Press, 1969, p. 1–20.

[AGV2] V. I. Arnol'd, S.M. Gusein-Zade, A.N. Varchenko, *Singularities of Differentiable Maps, Vol. II. Monodromy and Asymptotics of Integrals*, Translated from the Russian by Hugh Porteous, Translation revised by the authors and James Montaldi, *Monographs in Mathematics*, **83**, Boston, MA, Birkhäuser, 1988.

[Ar1] V. I. Arnold, Normal forms of functions near degenerate critical points, the Weyl groups A_k, D_k, E_k and Lagrangian singularities, *Funkcional. Anal. i Priložen.* **6** (1972), 3–25.

[Ar2] V. I. Arnold, Singularities of fractions and behaviour of polynomials at infinity, *Tr. Mat. Inst. Steklova* **221** (1998), 48–68.

[ACD] E. Artal-Bartolo, P. Cassou-Nogués, A. Dimca, Sur la topologie des polynomes complexes, *Singularities* (Oberwolfach, 1996), 317–343, *Progr. Math.* **162**, Basel, Birkhuser, 1998.

[Assi] A. Assi, Familles de corbes planes ayant une seule valeur irrégulière, *C. R. Acad. Sci. Paris* **322** (1996), 1203–1207.

[Be] K. Bekka, C-régularité et trivialité topologique, Singularity theory and its applications, Part I (Coventry, 1988/1989), 42–62, Lect. Notes Math. **1462**, Berlin, Springer-Verlag, 1991.

[BM] A. L. Blakers, W. S. Massey, The homotopy groups of a triad. III, *Ann. of Math.* **58** (1953), 409–417.

236

[Bl] D. Blair, Contact manifolds in Riemannian geometry, *Lect. Notes in Math.* **509**, Berlin, Springer-Verlag, 1976.

[Bo1] A. Bodin, Classification of polynomials from $\mathbb{C}^2$ to $\mathbb{C}$ with one critical value, *Math. Z.* **242** (2002), 303–322.

[Bo2] A. Bodin, Non-reality and non-connectivity of complex polynomials, *C. R. Acad. Sci. Paris* **335** (2002), 1039–1042.

[Bo3] A. Bodin, Invariance of Milnor numbers and topology of complex polynomials, *Comment. Math. Helv.* **78** (2003), 134–152.

[BT] A. Bodin, M. Tibăr, Topological triviality of families of complex polynomials, *Adv. Math.* **199** (2006), 136–150.

[BrSc] J.-P. Brasselet, M. H. Schwartz, Sur les classes de Chern d'un ensemble analytique complexe *Asterisque* 82–83 (1981), 93–147.

[BLS] J.-P. Brasselet, Lê D. T., J. Seade, Euler obstruction and indices of vector fields, *Topology* **39** (2000), 1193–1208.

[BMPS] J.-P. Brasselet, D. B. Massey, A. J. Parameswaran, J. Seade, Euler obstruction and defects of functions on singular varieties, *J. London Math. Soc.* **70** (2004), 59–76.

[Bre] G. E. Bredon, *Topology and Geometry*, Corrected third printing of the 1993 original, *Graduate Texts in Mathematics*, **139**, New York, Springer-Verlag, 1997.

[BMM] J. Briançon, Ph. Maisonobe, M. Merle, Localisation de systèmes différentiels, stratifications de Whitney et condition de Thom, *Invent. Math.* **117** (1994), 531–550.

[BrSp1] J. Briançon, J.-P. Speder, La trivialité topologique n'implique pas les conditions de Whitney, *C. R. Acad. Sci. Paris*, Sér. A **280** (1975), 365–367.

[BrSp2] J. Briançon, J.-P. Speder, Les conditions de Whitney impliquent μ^* constant, *Ann. Inst. Fourier* (Grenoble) **26** (1976), 153–163.

[BK] E. Brieskorn, H. Knörrer, *Plane Algebraic Curves*, Translated from the German by John Stillwell, Basel, Birkhäuser Verlag, 1986.

[Br1] S. A. Broughton, On the topology of polynomial hypersurfaces, *Proc. A.M.S. Symp. in Pure. Math.* **40** (1983), 165–178.

[Br2] S. A. Broughton, Milnor numbers and the topology of polynomial hypersurfaces, *Invent. Math.* **92** (1988), 217–241.

[BDK] J. L. Brylinski, A. Dubson, M. Kashiwara, Formule d'indice pour les modules holonomes et obstruction d'Euler locale, *C. R. Acad. Sci. Paris* **293** (1981), 573–576

[BV] D. Burghelea, A. Verona, Local homological properties of analytic sets, *Manuscripta Math.* **7** (1972), 55–66.

[Cau] C. Caubel, Structures presque de contact et singularités isolées, *C. R. Acad. Sci. Paris*, Sér. I Math. **333** (2001), 339–342.

[CNP] C. Caubel, A. Némethi, P. Popescu-Pampu, Milnor open books and Milnor fillable contact 3-manifolds, *Topology* **45** (2006), 673–689.

[CT1] C. Caubel, M. Tibăr, The contact boundary of a complex polynomial, *Manuscripta Math.* **111** (2003), 211–219.

[CT2] C. Caubel, M. Tibăr, Contact boundaries of hypersurface singularities and of complex polynomials, *Geometry and topology of caustics – CAUSTICS '02*, 29–37, Banach Center Publ., 62, Polish Acad. Sci., Warsaw, 2004.

[Ca] A. Cayley, On the theory of linear transformations, *Cambridge Math. Journal* **4** (1845), 1–16. *Collected papers*, Vol. I, pp. 80–94, Cambridge Univ. Press, 1889.

[Ch1] D. Chéniot, Topologie du complémentaire d'un ensemble algébrique projectif, L'Enseign. Math. **37** (1991), 293–402.

[Ch2] D. Chéniot, Vanishing cycles in a pencil of hyperplane sections of a non-singular quasi-projective variety, *Proc. London Math. Soc.* **72** (1996), 515–544.

[CL] D. Chéniot, A. Libgober, Zariski-van Kampen theorem for higher homotopy groups, *J. Inst. Math. Jussieu* **2** (2003), 495–527.

[CE] D. Chéniot, C. Eyral, Homotopical variations and high-dimensional Zariski-van Kampen theorems, *Trans. Amer. Math. Soc.* **358** (2006), 1–10.

[ChL] S. S. Chern, R. K. Lashof, On the total curvature of immersed manifolds, I, II, *Amer. J. of Math.* **79** (1957), 306–318; and *Michigan Math. J.* **5** (1958), 5–12.

[Cl] C. H. Clemens, Degeneration of Kähler manifolds, *Duke Math. J.* **44** (1977), 215–290.

[Co] S. Cohn-Vossen, Kürzeste Wege and Totalkrümmung auf Flächen, *Compositio Math.* **2** (1935), 69–133.

[CP] M. Coste, M. J. de la Puente, Atypical values at infinity of a polynomial function on the real plane: an erratum, and an algorithmic criterion, *J. Pure Appl. Algebra* **162** (2001), 23–35.

[dJ] T. de Jong, Some classes of line singularities, *Math. Z.* **198** (1988), 493–517.

[De] P. Deligne, Groupes de monodromie en géométrie algébrique. II., SGA 7 II, in P. Deligne and N. Katz, *Lect. Notes in Math.*, **340**, Berlin, Springer-Verlag, 1973.

[Di1] A. Dimca, On the homology and cohomology of complete intersections with isolated singularities, *Compositio Math.* **58** (1986), 321–339.

[Di2] A. Dimca, Monodromy and Hodge theory of regular functions, *New developments in singularity theory* (Cambridge, 2000), 257–278; *NATO Sci. Ser. II Math. Phys. Chem.*, **21** Dordrecht, Kluwer Acad. Publ., 2001.

[DN1] A. Dimca, A. Némethi, On monodromy of complex polynomials, *Duke Math. J.* **108** (2001), 199–209.

[DN2] A. Dimca, A. Némethi, Thom-Sebastiani construction and monodromy of polynomials, *Proc. Steklov Inst. Math.* **238** (2002), 97–114.

[DP] A. Dimca, L. Păunescu, On the connectivity of complex affine hypersurfaces, II, *Topology* **39** (2000), 1035–1043.

[Dold] A. Dold, Lectures *on algebraic topology, Grundlehren der mathematischen Wissenschaften*, **200** New York and Berlin, Springer-Verlag, 1972.

[Don] S. K. Donaldson, Polynomials, vanishing cycles and Floer homology, *Mathematics: Frontiers and Perspectives*, 55–64, Providence, RI, Amer. Math. Soc., 2000.

[Du1] A. H. Durfee, Neighborhoods of algebraic sets, *Trans. Amer. Math. Soc.* **276** (1983), 517–530.

[Du2] A. H. Durfee, Five definitions of critical point at infinity, *Singularities* (Oberwolfach, 1996), 345–360, *Progr. Math.* **162**, Basel, Birkhäuser, 1998.

[Dub1] A. Dubson, Classes caractéristiques des variétés singuliéres, *C. R. Acad. Sci. Paris* **287** (1978), 237–240.

[Dub2] A. Dubson, Formule pour l'indice des complexes constructibles et des D-modules holonomes, *C. R. Acad. Sci. Paris, Sér. I Math.* **298** (1984), 113–116.

[Eb] W. Ebeling, The monodromy groups of isolated singularities of complete intersections, *Lect. Notes in Math.*, **1293** Berlin, Springer-Verlag, 1987.

[EN] D. Eisenbud, W. Neumann, Three-dimensional link theory and invariants of plane curve singularities, *Ann. of Math. Studies*, **110** Princeton, NJ, Princeton University Press, 1985.

[El1] Y. Eliashberg, Symplectic geometry of plurisubharmonic functions (Notes by Miguel Abreu), Jacques Hurtubise, et al. (ed.), *Gauge theory and symplectic geometry, NATO ASI Ser., Ser. C, Math. Phys. Sci.* **488** (1997), 49–67.

[El2] Y. Eliashberg, Invariants in contact topology *Doc. Math.* Extra Vol. Berlin, ICM, 1998, 327–338.

[Ey] C. Eyral, Tomographie des variétés singulières et théorèmes de Lefschetz, *Proc. London Math. Soc.* **83** (2001), 141–175.

[Fed] M. V. Fedoryuk, The asymptotics of the Fourier transform of the exponential function of a polynomial, *Docl. Acad. Nauk*, **227** (1976), 580–583; *Soviet Math. Dokl.* **17** (1976), 486–490.

[Fen] W. Fenchel, On total curvatures of Riemannian manifolds. *J. London Math. Soc.* **15** (1940), 15–22.

[Fer] M. Ferrarotti, G-manifolds and stratifications, *Rend. Istit. Mat. Univ. Trieste* **26** (1994), 211–232.

[Fo] L. Fourrier, Topologie d'un polynôme de deux variables complexes au voisinage de l'infini, *Ann. Inst. Fourier* (Grenoble), **46** (1996), 645–687.

[Fuk] T. Fukuda, Types topologiques des polynômes, *Inst. Hautes Études Sci. Publ. Math.* **46** (1976), 87–10.

[Ful] W. Fulton, On the topology of algebraic varieties, in *Algebraic Geometry, Bowdoin*, (Brunswick, Maine, 1985), 15–46; *Proc. A.M.S Symp. in Pure Math.* **46** (1987).

[FL] W. Fulton, R. Lazarsfeld, Connectivity and its applications in algebraic geometry, *Algebraic geometry* (Chicago, Ill., 1980), pp. 26–92, *Lect. Notes in Math.*, **862** and Berlin/New York, Springer-Verlag, 1981.

[GN] R. Garcia Lopez, A. Némethi, On the monodromy at infinity of a polynomial map, *Compositio Math.* **100** (1996), 205–231.

[Ga] L. Gavrilov, Abelian integrals related to Morse polynomials and perturbations of plane Hamiltonian vector fields, *Ann. Inst. Fourier* (Grenoble) **49** (1999), 611–652.

[Gi] V. Ginsburg, Characteristic cycles and vanishing cycles, *Invent. Math.* **84** (1986), 327–402.

[GM1] M. Goresky, R. MacPherson, Stratified Morse theory, *Singularities, Proc. A.M.S Symp. in Pure Math.* **40** (1983), 517–583.

[GM2] M. Goresky, R. MacPherson, *Stratified Morse Theory*, Berlin, Heidelberg, New York, Springer-Verlag, 1987.

[JGr] J. W. Gray, Some global properties of contact structures, *Ann. of Math.* **69** (1959), 421–450.

[BGr] B. Gray, Homotopy theory. an introduction to algebraic topology, *Pure and Applied Mathematics.* **64** Academic Press, 1975.

[Gri] Ph. A. Griffiths, Complex differential and integral geometry and curvature integrals associated to singularities of complex analytic varieties, *Duke Math. J.* **45** (1978), 427–512.

[G] A. Grothendieck, Cohomologie locale des faisceaux cohérents et théorèmes de Lefschetz locaux et globaux (*SGA* 2), Séminaire de Géométrie Algébrique du Bois-Marie, *Advanced Studies in Pure Mathematics*, Vol. II, Amsterdam, North-Holland, 1962; Paris, Masson & Cie, 1968.

[GLM1] S. Gusein-Zade, I. Luengo, A. Melle, Partial resolutions and the zeta-function of a singularity, *Comment. Math. Helv.* **72** (1997), 244–256.

[GLM2] S. Gusein-Zade, I. Luengo, A. Melle, Zeta functions for germs of meromorphic functions and Newton diagrams, *Funct. Anal. Appl.* **32** (1998), 93–99.

[GLM3] S. Gusein-Zade, I. Luengo, A. Melle, On the zeta function of a polynomial at infinity, *Bull. Sci. Math.* **124** (2000), 213–224.

[GLM4] S. Gusein-Zade, I. Luengo, A. Melle, On atypical values and local monodromies of meromorphic functions, Dedicated to S.P. Novikov on the occasion of his 60th birthday, *Tr. Mat. Inst. Steklova* **225** (1999), 168–176.

[GS] S. Gusein-Zade, D. Siersma, Deformations of polynomials and their zeta functions, math.AG/0503450.

[GWPL] C. G. Gibson, K. Wirthmüller, A. A. du Plessis, E. J. N. Looijenga, Topological Stability of Smooth Mappings, *Lect. Notes in Math.* **552** Berlin, Springer-Verlag, 1976.

[Hà] Hà H.V., Nombre de Łojasiewicz et singularités à l'infini des polynômes de deux variables complexes *C. R. Acad. Sci. Paris* **311** (1990), 429–432.

[HàL] Hà H. V., Lê D. T., Sur la topologie des polynômes complexes, *Acta Math. Vietnamica* **9** (1984), 21–32.

[HZ] Hà H.V., A. Zaharia, Families of polynomials with total Milnor number constant, *Math. Ann.* **304** (1996), 481–488.

[Hm1] H. A. Hamm, Lokale topologische Eigenschaften komplexer Räume, *Math. Ann.* **191** (1971), 235–252.

[Hm2] H. A. Hamm, Zur Homotopietyp Steinscher Räume, *J. Reine Angew. Math.* **338** (1983), 121–135.

[Hm3] H. A. Hamm, Lefschetz theorems for singular varieties, *Proc. A.M.S Symp. Pure Math.* **40** (1983), 547–557.

[HmL1] H. A. Hamm, Lê D. T., Un théorème de Zariski du type de Lefschetz, *Ann. Sci. École Norm.*, Sup. **6** (1973), 317–355.

[HmL2] H. A. Hamm, Lê D. T., Local generalizations of Lefschetz–Zariski theorems, *J. Reine Angew. Math.* **389** (1988), 157–189.

[HmL3] H. A. Hamm, Lê D. T., Rectified homotopical depth and Grothendieck conjectures, in P. Cartier *et al.* (eds), *Grothendieck Festschrift*, Vol. II, pp. 311–351, Birkhäuser, 1991.

[Har] J. Harris, *Algebraic Geometry. A First Course*, *Graduate Texts in Mathematics*, **133**, Berlin, Springer-Verlag, 1992.

[HM] J.-P. Henry, M. Merle, Limites de normales, conditions de Whitney et éclatement d'Hironaka, Singularities, *Proc. A.M.S. Symp. in Pure Math.* **40** (1983), 575–584.

[HMS] J. P. Henry, M. Merle, C. Sabbah, Sur la condition de Thom stricte pour un morphisme analytique complexe, *Ann. Sci. Norm.*, Sup. **17** (1984), 227–268.

[Hiro] H. Hironaka, Stratifications and flatness, in *Real and Complex Singularities*, Oslo, 1976, Sijhoff en Norhoff, Alphen a.d. Rijn, 1977.

[Hirz] H. Hirzebruch, Topological methods in algebraic geometry, trans. from the German and Appendix One by R. L. E. Schwarzenberger, with a preface to the third English edition by the author and Schwarzenberger; Appendix Two by A. Borel; Reprint of the 1978 edition, *Classics in Mathematics*, Berlin, Springer-Verlag, 1995.

[JT] G. Jiang, M. Tibăr, Splitting of singularities, *J. Math. Soc. Japan* **54** (2002), 255–271.

[Ka] K. K. Karčjauskas, Homotopy properties of algebraic sets, *Studies in Topology, Vol. III*, Zap. Nauchn. Sem. Leningrad, Otdel, Mat. Inst. Steklova, **83** (1979), 67–72.

[KPS] L. Katzarkov, T. Pantev, C. Simpson, Density of monodromy actions on non-abelian cohomology, *Adv. Math.* **179** (2003), 155–204.

[Ken] G. Kennedy, MacPherson's Chern classes of singular varieties, *Commun. Algebra.* **9** (1990), 2821–2839.

[Ki] H. C. King, Topological type in families of germs, *Invent. Math.* **62** (1980/81), 1–13.

[Kl] S. L. Kleiman, The transversality of a general translate, *Compositio Math.* **28** (1974), 287–297.

[Ku] N. H. Kuiper, Minimal total curvature for immersions, *Invent. Math.* **10** (1970), 209–238.

[Lam1] K. Lamotke, The topology of complex projective varieties after S. Lefschetz, *Topology* **20** (1981), 15–51.

[Lan1] R. Langevin, Courbure et singularités complexes, *Comment. Math. Helv.* **54** (1979), 6–16.

[Lan2] R. Langevin, Courbures, feuilletages et surfaces, Dissertation, Université Paris-Sud, Orsay, 1980; *Publ. Math. d'Orsay* **80** 3. Université de Paris-Sud, Département de Mathématique, Orsay, 1980.

[Lau] G. Laumon, Degré de la variété duale d'une hypersurface á singularités isolées, *Bull. Soc. Math. France* **104** (1976), 51–63.

[Laz] F. Lazzeri, A theorem on the monodromy of isolated singularities, in *Singularités à Cargèse*, 1972, 269–275. Asterisque, Nos. 7 and 8, *Soc. Math. France*, Paris, 1973.

[Lê1] Lê D. T., Une application d'un théorème d'A'Campo à l'équisingularité, *Nederl. Akad. Wet., Proc., Ser.* A **76** (1973), 403–409.

[Lê2] Lê D. T., La monodromie n'a pas de points fixes, *J. Fac. Sci. Univ. Tokyo,* Ser. 1A, **22** (1973), 409–427.

[Lê3] Lê D. T., Some remarks on the relative monodromy, in *Real and Complex Singularities*, Oslo, Sijhoff en Norhoff, 1976, Alphen a.d. Rijn, 1977, pp. 397–403.

[Lê4] Lê D. T., The geometry of the monodromy theorem, *C.P. Ramanujam – a Tribute*, Tata Institute, Springer-Verlag, 1978, pp. 157–173.

[Lê5] Lê D. T., Complex analytic functions with isolated singularities, *J. Algebraic Geometry*, **1** (1992), 83–100.

[LM] Lê D. T., Z. Mebkhout, Variétés caractéristiques et variétés polaires, *C. R. Acad. Sci. Paris*, Sér. I Math. **296** (1983), 129–132.

[LêRa] Lê D. T., C. P. Ramanujam, The invariance of Milnor's number implies the invariance of the topological type, *Amer. J. of Math.* **98** (1976), 67–78.

[LêSa] Lê D. T., K. Saito, La constance du nombre de Milnor donne des bonnes stratifications, *C. R. Acad. Sci. Paris*, Sr. A-B **277** (1973), A793–A795.

[LêTe] Lê D. T., B. Teissier, Varits polaires locales et classes de Chern des varits singulires, *Ann. of Math.* **114** (1981), 457–491.

[LêWe1] Lê D. T., C. Weber, A geometrical approach to the Jacobian conjecture for $n = 2$, *Kodai Math. J.* **17** (1994), 374–381.

[LêWe] Lê D. T., C. Weber, Polynômes à fibres rationelles et conjecture Jacobienne à 2 variables, **320** (1995), 581–584.

[Lef] S. Lefschetz, *L'analysis situs et la géométrie algébrique*, Paris, Gauthier-Villars, 1924, (nouveau tirage 1950).

[Li1] A. Libgober, Alexander invariants of plane algebraic curves, *Proc. A.M.S.Symp. in Pure Math.* **40** (1983), 135–143.

[Li2] A. Libgober, Homotopy groups of the complements to singular hypersurfaces, *Ann. of Math.* **139** (1994), 117–144.

[LS] A. Libgober, S. Sperber, On the zeta function of monodromy of a polynomial map, *Compos. Math.* **95** (1995), 287–307.

[LiTi] A. Libgober, M. Tibăr, Homotopy groups of complements and nonisolated singularities, *Int. Math. Res. Not.*, 2002, 871–888.

[Łoj1] S. Łojasiewicz, Ensembles semi-analytiques, prépublication *I.H.E.S.* (1964).

[Loo] E. J. N. Looijenga, *Isolated Singular Points on Complete Intersections, London Mathematical Society Lecture Note Series*, **77**, Cambridge University Press, 1984.

[MP] R. MacPherson, Chern classes for singular varieties, *Ann. of Math.* **100** (1974), 423–432.

[Mal] B. Malgrange, Méthode de la phase stationnaire et sommation de Borel, Complex analysis, microlocal calculus and relativistic quantum theory (Proc. Internat. Colloq., Centre Phys., Les Houches, 1979), pp. 170–177, *Lecture Notes in Phys.* **126** Berlin and New York, Springer-Verlag, 1980.

[Mas] D. B. Massey, Numerical invariants of perverse sheaves, *Duke Math. J.* **73** (1994), 307–369.

[Mat] J. Mather, Notes on topological stability, Harvard University (1979).

[Me] M. Merle, Invariants polaires des courbes planes, *Invent. Math.* **41** (1977), 103–111.

[MW] F. Michel, C. Weber, On the monodromies of a polynomial map from $\mathbb{C}^2$ to $\mathbb{C}$, *Topology* **40** (2001), 1217–1240.

[Mi1] J. Milnor, On the total curvature of knots, *Ann. of Math.* **52** (1950), 248–257.

[Mi2] J. W. Milnor, *Singular Points of Complex Hypersurfaces, Ann. of Math. Studies* **61**, Princeton, 1968.

[Mo] S. Morita, A topological classification of complex structures on $S^1 \times \Sigma^{2n-1}$, *Topology* **14** (1975), 13–22.

[Ne1] A. Némethi, Théorie de Lefschetz pour les variétés algébriques affines, *C. R. Acad. Sci. Paris*, Ser. I Math. **303** (1986), 567–570.

[Ne2] A. Némethi, Lefschetz theory for complex affine varieties, *Rev. Roumaine Math. Pures Appl.* **33** (1988), 233–250.

[Ne3] A. Némethi, The Milnor fiber and the zeta function of the singularities of type $f = P(h, g)$, *Compositio Math.* **79** (1991), 63–97.

[NZ] A. Némethi, A. Zaharia, On the bifurcation set of a polynomial and Newton boundary, *Publ. RIMS* **26** (1990), 681–689.

[Neu] W. D. Neumann, Complex algebraic plane curves via their links at infinity, *Invent. Math.* **98** (1989), 445–489.

[NN1] W. D. Neumann, P. Norbury, Vanishing cycles and monodromy of complex polynomials, *Duke Math. J.* **101** (2000), 487–497.

[NN2] W. D. Neumann, P. Norbury, Unfolding polynomial maps at infinity, *Math. Ann.* **318** (2000), 149–180.

[Oka1] M. Oka, On the bifurcation of the multiplicity and topology of the Newton boundary, *J. Math. Soc. Japan* **31** (1979), 435–450.

[Oka2] M. Oka, Flex curves and their applications, *Geom. Dedicata* **75** (1999), 67–100.

[Oka3] M. Oka, On Fermat curves and maximal nodal curves, *Michigan Math. J.* **53** (2005), 459–477.

[PaSm] R. Palais, S. Smale, A generalized Morse theory, *Bull. Amer. Math. Soc.* **70** (1964), 165–172.

[Pa1] A. Parusiński, On the bifurcation set of a complex polynomial with isolated singularities at infinity, *Compositio Math.* **97** (1995), 369–384.

[Pa2] A. Parusiński, A note on singularities at infinity of complex polynomials, in Simplectic Singularities and Geometry of Gauge Fields, *Banach Center Publ.* **39** (1997), 131–141.

[Pa3] A. Parusiński, Topological triviality of μ-constant deformations of type $f(x) + tg(x)$, *Bull. London Math. Soc.* **31** (1999), 686–692.

[PZ1] L. Păunescu, A. Zaharia, On the Łojasiewicz exponent at infinity for polynomial functions, *Kodai Math. J.* **20** (1997), 269–274.

[PZ2] L. Păunescu, A. Zaharia, Remarks on the Milnor fibration at infinity, *Manuscripta Math.* **103** (2000), 351–361.

[Ph] F. Pham, Vanishing homologies and the *n* variable saddlepoint method, *Arcata Proc. of Symp. in Pure Math.* **40** (1983), 319–333.

[PS] E. Picard, G. Simart, Théorie des fonctions algébriques de deux variables indépendantes, Vols. I, II., 1897 and 1906 Reprinted in one volume Bronx, NY, Chelsea Publishing Co., 1971.

[Pi1] R. Piene, Polar classes of singular varieties, *Ann. Sci. Ecole Norm., Sup.* **11** (1978), 247–276.

[Pi2] R. Piene, Cycles polaires et classes de Chern pour les variétés projectives singulières, *Travaux en cours* **37** Paris, Hermann, 1988, pp. 7–34.

[Rud] L. Rudolph, Embeddings of the line in the plane, *J. Reine Angew. Math.* **337** (1982), 113–118.

[Sab] C. Sabbah, Monodromy at Infinity and Fourier Transform, *Publ. RIMS, Kyoto Univ.* **33** (1997), 643–685.

[Sae] O. Saeki, Topological type of complex isolated hypersurface singularities, *Kodai Math J.* **12** (1989), 23–29.

[Sche] J. Scherk, CR structures on the link of an isolated singularity, Canadian Math. Soc. Conf. Proc. **6** (1986), 397–403.

[Sch] J. Schürmann, *Topology of Singular Spaces and Constructible Sheaves, Mathematics Institute of the Polish Academy of Sciences, Mathematical Monographs (New Series)* **63**, Basel, Birkhäuser Verlag, 2003.

[Sch1] J. Schürmann, A short proof of a formula of Brasselet, Lê and Seade, math. AG/ 0201316.

[Sch2] J. Schürmann, A generalized Verdier-type Riemann-Roch theorem for Chern–Schwartz–MacPherson classes, math.AG/0202175

[Sch3] J. Schürmann, A general intersection formula for Lagrangian cycles, *Compositio Math.* **140** (2004), 1037–1052.

[Sch4] J. Schürmann, Lecture on characteristic classes of constructible functions, in P. Pragacz (ed.), *Topics in Cohomological Studies of Algebraic Varieties, Trends in Mathematics*, Birkhäuser, 2005, pp. 175–201.

[ScTi] J. Schürmann, M. Tibăr, MacPherson cycles and polar varieties of affine algebraic spaces, math.AG/0603338.

[Sc] M.-H. Schwartz, Champs radiaux sur une stratification analytique complexe, *Travaux en cours* **39**, Hermann, 1991.

[STV1] J. Seade, M. Tibăr, A. Verjovsky, Global Euler obstruction of affine hypersurfaces, *Math. Ann.* **333** (2005), 393–403.

[STV2] J. Seade, M. Tibăr, A. Verjovsky, Milnor numbers and Euler obstruction, *Bull. Brazilian Math. Soc.* **36** (2005), 275–283.

[Si1] D. Siersma, Vanishing cycles and special fibres, in: Singularity theory and its applications, Part I (Coventry, 1988/1989), 292–301, *Lect. Notes in Math.* **1462** Berlin, Springer-Verlag, 1991.

[Si2] D. Siersma, Variation mappings on singularities with a 1-dimensional critical locus, *Topology* **30** (1991), 445–469.

[Si3] D. Siersma, The vanishing topology of non isolated singularities, in *New Developments in Singularity Theory* (Cambridge, 2000), pp. 447–472; *NATO Sci. Ser. II Math. Phys. Chem.* **21**, Dordrecht, Kluwer Acad. Publ., 2001.

[SiSm] D. Siersma, J. Smeltink, Classification of singularities at infinity of polynomials of degree 4 in two variables, preprint 945 (1996), University of Utrecht.

[ST1] D. Siersma, M. Tibăr, Is the polar relative monodromy of finite order? An example, Proceedings of the 1994 Workshop on Topology and Geometry (Zhanjiang), *Chinese Quart. J. Math.* **10** (1995), 78–85.

[ST2] D. Siersma, M. Tibăr, Singularities at infinity and their vanishing cycles, *Duke Math. Journal* **80** (1995), 771–783.

[ST3] D. Siersma, M. Tibăr, Topology of polynomial functions and monodromy dynamics, *C. R. Acad. Sci. Paris, Ser. I* **327** (1998), 655–660.

[ST4] D. Siersma, M. Tibăr, Singularities at infinity and their vanishing cycles, II. Monodromy, *Publ. Res. Inst. Math. Sci.* **36** (2000), 659–679.

[ST5] D. Siersma, M. Tibăr, Deformations of polynomials, boundary singularities and monodromy, Mosc. Math. J. **3** (2003), 661–679.

[ST6] D. Siersma, M. Tibăr, On the vanishing cycles of a meromorphic function on the complement of its poles, Real and complex singularities, 277–289, *Contemp. Math.* **354** Providence, RI, Amer. Math. Soc. 2004.

[ST7] D. Siersma, M. Tibăr, Singularity exchange at the frontier of the space, in Real and Complex Singularities, 327–342, *Trends Math.*, Birkhäuser, 2006.

[ST8] D. Siersma, M. Tibăr, Curvature and Gauss-Bonnet defect of global affine hypersurfaces, *Bull. Sci. Math.* **130** (2006), 110–122.

[Sp] E. H. Spanier, Algebraic topology, New York-Toronto, McGraw-Hill Book Co., 1966.

[Stb] J. H. M. Steenbrink, The spectrum of Hypersurface Singularities, *Théorie de Hodge* (Luminy, 1987), Astérisque No. 179–180 (1989), 163–184.

[Str] N. Steenrod, The Topology of Fiber Bundles, Princeton Univ. Press, 1951.

[Sul] D. Sullivan, Combinatorial invariants of analytic spaces, in *Proceedings of Liverpool Singularities Symposium*, I, *Lect. in Notes Math.* **192** 165–168. Springer-Verlag, Berlin, 1971.

[Suz] M. Suzuki, Propriétés topologiques des polynômes de deux variables complexes, et automorphismes algébriques de l'espace C^2, *J. Math. Soc. Japan* **26** (1974), 241–257.

[Sw] R. Switzer, Algebraic Topology – Homotopy and Homology, Berlin, Heidelberg and New York, Springer-Verlag, 1975.

[Sz] A. Szpirglas, Singularités de bord: dualité, formules de Picard-Lefschetz relatives et diagrammes de Dynkin, *Bull. Soc. Math. France* **118** (1990), 451–486.

[Te1] B. Teissier, Cycles évanescents, sections planes et conditions de Whitney, *Singularités à Cargèse* (1972), pp. 285–362. Asterisque, Nos. 7 and 8, Soc. Math. France, Paris, 1973.

[Te2] B. Teissier, Variétés polaires, I. Invariants polaires des singularités d'hypersurfaces, Invent. Math. **40** (1977), 267–292.

[Te3] B. Teissier, Varietés polaires, II. Multiplicités polaires, sections planes et conditions de Whitney, *Géométrie Algébrique à la Rabida, Lect. Notes in Math.* **961** Berlin, Springer-Verlag, 1981, pp. 314–491.

[Te4] B. Teissier, Séminaire sur les Singularités des Surfaces, held at the Centre de Mathématiques de l'École Polytechnique, Palaiseau, 1976–1977. Edited by M. Demazure, H. C. Pinkham and B. Teissier, pp. 136–141, *Lect. Notes in Math.* **777** Berlin, Springer-Verlag, 1980.

[Th1] R. Thom, Les structures différentiables des boules et des sphères, *Colloque Géom. Diff. Globale* (Bruxelles, 1958), pp. 27–35, Centre Belge Rech. Math., Louvain, 1959.

[Th2] R. Thom, Ensembles et morphismes stratifiés, *Bull. Amer. Math. Soc.* **75** (1969), 249–312.

[Tho] A. Thorup, Generalized Plücker formulas, in *Recent Progress in Intersection Theory, Bologna, 1997*, 299–327, *Trends Math.*, Boston, MA, Birkhäuser, 2000.

[Ti1] M. Tibăr, Carrousel monodromy and Lefschetz number of singularities, *L'Enseignement Math.* **39** (1993), 233–247.

[Ti2] M. Tibăr, Bouquet decomposition of the Milnor fibre, *Topology* **35** (1996), 227–242.

[Ti3] M. Tibăr, On the monodromy fibration of polynomial functions with singularities at infinity *C. R. Acad. Sci. Paris, Ser.* I **324** (1997), 1031–1035.

[Ti4] M. Tibăr, Topology at infinity of polynomial mappings and Thom regularity condition, *Compositio Math.* **111** (1998), 89–109.

[Ti6] M. Tibăr, Asymptotic Equisingularity and Topology of Complex Hypersurfaces, *Int. Math. Research Notices* **18** (1998), 979–990.

[Ti7] M. Tibăr, Regularity at infinity of real and complex polynomial functions, *Singularity Theory* (Liverpool, 1996), xx, 249–264, *London Math. Soc. Lecture Note Ser.* **263**, Cambridge, Cambridge Univ. Press, 1999.

[Ti8] M. Tibăr, Topology of Lefschetz Fibrations in Complex and Symplectic Geometry, Newton Institute preprint NI01029, 2001.

[Ti9] M. Tibăr, Connectivity via nongeneric pencils, *Internat. J. Math.* **13** (2002), 111–123.

[Ti10] M. Tibăr, On higher homotopy groups of pencils, math.AG/0207108.

[Ti11] M. Tibăr, Singularities and topology of meromorphic functions, in *Trends in Singularities*, 223–246, *Trends Math.*, Birkhäuser, 2002.

[Ti12] M. Tibăr, Vanishing cycles of pencils of hypersurfaces, *Topology* **43** (2004), 619–633.

[Ti13] M. Tibăr, The vanishing neighbourhood of non-isolated singularities, *Israel J. Math.* **157** (2007), 309–322.

[Ti14] M. Tibăr, Duality of Euler data for affine varieties, math.CV/0506433. to appear in *Adv. Studies in Pure Math.* Math. Soc. Japan (2006).

[TZ] M. Tibăr, A. Zaharia, Asymptotic behaviour of families of real curves, *Manuscripta Math.* **99** (1999), 383–393.

[Tim] J. G. Timourian, The invariance of Milnor's number implies topological triviality, *Amer. J. Math.* **99** (1977), 437–446.

[To] J. A. Todd, The arithmetical invariants of algebraic loci, *Proc. London Math. Soc.* **43** (1937), 190–225.

[vK] E. R. van Kampen, On the fundamental group of an algebraic curve, *American J. of Math.* **55** (1933), 255–260.

[Var1] A. N. Varchenko, Theorems on topological equisingularity of families of algebraic varieties and families of polynomial maps, *Izvestiya Akad. Nauk* **36** (1972), 957–1019.

[Var2] A. N. Varčenko, The integrality of the limit of the integral of the curvature along the boundary of an isolated singularity of a surface in $\mathbb{C}^3$, *Uspekhi Mat. Nauk* **33** (1978), 199–200.

[Var3] A. N. Varchenko, Contact structures and isolated singularities, *Mosc. Univ. Math. Bull.* **35** (1980), 18–22.

[Ve] J.-L. Verdier, Stratifications de Whitney et théorème de Bertini-Sard, *Inventiones Math.* **36** (1976), 295–312.

[Vi] O. Ya. Viro, Some integral calculus based on Euler characteristic, Topology and geometry – Rohlin Seminar, 127–138, *Lect. Notes in Math.* **1346**, Berlin, Springer-Verlag, 1988.

[Za] A. Zaharia, On the bifurcation set of a polynomial function and Newton boundary. II, *Kodai Math. J.* **19** (1996), 218–233.

[ZL] M. G. Zaidenberg, V. Ya. Lin, An irreducible simply connected algebraic curve in $\mathbb{C}^2$ is equivalent to a quasihomogeneous curve, *Soviet Math. Dokl.* **28** (1983), 200–204.

Bibliography

V. I. Arnol'd, S.M. Gusein-Zade, A.N. Varchenko, *Singularities of Differentiable Maps, Vol. I. The Classification of Critical Points, Caustics and Wave Fronts*, Translated from the Russian by Ian Porteous and Mark Reynolds, *Monographs in Mathematics* 82, Boston, MA, Birkhäuser, 1985.

V. I. Arnol'd, S.M. Gusein-Zade, A.N. Varchenko, *Singularities of Differentiable Maps, Vol. II. Monodromy and Asymptotics of Integrals*, Translated from the Russian by Hugh Porteous, Translation revised by the authors and James Montaldi, *Monographs in Mathematics*, 83, Boston, MA, Birkhäuser, 1988.

G. E. Bredon, *Topology and Geometry*, Corrected third printing of the 1993 original, *Graduate Texts in Mathematics*, 139, New York, Springer-Verlag, 1997.

E. Brieskorn, H. Knörrer, *Plane Algebraic Curves*, Translated from the German by John Stillwell, Basel, Birkhäuser Verlag, 1986.

G. Fischer, *Plane Algebraic Curves*, Translated from the 1994 German original by Leslie Kay, *Student Mathematical Library*, 15, Providence, RI, American Mathematical Society, 2001.

M. Goresky, R. MacPherson, *Stratified Morse Theory*, Berlin, Heidelberg, New York, Springer-Verlag, 1987.

J. Harris, *Algebraic Geometry. A First Course*, *Graduate Texts in Mathematics*, 133, Berlin, Springer-Verlag, 1992.

K. Lamotke, *Regular Solids and Isolated Singularities*, *Advanced Lectures in Mathematics*, Braunschweig, Friedr. Vieweg & Sohn, 1986.

S. Łojasiewicz, *Introduction to Complex Analytic Geometry*, Translated from the Polish by Maciej Klimek, Basel, Birkhäuser Verlag, 1991.

E. J. N. Looijenga, *Isolated Singular Points on Complete Intersections*, *London Mathematical Society Lecture Note Series*, 77, Cambridge University Press, 1984.

J. W. Milnor, *Singular Points of Complex Hypersurfaces*, *Ann. of Math. Studies* 61, Princeton, 1968.

J.W. Milnor, *Topology from the Differentiable Viewpoint*, Based on notes by David W. Weaver., Revised reprint of the 1965 original, *Princeton Landmarks in Mathematics*, Princeton, NJ, Princeton University Press, 1997.

V. Guillemin, A. Pollak, *Differential Topology*, Englewood Cliffs, NJ, Prentice-Hall, 1974.

J. Schürmann, *Topology of Singular Spaces and Constructible Sheaves, Mathematics Institute of the Polish Academy of Sciences, Mathematical Monographs (New Series)* **63**, Basel, Birkhäuser Verlag, 2003.

V. A. Vassiliev, *Stratified Picard–Lefschetz theory, Selecta Math*, (N.S.) **1** (1995), 597–621.

V. A. Vassiliev, *Applied Picard-Lefschetz theory, Mathematical Surveys and Monographs*, **97** Providence, RI, American Mathematical Society, 2002.

Index